NATIVE PLANTS
OF THE
MIDWEST

NATIVE PLANTS OF THE MIDWEST

A Comprehensive Guide to the Best 500 Species for the Garden

Alan Branhagen

Timber Press
Portland, Oregon

To parents, family, teachers, friends, colleagues, and mentors who give added time and resources to inspire learning and open doors of opportunity.

Published in 2016 by Timber Press, Inc.

The Haseltine Building
133 S.W. Second Avenue, Suite 450
Portland, Oregon 97204-3527
timberpress.com

Printed in China
Fourth printing 2021

Text and jacket design by Laken Wright

Library of Congress Cataloging-in-Publication Data

Names: Branhagen, Alan, author.
Title: Native plants of the Midwest: a comprehensive guide to the best 500
    species for the garden / Alan Branhagen.
Description: Portland, Oregon: Timber Press, 2016. | Includes
    bibliographical references and index.
Identifiers: LCCN 2016009571 | ISBN 9781604695939 (hardcover)
Subjects: LCSH: Native plants for cultivation–Middle West. | Endemic
    plants–Middle West.
Classification: LCC SB439 .B63 2016 | DDC 635.9/5177–dc23 LC record
    available at http://lccn.loc.gov/2016009571

A catalog record for this book is also available from the British Library.

Contents

Foreword

I first met the author in the winter of 1980 when I was a graduate student teaching assistant for a plant taxonomy lab in the Botany Department at Iowa State University. I noticed Alan would always leave early, which usually is the sign of a mediocre student. A brief conversation with him proved that just the opposite was the case—Alan was a great student who already knew much of what we were teaching! He was leaving early to visit the university library or nearby Ledges State Park to learn more about plants. In subsequent years Alan was very generous in sharing with me his knowledge of the native plants of his beloved Upper Iowa River valley in northeast Iowa and elsewhere in the Midwest.

Generations of gardeners in the Midwest have focused on cultivated plants from Europe and Asia. This book is part of a wonderful growing trend, the return to gardens of native plants that are part of our natural heritage. Alan has learned a lot about native plants, gardening, and landscape architecture over the years and shares his knowledge with us. He reveals in these pages that the native plants of the Midwest's prairies, forests, and wetlands have much to offer—wonderful colors, shapes, sizes, textures, blooming and producing fruits at different times of the year, and locally grown for food.

Alan emphasizes that native plants not only have much to offer to gardens from a horticultural perspective, but they can also greatly add to our enjoyment by attracting native birds and insects. An iridescent ruby-throated hummingbird, bill deep, sips nectar as it hovers in front of a cardinal flower; black and orange banded Baltimore checkerspot larvae feed communally on turtlehead inside an envelope of silk; leaf miners trace intricate, meandering patterns while feeding within the thin leaves of columbine. These animals add color, a range of shapes and sizes plus the element of motion that enhance a garden's potential to be a more diverse, interesting, educational, and vibrant part of our lives. With the loss of so much native habitat in the Midwest to agriculture and development, gardens can even become a way to increase the chance that some of these species will be present for future generations to enjoy. The native plants in our gardens can also inspire us to assist public and private conservation organizations in preserving native habitat, now more important than ever in our fast changing world.

Read this book and be inspired by Alan's advice and photographs to return native plants to your garden. Follow these plants through the growing season to appreciate the changes in color and form as they progress from buds to flowers to fruits before resting in their winter mode. Learn to enjoy and share them with your neighbors, not just the humans, but also the animals native to your portion of the Midwest.

Mark Leoschke
Botanist, Wildlife Bureau, Iowa Department of
Natural Resources

Preface

As a young child, native plants grabbed my attention with their uniqueness; I found them more special and intriguing than popular garden flowers. The very first flower I remember was Dutchman's breeches on a walk at a local park. My mom pointed out that each flower looked like little pants. I inspected them closely: they were upside down, inflated, sparkling white, and even-layered. Dutchman's breeches remains one of my favorite plants to this day and I can't imagine spring without a romp to experience it in the glory of its native haunts. Now I know its natural history and taxonomy and that adds to its appeal.

As a budding young naturalist interested in all things living wild, I scoured our Iowa neighborhood collecting butterflies, moths, old bird nests (which I now know is not legal), and tree leaves. At age 8, I walked all over the neighborhood on a mission to collect the leaf of every tree I knew, which included bur oak, American elm, sugar maple, and "rarities" like Ohio buckeye, sycamore, and tulip tree. Our *Encyclopedia Britannica* helped with identifications.

Keenly aware of the living things around me, I remember the plants at my maternal grandparents' home more than anything else about the farm. Norway spruce created the windbreak. Eastern white pines, an Ohio buckeye, and black cherry grew around the house. Thickets of American plum and chokecherry marched along the driveway's hedgerow where I found prized cecropia cocoons, North America's largest moth. Plum blossoms perfumed the spring air while their fruit tasted sweet in late summer. To this day, the scent of plum blossoms recalls the laugh of my grandmother and the delicious apple-plum preserves she made.

Native plants taught me to live in the moment by enticing me to closely inspect their architecture and the goings-on around them through the seasons. I learned by observation what plants hosted various caterpillars and knew where to find milkweed for treasured monarch caterpillars. I observed that some plants were no place to look for wildlife treasures.

Whorled milkweed attracted so many bugs while it bloomed that it became my summertime favorite. I caught various insects on its strikingly scented white flowers; wasps were a daring thrill to catch without getting stung. I vividly remember observing a nectaring Gulf fritillary—a rare vagrant butterfly that far north. The resident fritillaries—meadow, Aphrodite, and great spangled—were familiar butterflies. Whorled milkweed inspired me to put together a list of flowers that attracted butterflies way before I found

Dicentra cucullaria (Dutchman's breeches) flowers resemble tiny pants.

any book describing butterfly nectar plants. Whenever I smell a blooming whorled milkweed, I am transported back into those childhood days.

When I was an early teenager, my world expanded from an edge-of-town neighborhood to a rural golf course as my parents changed careers to manage it. You would think that would have made me interested in turf grass and landscape maintenance, but this was no typical golf course. Its fairways sat on a ridge top crest surrounded by stands of native paper birch and the course's periphery was wild woodlands like nothing I had seen before. I encountered a diversity of native plants and animals that were simply astounding, as neither plow nor cow had touched it, and humans stayed on the links.

A field trip to mentor Elizabeth Lorentzen's acreage introduced me to the native wildflower blue-eyed grass (*Sisyrinchium* sp.). Here was a "grass" with exquisite blue flowers at the blade tips, not a true grass but an iris relative. From that moment, I was hooked on native wildflowers and literally budded into a botanist. Thanks to the publication of *Newcomb's Wildflower Guide*, I could finally identify these natives. Blue-eyed grass inspired me to expand my search for other unique flora and I quickly picked up on the different types of woodlands around me as each hosted specific trees and wildflowers.

Around the golf course, I discovered yellow lady's slippers, showy orchis, lily-leaved and Loesel's twayblades, late coralroot, and three birds orchid. Even the names were intriguing. Then came the day I found my first closed gentian, the color blue that can't be captured in any other way besides our own vision. Gentian indicated there were actual bits of prairie remnants around the golf course along with hoary puccoon, downy phlox, alumroot, and Culver's root amid big bluestem and Indian grass. To keep track of all these wild plants, I put together a series of loose-leaf notebooks—one page per species found, in taxonomic order. This led to explorations in the field and in literature for the flora (and fauna) found throughout my home county and the counties on either side.

Mrs. Lorentzen introduced me to the writings of Hal Bruce, which sealed my interest in gardening with native plants as it gave eloquent reasons why to plant them. Hal's book encouraged me to bring some of the native plant bounty back to our home in town. I carefully transplanted a sapling hophornbeam, delicately excavating each root with my bare hands as I read they were difficult to transplant. The tree is now over 40 feet tall and over 1 foot in diameter. The conservationist in me knew not to dig the orchids, but I did try to transplant a few puccoon divisions from the edge of a fairway. They died. I now know puccoons defy captivity. To this day, my parents' backyard is surrounded by wild-sourced natives: pagoda dogwood, dotted hawthorn, nannyberry, and American plum with added nursery-purchased shadblow serviceberry, red maple, and Kentucky coffeetree, plus redbuds and yellowwoods I grew from seed. All inspired by the writings of Hal Bruce.

My long-time love of birds, butterflies, and botany played an important role as landscape architecture and gardening became my vocation. I earned a bachelor's degree in landscape architecture from Iowa State University where I was introduced to the writings of legendary conservationist Aldo Leopold and Chicagoland native plant landscape architect Jens Jensen. Their classic writings resonated with my youthful experiences.

Robert W. Dyas, professor of landscape architecture, became a mentor and taught that the Midwest native flora landscape is simply spectacular and as beautiful as any on the planet. His plant materials classes instantly gelled how to use these plants as a designer as I was already familiar with their identification and already knew their native haunts. He taught plants by adaptability based on their native habitat and we visited remnants of woods and prairies all around central Iowa.

My favorite plant materials project was drawing a landscape and locating where it was in North America based on a single big sheet of paper covered in handwritten plant botanical names. You could figure out the topography, soils, and habitat based on the requirements of the listed plants. The project taught that every local region has a different mix of plants unique to it. My assigned landscape was from prairie-covered glacial ridges around a lake and bog in Riding Mountain National Park in Manitoba.

Mr. Dyas's field trips around the Midwest brought the flora of Missouri and Wisconsin to me. He gave me *Flora of Missouri* by Julian Steyermark, which is a monumental book, subsequently updated into three volumes. *Vegetation of Wisconsin* was required as a textbook; its author, John T. Curtis, was ahead of his time and instigated the first restored prairie that still graces the University of Wisconsin Arboretum in Madison. Mr. Dyas corresponded with me for more than 20 years since graduation and I cherish his inspirations to this day.

Richard W. Pohl, botany professor and agrostologist (grass expert), also inspired beyond words as he taught plant taxonomy and how important plant families are to humanity. When I walked into his graduate level agrostology (study of grasses) class, Dr. Pohl asked what an undergraduate landscape architecture student was doing there. I knew grasses were the spirit of the Midwest and the most important family of plants to humanity. I wanted to learn more and he taught it better than anyone.

Graduate student teaching assistants in the lab portion of my plant taxonomy classes, Kay Klier (lead) and Mark Leoschke (assistant), also were extraordinary. As a great teacher, Kay had a relative send her a plant (*Ephedra* sp.) from Utah to challenge me on a test. Kay and Mark are friends to this day and Mark is currently the botanist for the Wildlife Bureau of the Iowa Department of Natural Resources. Because botany is not in the College of Design, I could not have a minor in it though I feel I am a botanist. Those of us who can actually identify thousands of plants are a "dying breed" as botany classes have disappeared or changed to be more about the molecular aspect of plants.

A master's degree in landscape architecture from Louisiana State University led oddly enough to where my career began working with development of forest preserves in Illinois. I knew I wanted to work with native plants and land planning, so when I saw the job notice for the Winnebago County Forest Preserve District while attending LSU, I visited the campus library and lo and behold they had a copy of *The Flora of Winnebago County, Illinois*. The botanical diversity of that county made me even more intrigued with the job opportunity and I was hired from afar.

The conservation community of Rockford was just what I needed to embrace my public service position with the Forest Preserve District whose mission is "to preserve and protect the flora and the fauna." Rockford resident George Fell, one of the founders of the Nature Conservancy and father of the Illinois Nature Preserve system (a first for the United States) as well as the son of the writer of *The Flora of Winnebago County*, reflected the citizens of Illinois's commitment to preserve, protect, restore and manage their native landscapes. This concept is still applauded and what an opportunity to be a part of that.

On a trip to Morton Arboretum outside Chicago, I met renowned prairie expert Ray Schulenberg. He walked us through "his" prairie restoration (second oldest in the world), which remains one of the best and open to all who visit. It was a vision for my first big project: to design and landscape the new district headquarters. We used solely native plants and that preserve's restored prairie core surrounded by native trees inspires me to this day. Then-director Mark Keister and I laid out the design in the stubble of corn. Now it's a place of unique butterflies and bird songs amid prairie grasses and native trees.

My love for all plants, with a primary focus on natives, and the visceral role public gardens play in our society landed me the position of director of horticulture at Powell Gardens, a major midwestern botanical garden east of Kansas City, sitting on spacious ground with native trees and prairie remnants. Powell Gardens' mission as "an experience that embraces the Midwest's spirit of place" is a great fit for me. The palette of plants to work with is practically unlimited and professional gardeners are on hand to maintain the 970-acre site. Here native plants and native plant (and animal) communities are woven through a plant collection that embraces the importance of all plants (both native and exotic) in our lives. From food for the body (including America's largest edible landscape) to food for the soul (beauty), they fulfill part two of Powell Gardens' mission to "inspire an appreciation for the importance of plants in our lives."

At middle age now, I have witnessed many changes in the midwestern landscape. My grandparents' farm is no more, the house, barn, outbuildings, hedgerows, windbreak, and sloughs bulldozed and drained for industrial crop production. The grassland meadowlarks and bobolinks are long gone from my old neighborhood and just a thread of the Aphrodite and great spangled fritillary butterflies hang on even though the habitat is now a masterful prairie restoration: Decorah Prairie. The Dutchman's breeches still bloom in a nearby park but nonnative invasive bush honeysuckles and buckthorns plus garlic mustard and an out-of-control, in-town deer herd have seen the ranks of many native plants dwindle. It is a clarion call to plant what supports the web of life.

My favorite current author is Doug Tallamy, who wrote *Bringing Nature Home*. One of Doug's lectures begins with Aldo Leopold's writing: "There are those who can live without wild things and those who cannot," yet Doug reminds all present, "If you think you can live without wild things, you should know that you cannot." Knowing where your food comes from, what pollinates it, and what keeps the pests in check by understanding the web of life in Nature around us is important. Nothing brings back wild things better than growing as many native plants as possible.

The lack of diversity in planting projects and the use of low or no value, nonnative and hybrid plants has contributed to the dearth of beneficial insects, and up the food chain to birds and bees. The system is becoming more out of balance. Herbicides and pesticides can only work for so long until their living targets evolve resistance—not to mention the toll misuse of these toxins puts on our environment and our personal health. The honeybees are in trouble, along with the iconic monarch, meadowlarks, and bobwhite. And with ever-increasing cancer rates, are we too?

May this book inspire you to learn about your local and midwestern native plants. Observe them through the seasons and marvel at the creatures they engage. Visit and support local public parks, nature preserves, and gardens. Bring nature home by incorporating appropriate native plants into your home landscape. Know that you can create a sustainable, healthy, and environmentally sound place full of wonder for yourself and for all of life.

Introduction

This is a book about the plants indigenous or native to the heartland of North America. No place else on earth has such an extreme continental climate, yet it is a place filled with plants of every size and in every hue. This book aims to inspire readers to plant native plants while learning how and where to grow them successfully. There is no perfect plant so understanding the strengths and limitations of each species is a critical component. I also aim to explain why it is important to utilize native plants in a landscape wherever possible.

Humans have manipulated the landscapes of the Midwest since arriving in the region. The first English-speaking settlers described a forest that stretched from the Appalachians to the Mississippi. The prairies were celebrated as a sea of grass, so vast it stretched to the horizon in many places. Much of the forest was portrayed as open woodland, with a parklike appearance of scrub and gnarly trees interspersed with grass. We like to think of these early descriptions as depicting a pristine place, but we know that the bison, elk, and other creatures along with Native Americans and their use of fire created that landscape.

When settlers arrived, the region was already changing as the great glacial ice had melted not that long ago (in the big scheme of things). Northern forest trees were retreating northward and southern species advancing as the climate changed. Grasslands had periodically advanced eastward and northward through periods of heat and drought, the habitat they required maintained by natural and man-made fires that burned through entire landscapes. Plants filled every niche, segregated by their adaptations to all the various conditions from wet to dry, muck to sand, sun to shade, and hot to cool microclimates.

Today the Midwest is one of the most human-manipulated landscapes: the seas of prairies are now a vast expanse of farmland while the forest has been fragmented into smaller tracts. Once open woodlands and savannas are now dense forests. The region's great herbivores, bison and elk, no longer roam, while wildfires no longer burn. Some native animals like white-tailed deer and some imported plants like bush honeysuckles and reed canary grass have gone awry, usurping indigenous plants in remnant wildlands. With the forces that shaped the original landscape now gone, the remaining natural areas must be managed almost like gardens to protect their inhabitants.

The indigenous plants are important because they sustain all of life in this landscape. Could you live solely off native plants? Many species, mainly insects, through millennia of adaptations and evolution, are viscerally linked to a specific plant. Two butterflies are a good example of these links: zebra swallowtail can only survive where its host, the pawpaw, grows, the Karner blue where the wild lupine grows. We know a healthy environment for humans includes a diversity of life around us. Aldo Leopold's saying still holds true: "The first part of intelligent tinkering is to save all the parts." By including native plants in our landscape, we are

The prairie restoration that was the late Doug and Dot Wade's backyard in Ogle County, Illinois, also was the site of Windrift Prairie Nursery, one of the first native plant nurseries.

helping to save this diversity, especially important in the manipulated and fragmented Midwest.

The typical suburban landscape includes a home, expansive (and rarely used) lawn, foundation plantings, plus token shade and evergreen trees with various adornments of ornamental plantings. In many cases, the shade trees are native but most of the smaller trees, shrubs, and groundcovers including turf are not. In 1976 Hal Bruce wrote: "Americans simply do not utilize their wildflower resources. Yet there is still time to begin." After almost 40 years since his book was published, we have begun, but we can do even better.

What Defines a Native Plant?

In North America, a native plant is typically defined as one that grew wild in a particular or defined area prior to settlement by Europeans. Missouri's award-winning "Grow Native!" program provides an excellent example and defines a native plant as a species that grew wild within the boundaries of the state of Missouri before settlement. (Grow Native! is now a nongovernmental program promoting plants native to the Lower Midwest.)

What defines the Midwest? It's the central hardwood forest and tallgrass prairie bioregions that comprise the core of the midwestern states and a bit into Canada. Thus, Midwest native plants are those that originally grew in the hardwood forests, prairies, and associated wetlands that so beautifully interface here in the heartland of North America.

Native plants and wildflowers are not necessarily the same. A wildflower can be any plant that flourishes in a particular area on its own including those that have naturalized from foreign shores, usually aided by humans, to reach a new region. Queen Anne's lace (*Daucus carota*), chicory (*Cichorium intybus*), Deptford pink (*Dianthus armeria*), and oxeye daisy (*Leucanthemum vulgare*) are examples of widespread midwestern wildflowers that are native to Europe and so are *not* Midwest native plants.

Ranges of all plants are not static, but dynamic, and that creates a challenge to define native plant boundaries. Recently, Arkansas native plants have crossed the border north into Missouri whether from human aid or dispersed by wild creatures such as birds. Botanists have documented recent or an increasing number of site collections in Missouri of trumpet honeysuckle (*Lonicera sempervirens*), American beautyberry (*Callicarpa americana*), and cedar elm (*Ulmus crassifolia*). The changing range of plants is happening everywhere. At Powell Gardens, there are no fruiting plantings of sassafras (*Sassafras albidum*), but the species

is "native" one county to the east. A few maturing saplings have shown up on its wild lands. Birds disperse sassafras and there is no sign that says: "Don't poop beyond this line!" for them. There is the caveat that herbarium specimens documenting what grew native historically can be incomplete, and that more recent collections may have been present in the past.

Osage-orange (*Maclura pomifera*) is a prime example of dynamic native range. At settlement, it was growing wild only in small areas of Arkansas, Oklahoma, and Texas. Native Americans cultivated the plant elsewhere as it was a valuable commodity. The French name *bois d'arc* suits it, as the wood was valued for bows. Lewis and Clark "discovered" the plant at the beginning of their historic journey, though their collection originated from plants cultivated at a Native American Osage village. Archaeologists have found Osage-orange remains all over the heartland and into Canada. Paleobotanists look at its large fruit and wonder why the tree nearly died out. The scientific conclusion is that extinct fauna, including prehistoric horses and mastodons, ate its large fruit and dispersed its seeds. The tree was on its way to oblivion when settlers found a use for it. Prior to the invention of barbed wire, settlers widely planted it as a living hedge across much of the Midwest and it has naturalized extensively from those plantings. Is it native? My botanist friend Mark Leoschke gives a resounding "no," but I include it.

Land use, climate, and soils determine the range of native plants through time. Prior to settlement of the Midwest, Native Americans and lightning set widespread fires that burned across the landscape. Settlers suppressed these fires, allowing many plants formerly limited by wildfire to spread. When rural populations peaked, farmers grazed, burned, and cut wood from woodlands, keeping them relatively open. Now such places are pretty much abandoned, creating dense forests where only shade-tolerant species thrive.

Climate also is changing with the Midwest experiencing increases in average temperatures, thus allowing more southerly plants to spread northward. Northern plants, especially those that were relicts in more southern areas, are dying out.

Soil type remains constant in any given place and is based on its parent material, pH, and moisture availability. Many plants are adapted to specific soil conditions and thus have restricted ranges defined by those soils. Sandy, loess, or more acidic, igneous bedrock-based soils harbor many range-restricted midwestern plants.

Native plants depicted and described in this book are mainly widespread midwestern species. Some species from relict populations or peripheral areas are included if they are currently widely cultivated and perform well in gardens across most of the Midwest. Species also must support a component of the native fauna to be considered and are included if insects and birds readily utilize the plant and it's not invasive. Most Northwoods, or northern relict, species are not included as they are often very challenging to grow outside of their uniquely cool relict habitats. I find it is acceptable for each reader to have a more conservative or widespread view of what's native—whatever works for you in your own garden and community.

Why Cultivate Native Plants?

Pondering why to cultivate native plants before a talk I gave at the Dyck Arboretum of the Plains in Hesston, Kansas, reminded me of Dr. Seuss's Grinch pondering the true meaning of Christmas. That thought led me to say, "They grew without irrigation, they grew without hoes, / they grew without fertilization, pesticides, or Lowe's."

What an amazing garden was here before us, surviving on native soils, with seasonal rainfall, fertilized by creatures and by their own recycling program. Growing, flowering, and fruiting in the region's growing season, in balance with a web of life that included a nearly infinite number of soil microfauna, insects, birds, and mammals, native plants were the epitome of sustainability.

Cultivating native plants in the types of soils in which they grow wild makes them sustainable and makes amending the soil unnecessary. Planting them where they are able to thrive and watering them only until established makes extra irrigation, saving our most precious commodity—clean fresh water. Native soils also have the proper fertility—making purchasing of fertilizers unnecessary, saving that resource for other more important uses and reducing polluted run-off into local streams and wetlands. All this reduces maintenance required.

There are caveats. Many soils in our manipulated landscapes have construction-damaged or otherwise destroyed structure and profiles, so that these soils must be amended unless species that tolerate disturbed soils are utilized. Plants must be irrigated after transplanting and often during periods of severe drought, but they may require extra watering in altered soils because the microfauna and other soil life are not fully present to help a plant obtain water. (Soil mycorrhizae are important fungi because they bridge the gap between soil and root, helping plants absorb water and nutrients.) But for the most part, planting native plants reduces all these needs.

Quercus macrocarpa (bur oak) is a widespread and iconic midwestern tree.

My childhood discoveries of how native plants sustain native wildlife taught me the integral role that plants play in the web of life. The wasps and fritillaries on the milkweed flowers collect nectar for food, but in so doing fill an important role in pollinating the flowers. Pollination brings about fruits with seeds and ultimately more milkweeds so each benefits from the other. Monarch caterpillars are adapted to only eat milkweed as is a whole "milkweed village" (there's a t-shirt) including aphids, several species of true bugs, and a moth caterpillar with a gorgeous coat of hairs—all able to feed on no other plants. A midwestern monarch needs not only a cool, montane oyamel fir (*Abies religiosa*) forest in Mexico in which to overwinter, but also a wasp or other pollinator—as does a whole cast of other creatures.

A similar scenario is true for every plant and creature we see. We need this cast of characters, too, as a lot of our food requires pollinators.

If you embrace native plants and their accompanying balance with the web of life, you not only attract a plethora of entertaining creatures to your landscape, you also work with them as predator and prey to help keep them all in balance. Pesky bugs rarely reach populations beyond a threshold that damages your prized plants because other predatory and beneficial insects and other wildlife like hungry birds are there too. This will reduce and often eliminate the need for any use of pesticides. Many pesticides are misapplied and cause damage to the health of our environment beyond intended usage—often killing the very beneficial creatures we need. The conclusion is that native plants are a visceral component of a healthy landscape.

Native plants were the original flora that decorated this land. They define the aesthetics of the Midwest and help differentiate it from the Northwoods, the East Coast, the South, or the Western Mountains. The unique aesthetic and inexplicably linked web of life is what "spirit of place" is all about. Do you want to live in Anywhere, U.S.A., on planet Earth, or do you want to bloom where you're planted and be a part of the nature of your place?

The problem is midwestern native plants have a P.R. problem. Compared to the coasts, the Midwest is fly-over country. Turn on the TV and virtually all commercials show mountains, deserts, or tropical shores. Any midwestern-themed commercial, say for agriculture, shows vast fields of perfect, weed-free crops and never a single native plant or field in context with the native landscape. Aldo Leopold commented on this in his *Sand County Almanac* essay—"Illinois and Iowa"—a long time ago, lamenting a farmer not knowing why his soil is better and passersby not recognizing native plants clinging to life along the roadside.

I've known native plants and native landscapes all my life and love their special beauty, but I know it's an acquired taste. With exposure and understanding, I think native plants would be more embraced.

I recall a late friend, Rod Myers, disabled with muscular dystrophy. He loved nature but had never experienced a prairie. Is there such a thing as an accessible prairie? Three of us grabbed his wheelchair and lugged him to a hillock in

TOP This private native prairie remnant overlooks verdant farmland in Winneshiek County, Iowa.
BOTTOM Flowering forbs on Paintbrush Prairie, Pettis County, Missouri.

Illinois's Nachusa Grasslands and left him there for an afternoon while we explored the prairie. Afterward, he simply gushed about the experience immersed in a real prairie—of watching the grasses in the winds, the wildflowers, insects, and birds. Something we took for granted was the experience of a lifetime for a disabled person. He never forgot and often reminded me about the experience. It inspired him to install a native plant landscape by his assisted living patio where I was amazed by the creatures that showed up in densely developing suburbia. That planting enriched his life, and he actually was on the garden tour of the local chapter of Wild Ones. If you plant it, they will come.

I recall my first indelible experience witnessing the tallgrass prairie stretching horizon-to-horizon. It happened on the Kansas Turnpike where it crosses the Flint Hills of Kansas. It was life changing and to think the whole tallgrass prairie component of the Midwest once looked like that. I know it will never look that way again, but I sleep better knowing there is a place where it still exists. Thank you to many ranchers and conservationists who maintain this landscape. Take time to visit the Tallgrass Prairie National Preserve, Kansas State University's Konza Prairie, or the Nature Conservancy's Tallgrass Prairie Preserve in northeast Oklahoma.

It's still important to visit, support, and manage local remnants of tallgrass prairie, savannas, glades, woodlands, wetlands or whatever you have locally that preserves vestiges of native flora. Each serves as a reservoir of the plants most adapted to the particular conditions of your region. Although virtually all of these remnants are small and can no longer support keystone species like bison and prairie chickens, they may harbor other treasures of our fauna from insects, lizards, and birds. Your local remnants provide the ideal reservoir of plants and animals that ultimately colonize your own garden.

Acquiring Native Plants

Growing native plants from seed and seedlings is the most economical but likewise most time-consuming method of acquiring plants. Most states have a state nursery that provides low cost tree and shrub seedlings. There are also many local and midwestern regional nurseries that sell native plant seed and seedlings. Many native plants are quite difficult to grow from seed so purchasing a plant is the only choice.

When acquiring plants, make sure they are nursery propagated or divisions from gardens and not pillaged from the wild.

Pedicularis canadensis (wood betony)

Houstonia minima (little bluets)

Lycopodium lucidulum (shining clubmoss)

Polypodium vulgare (common polypody fern)

Gentianopsis crinata (fringed gentian)

If a plant is inordinately cheap—especially if it is one of more slow growing and slow-to-propagate spring wildflowers—that is often a warning that the plant may have been collected from the wild. Local plant societies and organizations occasionally salvage wildflowers from construction projects but always after obtaining permission from the landowner.

Many Midwest native plants do not conform to nursery mass production. They are programmed to grow roots with little top growth as an adaptation to the climate. This makes them more expensive than standard nursery items, but I guarantee they will be worth the wait and added expense in the long run. Many upland oaks, hickories, pecan, buckeyes, hophornbeam, leadplant, New Jersey tea, trilliums, yellow stargrass, and other spring wildflowers grow that way.

I dream of a time when native bulbs are readily available and their purchase as popular as tulips and daffodils. I relish the thought of lawns filled with spring beauty and trout-lilies rather than (or at least along with) Siberian squill (*Scilla siberica*) and crocus (*Crocus* spp.). Dedicated native plant horticulturists are slowly cracking the code to produce many of our more difficult native plants including orchids. Many native plants are not grown in quantity because the demand for them is simply not there. Every native plant purchase you make helps towards the cause.

I also hope to see the day when drought-tolerant, difficult-to-transplant trees and shrubs will be as readily available as red maples and invasive Callery pears. Dedicated nursery professionals are working out these production challenges. Resist the temptation to buy cheap, easy-to-produce trees that usually are shorter lived. Many of our communities' native trees are mature and not being replaced. We owe it to future generations to know what the real midwestern flora was. Create your own little episode in the rebirth of our priceless native flora.

Gentiana puberulenta (prairie gentian)

Spiranthes cernua (nodding ladies' tresses)

Aplectrum hyemale (puttyroot orchid)

Lithospermum canescens (hoary puccoon)

Plants That Defy Captivity

For whatever reason, many native wildflowers require natural lands to live out their lives. They may grow in a garden for a bit but rarely persist though they may be inherited in gardens built into remnant natural areas. These plants are poster children for why we need to protect and manage our remaining natural lands whether wetlands, forests, or prairies.

Perhaps no other ecosystem on earth has been so severely impacted by mankind as North America's tallgrass prairie. Watch Carol Davit's (Director of the Missouri Prairie Foundation) Ted-X talk in St. Louis for a succinct, though thorough and inspiring call to preserve and protect the remaining remnants of native prairies: youtube.com/watch?v=Gl5wzHjzvMk.

Lilium philadelphicum (wood lily)

The Midwest Spirit of Place

How does one define the spirit of the Midwest? It's the blend between the verdant lush forests of the Appalachians on the east and the dry, short grasses of the Great Plains on the west. It lies below the cold boreal evergreen northern woods and above the great steamy southern swamps and pinelands. It is a land defined by a prevalence of open woods and savannas and tallgrass prairie.

The Midwest encompasses the core of the Great Lakes and the amazingly rich soils of the glaciated Corn Belt Plains that sweep from central Ohio westward to the Central and Northern Great Plains. Here lies one of the world's richest swaths of soils—Iowa and Illinois comprising nearly half of the prime farmland in the United States. It's now the supreme agricultural area on earth and drained by five of the continent's greatest rivers: Mississippi, Missouri, and Ohio to the south, Red to the north, and Saint Lawrence to the east.

It's a landscape that is relatively flat as glaciers once scoured this land more than once across the northern regions. Here the glaciers left a landscape of prairie potholes rich with waterfowl in the Northern Glaciated Plains, marvelous moraines in the Southeast Wisconsin Drift Plains, and remnant bogs in the southern Michigan–northern Indiana Drift Plains. Only the Flint Hills, Osage Plains, and Ozark Highlands in the west, Interior and Allegheny Plateaus in the east, and, like an island, the Driftless Area escaped. There are no mountains, lest ancient mini-remnants in the Baraboo Range in Wisconsin and St. Francois in Missouri. Marvelous rivers have cut into its flat plains and plateaus; even the Ozarks are flat on top. The coulees of the Driftless

Area and bluffs along the major rivers also testify to this. Great dunes of loess flank the eastern valley of the Missouri, the marvelous Loess Hills, while great dunes of sand rim the eastern and southern shore of Lake Michigan.

I never knew how much this vast space meant until driving back from a trip to the West Coast. When descending the front range of the Rockies on I-70, I could see the vast plains ahead. My thought instantly was that "home" lies ahead. No offense to mountain lovers, but I believe I mumbled, "I'm so glad to be leaving the mountains!"

Several plants are iconic to the Midwest. One tree's range covers this entire area and it is symbolic to me, our place's mascot: bur oak (*Quercus macrocarpa*). This tree usually grows wider than tall with rugged, sturdy branches adapted to the wild weather and thrives in our climate of extremes. On the prairies and in open woods, two grasses stand tall: big bluestem (*Andropogon gerardii*) and Indian grass (*Sorghastrum nutans*). A stand of tall native grasses flowing in the currents of wind is a purely midwestern experience. Their warm colors in autumn bleach out some by winter, but nonetheless add to a warm-colored winter landscape on the prairie. On forested floodplains, two of the largest trees—the eastern cottonwood (*Populus deltoides*) and silver maple

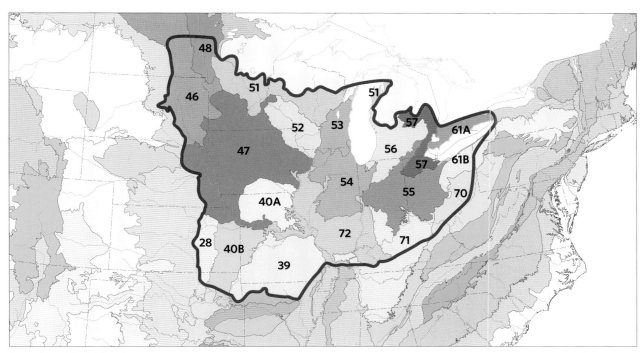

Ecoregions of the Midwest

28 Flint Hills
39 Ozark Highlands
40A Southern Iowa/Northern Missouri Drift Plains
40B Osage Plains
46 Northern Glaciated Plains
47 Western Corn Belt Plains
48 Southern Lake Agassiz Plains

51 North Central Hardwood Forests
52 Driftless Area
53 Southeast Wisconsin Drift Plains
54 Central Corn Belt Plains
55 Eastern Corn Belt Plains
56 Southern Michigan/Northern Indiana Drift Plains
57 Huron/Erie Lake Plains

61A Carolinian Forest
61B Western Erie Drift Plains
70 Eastern Ohio Allegheny Plateau
71 Interior Plateau: Shawnee Hills & Bluegrass Region
72 Interior River Valleys and Hills

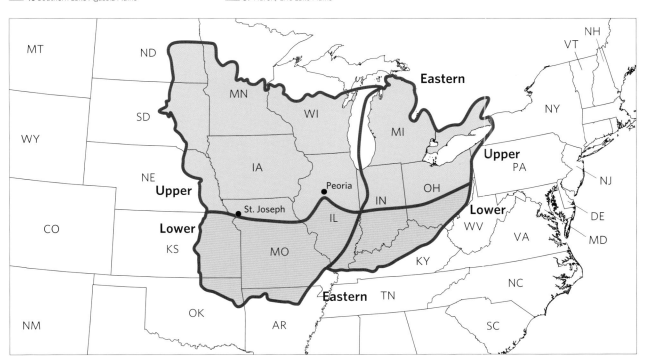

Subregions of the Midwest: Upper, Lower, and Eastern

(*Acer saccharinum*)—reign region-wide. Cottonwoods readily tower well over 100 feet high while silver maples are not far behind. Treeless wetlands contain prairie cordgrass (*Spartina pectinata*) as a signature—its roots held the prairie sod together like cord, making it a favored material in the sod houses of the region's pioneers.

Subregions

From a gardener's perspective, the Midwest can be divided into three subregions: the warmer Lower Midwest, the milder Upper Midwest, and the moister Eastern Midwest.

The Upper Midwest contains some elements and remnants of the Northwoods. Some Lower Midwest trees and shrubs are not hardy in this subregion. Snow cover is reliable almost every winter.

The Lower Midwest actually has a high temperature above freezing on most days all through winter. Snow cover is not reliable and rarely lasts for more than a week. Summer heat is more steadfast so some Upper Midwest plants can't tolerate that aspect.

The Eastern Midwest lies east of Lake Michigan and has more reliable rainfall and higher humidity. It includes the lake-effect snow lands where blankets of clouds and snow snuggle plants through the coldest winter nights.

Today's Midwest is organized around major cities radiating out to farms wherever the soil suits cultivation. The wettest floodplains of the rivers remain intact, and areas too steep or of poorer soils still conserve relicts of the native flora in both woodlands and rare prairie and savanna remnants. The deep soil tallgrass prairie and its accompanying wetlands are probably the rarest of originally widespread habitats on earth. Most only exist as tiny postage stamps of what once was set like an island in a vast sea of farmland.

Natural Vegetation

The Midwest embodies the diverse blend between the eastern deciduous forest and tallgrass prairie. The actual interface can be called *savanna*, where scattered trees, shrub thickets, and woodland plants grow among prairie flora. Trees cover only 10–30 percent of the savanna, and this habitat has become extremely rare, as trees either have reclaimed such areas, making them woodlands, or have been destroyed by agriculture and development.

The term *woodland* refers to a more open forest with 30–80 percent tree cover, while *true forest* has dense tree

Tallgrass Prairie National Preserve in Chase County, Kansas, epitomizes what the Midwest landscape looked like before settlement.

cover over 80 percent. Even once-widespread woodlands have become solid forests now with lack of fire, grazing, and wood harvest. *Grassland* is the treeless province usually dominated by grasses and termed *prairie* in North America.

GRASSLANDS

The Midwest's grassland province is easy to define as the tallgrass prairie component: bookended on the west by the Flint Hills of Kansas and on the east by remnant prairies of Adams County, Ohio, on the edge of the Appalachian Highlands. These are the prairies of the eastern part of the continent where rainfall was sufficient so that tall grasses were dominant.

There were four main species: big bluestem, Indian grass, switchgrass (*Panicum virgatum*), and, defining wet soils, prairie cordgrass. The drier mixed and short-grass prairie region of the Great Plains is not included here because it is such a different gardening region with many specific species plus climate and soils that won't allow gardening with plants from much of the core of the Midwest.

It must be stated that grasses were a major component of the tallgrass prairie, yet a simply spectacular accompanying suite of forbs (broad-leaved flowering plants), "wildflowers," flowered in sequence from spring through fall between the grasses. If you have not witnessed these floral displays, get out there and immerse yourself in them.

Many of our midwestern native plants are currently very popular as garden perennials especially across the pond in Europe. There many of them have "gone to finishing school" where European gardeners have selected new cultivars of our midwestern native plants. Through the international garden trade, these cultivars eventually find their way back into our gardens. What does that say about our understanding and appreciation of our local flora?

Ha Ha Tonka State Park's savanna in Camden County, Missouri, is a premier and rare extant example of this once-common habitat.

Lynx Prairie represents some of the easternmost tallgrass prairie remnants in Adams County, Ohio.

The tallgrass prairie can be divided into many diverse habitats, but I'll keep it relatively simple here based on whether the prairie is dry, mesic (moist or medium), or wet. Dry prairies occur on thin, gravelly or sandy soils or on steep slopes that face hot south and west afternoon sun and summer winds. Where bedrock is at the surface, these areas could be called *glades*, but there are also specific loess, sand, and gravel prairie communities and even hill or "goat" prairies—so named because they are so steep as to only be grazed by goats, though cattle still graze many of these remnants. Many shorter grasses like little bluestem (*Schizachyrium scoparium*) and sideoats grama (*Bouteloua curtipendula*) thrive in these types of prairie, and special wildflowers are able to survive in the harsher environment.

Mesic prairie is rare. It's what made the prime farmland of the Midwest and nearly all of it has been plowed for cultivation. Tiny remnants survive, many as early cemeteries and along railroad right-of-ways. The phenomenally deep and rich soils produce tall grasses and robust forbs that once were ubiquitous. I find it ironic that these plants are now regarded as weedy looking and untidy. Were it not for the great fires that swept across the landscape, these prairies would have been woodland, so today constant management to remove woody native shrubs and trees from these

TOP This hill prairie in Root River Valley, Houston County, Minnesota, depicts a classic dry prairie.
BOTTOM Markham Prairie ablaze with blazingstars (*Liatris spicata*) is a classic remnant of mesic prairie in Chicago's suburbia in Cook County, Illinois.

tiny remnants is critical for their conservation. The magnificent mesic prairie soil will grow almost any tree.

Wet prairies had extensive prairie cordgrass and blue-joint (*Calamagrostis canadensis*), which, like most wetland native grasses, spread by underground rhizomes. Wet prairies included native strains of phragmites (*Phragmites australis*) and canary reed grass (*Phalaris arundinacea*), but "foreign" running forms of these latter two grasses were introduced and have now usurped native species and become omnipresent invasive weeds. Here, too, were many plants in the sedge and rush family. Areas dominated by sedges (mainly *Carex stricta*) are known as sedge meadows.

FORESTS

The forested province of the Midwest is composed of the deciduous central hardwood forest, and can be broken down into four basic types based on soil moisture and the dominant trees:

> the mesic maple-basswood woodlands of the Upper and Western Midwest
> the drier oak-hickory woodlands throughout
> the mesic beech-maple of the Eastern Midwest
> the elm-ash-cottonwood forests of the whole region's wet floodplain forests

The reality is more complex than this with the mix of plants varying from region to region, but it's a great way to categorize the various midwestern woodlands.

The midwestern woodlands are mainly found on calcareous (limestone-based) soils with a lower pH than many of the surrounding regions, but not so alkaline as west of the region. These soils, rich in calcium and magnesium, support a ground flora like almost no other, much of it known as prevernal flora comprised of spring ephemerals that bloom early before the trees leaf out, capturing enough sun to set seed, and wither away before trees cast midsummer's dense shade. To quote Michael Homoya, botanist with the Indiana Department of Natural Resources, from his book *Wildflowers and Ferns of Indiana Forests*: "Certainly Indiana forests harbor some of the best and most extensive displays of spring wildflowers in the country, if not the world." The

Wet prairie/sedge meadow wildflowers include white spires of *Veronicastrum virginicum* (Culver's root), white-flowering *Pycnanthemum virginicum* (Virginia mountain-mint), yellow-blooming *Lysimachia quadriflora* (prairie loosestrife), and the fuchsia flowers of *Lythrum alatum* (winged loosestrife) at Kieselburg Forest Preserve, Winnebago County, Illinois.

same could be said of much of the Midwest's untouched marvelous and underappreciated woodlands.

The Upper and western Midwest's maple-basswood woodlands are rich soils that hold moisture. The dominant trees in this habitat are sugar maple (*Acer saccharum*) and basswood (*Tilia americana*), sometimes called American linden. Northern red oak (*Quercus rubra*) and bitternut hickory (*Carya cordiformis*) also are prevalent tree components. Many woodland spring ephemeral plants thrive in this habitat and show their best here. Any plant liking moist woodlands will do well. The soils are well drained, not too wet, and only very rarely dry so this habitat is known as "mesic" or medium. This woodland type can be found in the oak-dominated areas on more moist and sheltered east- and north-facing slopes.

The oak-hickory forest covers well-drained soils that dry out more often in the summer. The oak component of this forest refers to various species but almost universally includes bur oak, white oak (*Quercus alba*), northern red oak, and black oak (*Q. velutina*). The hickory component references shagbark hickory (*Carya ovata*), but includes other hickories in the Lower Midwest. Oak-hickory forest can be found on warmer south- and west-facing slopes in otherwise more mesic areas. The deep coarse root systems of oaks and hickories and their open canopies allow for a more diverse ground and understory flora than other woodland types.

Beech-maple forest is confined to the Eastern Midwest—east of a line drawn north-to-south through Lake Michigan but including areas hugging the western shore of the lake in Wisconsin. Maple again refers to sugar maple and the beech is American beech (*Fagus grandifolia*) with tulip tree (*Liriodendron tulipifera*) and northern red oak common companions. These trees thrive on well-drained soils that stay moist year-round. Both sugar maple and American beech are very shade tolerant species with dense surface roots. Spring ephemeral wildflowers thrive here as well as ferns, but other species must have high tolerance for shade.

Elm-ash-cottonwood forest grows on the seasonally wet floodplains throughout the Midwest. The elm refers to American elm (*Ulmus americana*), which was killed off as a shade tree by Dutch elm disease (DED, *Ophiostoma novoulmi* and *O. ulmi*) but still reseeds and reaches maturity in wild woodlands. Ash refers to green ash (*Fraxinus pennsylvanica*) whose fate also hangs in danger by the importation of the lethal emerald ash borer (EAB, *Agrilus planipennis*), an exotic pest from northeastern Asia. In North America, EAB has no natural control beneficial insects and our native ash trees have no resistance to this insect. Cottonwood refers to the monstrous growing eastern cottonwood. All these trees grow in soils that may be seasonally inundated by floodwaters, usually during the dormant season in early spring following snowmelt.

The mesic maple-basswood woods at Seed Savers' Heritage Farm shows a carpet of prevernal flora in Winneshiek County, Iowa.

TOP The Ozark Highlands contain some of the largest remaining tracts of oak-hickory forest in the Midwest as shown in this view from Taum Sauk Mountain, Iron County, Missouri.
BOTTOM A premier, old-growth beech-maple forest can be seen at Warren Woods, Berrien County, Michigan.

WETLANDS

Wetlands, besides the floodplain forests, were widespread across the Midwest but many of them have been drained for agriculture. Lakes are abundant in the recently glaciated parts of the Upper Midwest but have been created by man across the whole region. How can we forget the importance of the gigantic Great Lakes and their unique shorelines?

Swamps are forested wetland where water stands much of the year and are mainly confined to floodplains and low lands surrounding lakes, but were extensive in the Huron/Erie Lake Plains. Marshes are nonforested wetlands where water also stands most of the year. Fens are actually quite widespread across the Midwest and are unique peaty wetlands fed by mildly acidic to alkaline groundwater.

Bogs are acidic wetlands of sphagnum moss, which are raised above the influence of groundwater (receiving water only from rain) and are relicts of the retreating glaciers but a common component of the Northwoods. All of these wetland habitats contribute to the flora available to water gardening.

SUCCESSIONAL LANDS

Many lands are successional in nature: in transition from clearing or farming and comprised of plants that readily invade disturbed areas and set the stage for the habitat to change as it becomes more shaded and mature so that other plants can thrive. My own home is set in such a woodland–former pastureland.

Shingle oak (*Quercus imbricaria*) is the classic oak for such lands in much of the Lower Midwest, along with persimmon (*Diospyros virginiana*), honeylocust (*Gleditsia triacanthos*), American elm, hackberry (*Celtis occidentalis*), and eastern red cedar (*Juniperus virginiana*)—the only widespread native evergreen. In the Upper Midwest, northern pin oak (*Q. ellipsoidalis*) takes the place of shingle oak. Quaking aspen (*Populus tremuloides*), bigtooth aspen (*P. grandidentata*), and paper birch (*Betula papyrifera*) are signature species along with red cedar. In the Eastern and Lower Midwest, sassafras and persimmon readily form thickets to heal open ground.

Most of the Midwest's classic small trees and shrubs also thrive only in these successional "edge" habitats between the forest and the prairie. Hawthorns (*Crataegus* spp.), wild plums (*Prunus* spp.), wild crabapples (*Malus* spp.), and gray and roughleaf dogwoods (*Cornus racemosa* and *C. drummondii*) are prime examples. As successional areas mature, woodland trees shade out these species. We manage prairies to keep them out while their habitat in pastureland disappears in favor of row crops. Hedgerows between fields harbor these plants too, but are now cleared to farm fencerow-to-fencerow. I fear that the habitat for these once-abundant and iconic plants will soon be gone.

Learn what bioregion of the Midwest you live in and what your soils are. List the native plants adapted to your conditions as a palette from which to choose. These will be successful for you, contribute to a sustainable design, and celebrate the spirit of your place. If you want more detail, there are plenty of state and local flora books to help with the flora of your particular region.

Lincoln Memorial Garden in Springfield, Illinois, was designed by Jens Jensen and displays the Midwest's classic small trees and shrubs.

TOP The Upper Mississippi National Wildlife and Fish Refuge protects extensive stands of the Mississippi River's elm-ash-cottonwood floodplain forests. View from Mt. Hosmer, Lansing, Iowa.
BOTTOM Wetlands known as "prairie potholes" are common in the Northern Glaciated Plains as seen here at Ordway Prairie, McPherson County, South Dakota.

Inspirations

I am writing this during a winter that has brought back memories, more Arctic polar than I can remember for decades. The national media shrieks of polar vortex and calamity and yet I feel the thrill of the weather that defines the Midwest. It is not easy to get through these spells, but I always try to take a moment and see the spectacular spectacle of it all—the snow drifts like great desert dunes, the sparkling beauty of the snow itself, the otherworldly squeaky crunch of bitter cold. When gardening with native plants, it's comforting to know that they are adapted to this, no loss of sleep whether they will make it through bitter cold nights.

The wait for spring's arrival can sometimes be as painful as a child fidgeting before opening Christmas presents. I was that child complaining, "When will spring come?" I still remember my grandmother saying, "It'll come." You cherish the smell of earth again, of rains, and seeing earthworms and robins on the lawn. You again hear songs of birds along with the peeps and trills of frogs and toads.

The change to summer's warmth brings storms with dynamic, monumental clouds—more spectacular and higher than any of the world's mountain ranges. There is the terrifying threat of severe storms and tornadoes, but at age 52, I have yet to see one. I have stood in awe in a sunlit chasm between supercell thunderstorms that produced them and witnessed the wake of a tornado's destruction. Note to world: they are really isolated and mostly short-lived and more than 99.9 percent of the midwestern landscape remains intact.

The Midwest's greatest asset—its vast, rich soil—created an unprecedented opportunity for many an immigrant. My forefathers came here from Norway and harnessed that prairie soil to create wealth measured in family and community. The growth of the Midwest in the nineteenth century and its city-of-the-century Chicago, the economic heart of the region, saw the native landscapes and their plants almost completely converted for human use. Luckily, there were people who looked forward and started a conservation movement to preserve and protect the last remnants of wildlands before they were lost. The native plants were conserved and this book about them could be written.

Role Models

Many botanists, naturalists, architects, and landscape architects were inspired by native plants and understood their value in the local landscape. The places they preserved or designed and the writings they left behind are as inspiring as ever. Seek the works and writing of these now-gone role models and continue to spread their words. Conserve their legacy in your own community. Don't forget that every community has unsung heroes working and volunteering to protect local parks, gardens, and preserves. Embrace their efforts so that the unique flora of the Midwest is never lost.

Eloise Butler, from the turn of the last century in Minneapolis, was the most influential to me, and her writings are as relevant as ever. To this day, the Eloise Butler Wildflower Garden is sacred ground, a living jewel in the emerald

Thunderstorm clouds tower above Roscoe Prairie in Stearns County, Minnesota.

necklace of Minneapolis's parks. Go there—it is pure magic. Eloise's spirit resounds through the garden's native plants from woodland to prairie and wetland. Martha Hellander's book about Eloise, *The Wild Gardener* in 1992, made it all come together about what I was experiencing there. I hope Eloise would approve of this book and here's an excerpt from her unpublished "Early History of Eloise Butler Plant Reserve, 1926." Who could ever describe it better?

IN THE EARLY [18]80s Minneapolis was a place of enchantment—a veritable fairyland. Along the [Mississippi] river banks grew in profusion trillium, bloodroot, wild phlox, anemones, Dutchman's Breeches, and hepatica; the meadows were glorious with Indian paint brush, both red and yellow, with gentians, purple fringed orchids, and royal clumps of blue violets. In the tamarack swamps of the suburbs might be seen long vistas of our state flower, the showy lady's-slipper, together with the wild calla, and pitcher plants without number. And who could describe the outlying prairies, rioting in colors far exceeding the brilliancy of tropical flora. A long procession beginning with the pasque flower, "the crocus in chinchilla fur," the rosy three-flowered avens [prairie smoke], and the equally profuse bird's-foot violet, that gave way in turn to the more gorgeous blooms of midsummer and early autumn, as the purple blazing stars, giant sunflowers, goldenrods, and asters of many species and hues.

Eloise witnessed native plants like we never could have. Her writings described the woodlands, prairies, and wetlands out of which midwestern cities grew. Her words reflected my undergraduate college professor Dyas's words about the spectacular beauty of our own flora.

Regarding gardening, here's what Eloise Butler wrote in the same "Early History." She observed how our native landscape was usurped:

Cottagers on the suburban lake shores have fettered ideas of planting that are more appropriate for city grounds, and condemn their neighbors who strive to preserve the wildness, for lack of neatness in not using a lawnmower and in not pulling down the vine tangles in which birds nest and sing—apparently dissatisfied until the wilderness is reduced to a dead level of monotonous, songless tameness.

This strikes a chord on why we once again embrace our native plants and a natural style. We know the true price of the ubiquitous tame and songless landscape.

Aldo Leopold also captured this sentiment with his essay *Natural History* regarding farmer's replanting exterminated native plants on their land. "Perhaps they wish for their land what we all wish for our children—not only a chance to make a living but also a chance to express and develop a rich and varied assortment of inherent capabilities both wild and tame."

Natural landscaping is an "acquired taste" as one of my Rockford friends, Ruth Little, describes it. I have seen a big change in accepting ornamental grasses into the landscape, step one towards capturing the spirit of our place where prairie and woodland meet.

Sacred Places

The Midwest has no national park the likes of Yellowstone, Yosemite, or the Great Smoky Mountains. American preservationists of the time sought splendor in mountains and other unique geological features while the Midwest was relatively flat and economically productive. Patchworks of smaller preserves were saved over time and these can be experienced by every midwesterner as inspiration of native landscapes and places to witness native plants growing in the wild.

MIDWESTERN NATIVE PLANT ROLE MODELS

E. Lucy Braun (Ohio)	Robert H. Mohlenbrock (Illinois)
Eloise Butler (Minnesota)	Lorrie Otto (Wisconsin)
Alfred Caldwell (Illinois)	O. C. Simonds (Illinois)
John Curtis (Wisconsin)	Julian Steyermark (Missouri)
Charles Deam (Indiana)	Floyd Swink (Illinois)
Edgar Denison (Missouri)	Edward G. Voss (Michigan)
Jens Jensen (Illinois, Wisconsin)	Frank Lloyd Wright (Wisconsin, Illinois)
Aldo Leopold (Wisconsin)	

The Indiana Dunes on the south shore of Lake Michigan are the crossroads of the midwestern flora where species from all directions converge.

There is no better way to gain an understanding of and be inspired by native plants than by visiting your local parks, botanical gardens, and preserves that contain natural landscaping and native lands. When I was a young man, the Twin Cities were an omnipresent inspiration. The opening of the Minnesota Zoo in the 1970s with its design embracing nature and its wildlands certainly had an impact on me. My discovery of the Minnesota Landscape Arboretum was more inspiration from its incredible plant collections, but my favorite aspect of the garden is its native ancient oaks, sugar maples, and basswoods and its spacious design woven through glacial hills and dales of wetlands. There is beauty in the ever-present colorful canna and the new hybrid gee-gaw plant, but it takes lifetimes for a magnificent oak's branches to sweep to the ground. Thank you to my parents for trusting me to drive to these places as soon as I got a license—slightly over a three-hour drive from home. Would a parent be willing to let a 16-year-old do that today?

When I lived in Rockford, Illinois, day trips to the Mississippi River, Madison and Milwaukee, Lake Michigan and Chicagoland were common events. Illinois Beach, the world's first nature preserve, was and is a favorite wildland along with the premier midwestern national park: Indiana Dunes National Lakeshore (the fourth most biologically diverse national park in the United States).

Other favorites are Warren Dunes and Warren Woods in southwestern Michigan and Chiwaukee Prairie, named so as it lies halfway between Chicago and Milwaukee. As a birder, I found Chicago's Lincoln Park and its "magic hedge" at Montrose point to always be a thrill and its restoration with extensive native plantings make it even better today. The Milwaukee lakefront and its parks kept me in touch with beech trees.

With so few national parks and monuments, some of our best midwestern natural areas are registered national landmarks. Visit them. The plaque at one closest to my hometown reads:

Hayden Prairie has been designated a Registered National Landmark under the provisions of the Historic Sites Act of August 21, 1955. This site possesses exceptional value in illustrating the natural history of the United States.

—*U.S. Department of the Interior,*
National Park Service, 1966.

Selecting Native Plants

The first thing to think about when selecting plants is whether they can thrive under your growing conditions. Elements to consider include your ecoregion, the climate's hardiness and heat zone, your specific soils, and whether you have sun, shade, or combinations thereof. Most midwestern homeowners know a palm tree won't grow here and understand how winter hardiness is important in selecting plants, but making sure plants will grow under your particular soil and light conditions is just as important.

Once you know the soil, light, and moisture conditions—the underlying horticulture of your landscape—you can move on to part two of selecting plants: picking plants whose function works towards, solves, or ameliorates your needs. This is where you need to think about what each plant's role is in the landscape.

The third consideration is often our starting point, as we usually pick plants because we like the way they look and how they make us feel. It's all about the art and beauty of a plant. This is the most fun and artistic part of the plant selection process. I place it last because we must not forget that the plants we choose for beauty must be able to grow and thrive where we plant them and they must fulfill the functions we need to create a working landscape. A hodgepodge of plants selected solely for beauty and unable to survive a growing season or winter will not create a successful landscape. Aesthetics are but the icing on the cake of good planting design. In the end, we must choose and combine beautiful plants aesthetically, with seasonal colors and textures, and fragrances for a well-designed landscape.

The Whitmire Wildflower Garden at Shaw Nature Reserve in Franklin County, Missouri, offers an ideal place to learn how to select native plants for a landscape.

Horticultural Requirements

Native plant promoters have spread the word that native plants are easy to grow and adapted to their homeland. For many species, this is true only if they are cultivated in conditions that mimic those where they grow native. Each species has a specific set of habitats where it thrives. Under those conditions, yes, native plants grow well and require little or no input like extra watering, fertilizer, and pesticides.

WATER

A site's ability to meet the water needs of a plant is critical and the place to begin when selecting plants. Most wetland (water-loving) plants will thrive in wet conditions, grow in good soil that is not wet, but will simply struggle in a dry site. A plant adapted to dry conditions languishes and rots if it is too wet. There are always a few exceptions to this rule as, for example, with many floodplain trees that grow in flood-prone soils but, oddly enough, a flood is a physiological drought to them. Most floodplain trees are very tolerant of droughty sites, but this is rarely the case for the herbaceous plants beneath them.

SOIL

Soils dictate the availability of moisture and nutrients so growing natives in the same soil types that they grow wild in are also paramount. Clay soils are poorly drained and hold moisture, becoming wet in times of heavy rain, but because of their small particle size, retain moisture in a drought. Sandy soils are just the opposite as their large particle size does not hold moisture and drains any excess moisture away. Silty soils have medium "just right" soil size, holding moisture so that the soil is neither too wet nor too dry. The ideal three-way blend of clay, sand, and silt soil particles is called loam.

What makes soil grow plants best is organic matter or humus; the remains of plant, animal, and all life are what can be observed as the dark color in soils. The black soils of Iowa and Illinois are the remains of thousands of years of prairie plants (and animals), especially the plants' extensive and deep roots. Organic matter holds moisture and nutrients and also helps aerate soil and provide food for a wealth of soil microorganisms. Some fungal (mycorrhizae) organisms are critical in linking with plant roots and the uptake of moisture. Many wetland soils have very high organic matter because the remains of plants decompose slowly when continually wet. Without organic matter in our soil, the Earth's surface would simply look like Mars.

Peat and muck are organic soils formed under continually saturated conditions. Such soils can be found in marshes, sedge meadows, and fens where groundwater flows through bedrock and/or mineral soil. The groundwater rises to the surface and maintains a high water table, which keeps decomposing plant material saturated. This allows little oxygen to penetrate and for the accumulation of organic soils. Most peat as a soil amendment is actually mined from acidic fens (not bogs) of the Northwoods. A true bog is raised above contact with groundwater and stays wet due to rainfall and the ability of sphagnum moss to hold water. Bogs occur, rarely, southward into the Upper Midwest. The organic soils of fens can be acid to alkaline but commercially mined peat is acidic. Fens, though uncommon, can be found throughout the Midwest.

The pH of a soil also plays an important role. Many plants are adapted to a particular range of pH and this has to do with how pH changes the availability of nutrients in the soil. Many plants require acidic soils (low pH) to grow well and will become chlorotic (yellowish instead of green) when grown in soils too basic or alkaline (high pH). It also works the other way, as some plants prefer more alkaline soils. Across the Midwest, acidic soils are more apt to be found in areas of higher rainfall and in areas with bedrock substrate that is sandstone, metamorphic rock like chert, or igneous rock like granite. Limestone, on the other hand, is basic and found under much of the Midwest and contributes to the region's usually higher pH soils. Concrete also contains lime and contributes to higher pH soils in many urban areas.

LIGHT

Exposure to sunlight is critical to growing a successful plant. Some plants are adapted to full sun, others to varying degrees of shade, and still others are designed to thrive in full shade and will literally burn in sun. Plants requiring full sun will be weak and languish in shade.

It's important for gardeners to know that morning sun is cooler than afternoon sun and less harsh to sun-sensitive plants while the late afternoon sun is the hottest and harshest in much of the Midwest. This principle explains why a sun-demanding plant may do just fine with morning shade, as the midday and afternoon sun will make up for it. Likewise, a shade-demanding plant may do fine with morning sun whereas the afternoon sun would fry the same plant.

Full sun is also harsher the further south you go and under drier conditions of the west. A plant that thrives in full sun in the Upper Midwest may not be able to do so in the Lower Midwest, and likewise one doing fine in full sun in the Eastern Midwest may do better in some shade farther west.

COLD HARDINESS AND HEAT TOLERANCE

Hardiness also plays an important role and reflects many things but usually refers to a plant's tolerance of winter's cold temperatures. The Ohio and Lower Mississippi River valleys and Ozarks are the mildest with the coldest regions being the northern and western edges of the Midwest. The Great Lakes act as "hot water bottles" creating milder winter microclimates, especially on their leeward south and east sides where cloud cover caused by their lake effect moderates what would otherwise be colder, clear nights. I follow the USDA Hardiness Zone Map (http://planthardiness.ars.usda.gov) when describing hardiness zones.

Summertime heat is another consideration for a plant's hardiness. Summer heat stresses plants, especially cool-season grasses and plants of northern affinity. Summer heat varies from the hotter southern and western areas of the Midwest and decreases the more northeastward you go. The Great Lakes tend to moderate the effects of summer heat, creating a plant-benevolent climate in much of Michigan.

Summer heat is important for growth and hardening off (a plant's preparation for winter) for some Lower Midwest plants, reducing their hardiness rating in cool summer regions. The Midwest has warm summers and many plants from redbuds to hackberry require this and will languish in the cool summer, maritime climates of the Pacific Northwest and Northern Europe as examples. I follow the American Horticultural Society's Plant Heat Zone Map (ahs.org/gardening-resources/gardening-maps/heat-zone-map) when describing a plant's heat tolerance or heat requirements.

In summary, a native plant, in the proper soils and with appropriate moisture and sunlight, is more likely to be healthy and most able to survive extremes of heat and cold. This right plant in the right place also translates into pest and disease resistance and tolerance to drought. This concept is similar to our behavior of eating right and exercising to stay healthy. Most midwestern plants have a wide range and are tolerant of average garden conditions region-wide. Do research on a plant's cultural requirements to select and grow plants that will thrive over the long term with less input under your specific conditions.

Function in the Landscape

Several main functions come to mind. The first is to cover the ground and prevent erosion. Most landscapes utilize turf and/or lawn grasses for this role, but many low groundcover plants are available and actually all plants accomplish this by covering bare soil.

The second role of plants in a landscape is to screen. This function includes blocking winds and unsightly views, and parceling components of the landscape like walls of a room. The premier use here is evergreen windbreaks to lessen the impact of cold northwestern winds, but also includes functions like hedges.

The third main function of plants in a landscape is to provide important shade. Large shade trees are a very sound investment, especially when sited so that they will cool your home or outdoor seating space in summer.

In the plant profiles, the paragraph titled "Landscape use" describes how a plant functions in the landscape. For covering the ground, you want a plant that spreads and reduces maintenance; often you want to be able to walk or play on this plant, so the plants needs to be low-growing and tough.

For screening, we often want plants that are evergreen so they function through all seasons and aren't open or bare in the winter. Plants that are deciduous, but densely twiggy, also work, or maybe you are fine with the screen simply being a baffle, especially in winter when you may not be outdoors as much.

The size of a shade plant varies for your needs. If you want to shade the front stoop of your house, a large shrub or small tree will do, but to shade an entire house requires a large tree.

AS GROUNDCOVER, WINDBREAK, AND SHADE

Because it is important to conserve energy for both economic and ecologic reasons, two functional aspects of plants have been of foremost important in the Midwest: windbreak and shade. The farms originally laid out across the prairie reflected this from the time of settlement, and various state extension agencies provide information on how important this role is. Windbreaks slow the speed of the heartland's fierce winter winds, the biggest impact in creating a moderated microclimate for your home. Heat loss in a winter home is greatly reduced by a windbreak, saving a homeowner considerably on heating use and bills. Likewise, the Midwest's warm summers can be cooled by the appropriate placement of shade trees. One or more trees shading the southwest aspect of a home can reduce the heat gain on the home, lowering cooling needs and bills.

Windbreaks need to be evergreen or densely twiggy to function best, as winter is their primary time of utility. Shade trees, on the other hand, should be deciduous. Why? Because in the wintertime, the sun creates a passive solar gain on your home and you do not want to block the winter

Canada moonseed (*Menispermum canadense*) is
an excellent groundcover in rich soil.

Regional botanical gardens, such as Powell Gardens in
Johnson County, Missouri, offer a great place to visit and
see how native plants function in the landscape.

sun's warmth. Winter shadows created by evergreens are places where snow and ice remain; often creating problematic situations on walks and drives.

Today, the function of groundcovers to a sustainable landscape has become equally as important as windbreaks and shade. We know that live plant groundcovers provide much more air-cooling, rainfall retention, and more sustenance for nature than mulches and hardscape. Nonnative turf grasses have been the norm for our groundcover choices, but outdoor play has become rare and such spaces are rarely utilized; rainfall is not well absorbed; and there are no flowers and fruits for pollinators or birds. The inputs of routine maintenance (mowing) with noise, air pollution, and consumption of fossil fuels plus the high cost of labor to maintain turf has made their reduction in the landscape of paramount importance. Lower maintenance groundcovers, as a replacement of turf, is a wise, sustainable choice.

As Green Infrastructure and Wildlife Support

The environmental movement has brought to light another important function of plants: "ecosystem services." What's that? It's what a plant can do in the landscape to provide green infrastructure—how a live plant can provide services that were once accomplished by engineering. Examples include green roofs to deter heat gain on or from buildings; rain gardens to alleviate storm water runoff and improve groundwater recharge; and stream or waterway stabilization through using appropriate plants rather than concrete, gabions of stone, and other such man-made structures.

Green structure also includes the neglected roles plants play as host, food, and nectar to the web of life, which includes us—directly so with human-edible plants. Plants are the beginning of the food chain, but insects that feed on them are the next link that supports so much of the life around us. With population crashes of honeybees and monarchs, for example, we're reminded how important these creatures are to us—whether for pollination for fruits we eat or simply for our sense of wonder. Native plants, as the original resource with which this whole web evolved, play the most important role here bar none.

Gardens comprised of plants purposefully grown to attract appropriate insects into landscapes, gardens, or farms are called "insectaries gardens." There are three main components to plants selected for an insectaries garden: to attract beneficial predatory insects to help control pests, to attract pollinators for pollination services, and as trap plants that lure pests away from other plants you wish to protect. Insectaries gardens are the mainstay of an organic garden because they enhance working with Mother Nature not against her.

Beautiful fall color makes *Acer saccharum* (sugar maple) one of the most beloved midwestern native trees. Loose Park, Kansas City, Missouri.

Aesthetics: The Art and Beauty of the Plant

A beautiful plant is often called an ornamental. Because of that, I will call a plant's beautiful characteristics its ornamental attributes. These traits can arise from the overall shape or form of a plant, its foliage, bark, and stems, or from its flowers and fruit.

PLANT FORM

The form of a plant is something depicted over the life of a plant. For woody plants that don't die back in winter, form is a year-round attribute that is present in the landscape in all seasons. Plant forms range from prostrate (hugging the ground) to vase-shaped (growing upward yet spreading and weeping outward towards the top). Use plants with striking forms as focal plants, but for the most part the form of a plant translates into its functional qualities.

FOLIAGE

The foliage of a plant is one character that is present year-round on evergreens, but otherwise an ornamental attribute of the growing season. As foliage is present to provide food for the plant to live and grow, it is green—the color of its chlorophyll to make that whole marvelous process of photosynthesis happen. There are unusual plants with various foliage colors from yellow to purple, but almost always these are clones of anomalies where other pigments in a plant's leaf may mask the green chlorophyll. Plants with white-variegated leaves also fall in this category of cloned anomalies that many gardeners covet but others loathe. Fall color of deciduous plants is a major attribute gardeners consider in selection.

Some plants have bluish or silvery leaf surfaces because of waxy coatings or silky hairs present to help control a plant's water loss or as protection against harsh sun. Often this characteristic is just on one side of the leaf, creating a bi-colored effect. Swamp white oak is a great example: the upper side of the leaf is dark green whereas the underside is whitish. This characteristic can be seen from afar as the breeze blows the leaves to show their undersides and the plant is aptly named in Latin *Quercus bicolor*.

The size of leaves is an ornamental characteristic as plants with smaller leaves are considered to have "fine" texture while plants with large leaves are considered "coarse." You can actually create quite a striking landscape simply by harmonizing and contrasting leaf textures of plants. Leaf sizes also affect how we perceive a landscape. Plants with tinier, fine-textured leaves actually make an object look bigger because of the added detail, but large-leaved plants do the opposite and can be used to make an overwhelming space or object appear smaller. The most delightful spaces utilize a full range of leaf textures.

The shape of leaves also can be ornamental as unusual

The heart-shaped leaf of eastern redbud (*Cercis canadensis*) adds interest to the landscape.

The stunning blocky bark of persimmon (*Diospyros virginiana*) resembles an alligator's hide and shows best in winter.

outlines attract attention or cater to a particular design. Sassafras with its various mitten-shaped leaves, tulip tree with its tulip flower-shaped leaves, or redbud with its heart-shaped leaves can be sited to take advantage of these shapes. Plants with finely divided or intricately lobed leaves or with small leaflets are also unique design-wise. These plants also can display the same fine texture qualities as those with small leaves.

BARK AND STEMS

Characteristics of bark and stems of plants are aesthetically pleasing and can be highly ornamental. Paper birch or sycamore (*Platanus occidentalis*), with its striking white exfoliating bark, is what first comes to mind with midwesterners when asked about ornamental bark. Bark varies from smooth and gray to ruggedly furrowed and almost black. Some shrubs have unique red or otherwise colorful stems epitomized by the red-osier dogwood (*Cornus sericea*).

FLOWERS

Flowers are the favorite ornamental attribute of plants and the foremost characteristic by which most plants are chosen for the landscape. Colorful spring bloom on small trees

and shrubs is most often the first request nursery professionals hear at their garden centers. Herbaceous perennials and annuals are usually chosen for this aspect, at which they do excel. The challenge is how to select and utilize the sequence of colorful blooms on various plants through the season from spring through fall. Such a landscape provides beauty through the entire growing season.

FRUIT

The greatest ornamental character of many plants is their fruit. I will include cones on conifers here, as this is not a botany textbook (technically only angiosperms [flowering plants] produce fruit). The best fruiting season for many plants is in the fall and into winter when the fruits really stand out. The brilliant red berries on American highbush-cranberry (*Viburnum trilobum*) are so showy in a snowy landscape and the fruit are colorfast through the worst arctic blast. Herbaceous plants are highly ignored for their fruit, but this is one of the finest attributes to consider for the wintertime landscape. The interesting seed heads of various coneflowers, St. John's worts, mints, and grasses are really something to look at and are especially pretty when adorned with snow.

The colorful berrylike cones of eastern red cedar (*Juniperus virginiana*) are used in flavoring.

COLOR

Our most important sense in the garden is sight, and no part of this is more enjoyable than experiencing color. Blue may be the most widely loved color, but we all have a personal favorite(s). I also always work with color in landscape in how it makes us feel. We often overlook that aspect of color, but by making thoughtful choices, you can really improve your well-being and enjoyment of a landscape.

Want to sit and relax after a hard day of work—what colors should you choose? The cooler colors green, blue, and lavender along with white and off-whites including pale yellow and pink. The white and whitened color show up well in evening light too. Want an outdoor spot to sit that energizes you? Vibrant reds, oranges, and golden yellows including vivid purples energize. How about a winter's day and you look outside and what do you want to see? The warm colors of fruits, berries, and twigs really stand out in the drabber season.

Color is the epitome of a well-designed landscape and can make or break a successful place. There is a lot to color theory and there are entire books written on the subject— too much detail for this book.

Armed with fierce spines on its trunk and stems, wild honeylocust (*Gleditsia triacanthos*) is clearly not to be touched carelessly.

FRAGRANCE

Nothing evokes memories stronger than scent. I can't imagine spring without the scent of wild plums blooming—bringing back memories of my grandparents and their farm. The smell of Kentucky coffeetree's (*Gymnocladus dioicus*) flowers reminds me of a pleasant scented floor cleaner my mom used when I was a child.

All too often we don't even bother to smell the flowers anymore and this is sure a sad state when it comes to native plants. What does a swamp milkweed (*Asclepias incarnata*) smell like? How about a native sunflower (*Helianthus* spp.)? If you don't know, you are missing something special. Add that to your bucket list of things to embrace next time you see them.

We usually select plants with fragrances purely for their pleasantries and enjoyment. Consider fragrant plants near a front door, but especially around an outdoor deck or patio where people spend time outdoors.

Most plant scents are based on their flowers, but some plant fragrances come from foliage. The scent of pine (*Pinus* spp.) is one of the first that comes to mind, but I cannot help myself to sample many native mints by rubbing or picking a sprig of their leaves.

Other favorites come from scraping a twig or snapping dead twigs, including lemony sassafras, colognelike spicebush (*Lindera benzoin*), and oil of wintergreen in yellow birch (*Betula alleghaniensis*). Never underestimate plant fragrances for a successful garden, and be sure to include several to inspire your everyday experiences in the home landscape.

TASTE

My grandparents lived in a landscape where edible plants in the farmyard and fencerows were cherished for their tasty treats—beyond the bread-and-butter field crops that earned a living for them as farmers. Many of our midwestern native

Listen for the audible snap when Kentucky wisteria (*Wisteria frutescens*) fruits pop open and fling their seeds afar.

When giving garden tours, the native plant I make sure everyone touches is prairie dock silphium (*Silphium terebinthinaceum*). Its somewhat raspy surface holds a cool layer of air next to its huge leaves, creating quite a cool sensation on a hot, sunny day. No wonder this plant defies logic that plants with big leaves don't occur in such full sun situations—it has its own air conditioning.

And what better way of engaging children (or the young at heart) in the garden by introducing them to our native jewelweeds "touch-me-nots" (*Impatiens capensis, I. pallida*)? Pow! That's what happens when you touch their ripe or nearly ripe fruit seedpods in late summer or fall: they explode, dispersing the seeds. Also note that the mucilaginous juice from their stems is the anecdote to the sting of a nettle.

I find myself always touching plants in my garden. You get to know your plants better that way—a way to distinguish a roughleaf dogwood seedling from a gray or flowering dogwood (*Cornus florida*), a way to release the aroma of a mint, a way to tell if a persimmon fruit is ripe. Don't forget to touch.

PLANT SOUNDS

The state tree of both Kansas and Nebraska is the stately cottonwood. Have you ever listened to a cottonwood? Its leaves flutter in the wind (or even the slightest breeze) across the Midwest and create a pleasant sound that is hard to describe but somewhat like "light applause." The cottonwood's cousin, the quaking aspen, is more widely known for this attribute and named after that aspect of the plant. Both are great choices for plants where you want to muffle sounds. Beyond these two poplars, it is hard to name plants for their wind-made sounds other than the sounds blowing through pines (especially eastern white pines) and various grasses.

Some plants in my yard are audible because of the insects they attract. I have noticed the little white urn-shaped flowers on my persimmon trees from across the front yard only because of the audible buzz of bees attracted to them.

Two noisy plants sound off only at the perfect instant and it makes me laugh just thinking about them: native Kentucky wisteria (*Wisteria frutescens*) and witchhazels (*Hamamelis vernalis, H. virginiana*). These produce fruits that dry on the plant: a pendant, polished brown pea pod on the wisteria and an urnlike fruit against the stem of the witchhazel. Both audibly snap when they reach certain dryness and fling their seeds far and wide. Both are studies in natural engineering and materials. Open your ears in your garden or on your next visit to gardens and wildlands and listen if any plant makes a sound you like.

plants have delicious fruits from berries to nuts, but lesser-known natives have delicious tubers, stems, or flowers (or nectar-rich flowers highly desired for honey).

I can taste the fresh spear of greenbrier (*Smilax* sp.) in spring, followed by native red mulberry (*Morus rubra*), blackberry, and black raspberry (*Rubus occidentalis*) in mid to late summer, pawpaw (*Asimina triloba*) in September, persimmon after frost, and hazelnut (*Corylus americana*) later in autumn. Who said a landscape can't be tasty?

TACTILE PLANTS

We live in a society so engrained with "look, don't touch" that I think the tactile experience of the garden has largely been lost. Yes, we have a few harmful-to-the-touch plants such as poison-ivy (*Toxicodendron* spp.) various stinging nettles, thorny mature greenbrier stems, and fallen leaves of American holly (*Ilex opaca*), but we also have a phenomenal array of plants that invite us to touch them and that will enrich your life if you just take the time to feel their extraordinary surfaces.

Design with Native Plants

What comes to mind at the mention of the terms *Japanese garden*, *English garden*, or *tropical resort garden*? We have pretty good ideas of design principles for those styles, but what makes a midwestern garden? It's certainly one that reflects our native landscape and native plants.

Prairie-style architects like Frank Lloyd Wright and E. Fay Jones and landscape architects like Jens Jensen, O. C. Simonds, and Alfred Caldwell probably best capture what would be called a midwestern style. It is profoundly sad that the prairie-style genre is not currently popular. Of these notables, only Wright is a household name and his buildings are popular attractions with a timeless quality. Incidentally, Mr. Wright would also have objected to the term style or any label, as he sought inspiration from the study of nature—certainly reflecting the native plants and our landscape.

Most homeowners and gardeners think in terms of style set by real estate standards of curb appeal. It has lead to homogeneous designs that create order but do not reflect the inherent beauty of individual places. Read Andy and Sally Wasowski's *The Landscaping Revolution* with a chapter "Homogenize Milk, Not Landscapes." Simply by utilizing more native plants, the styles in their book reflect the location and local style while maintaining a home landscape with mass-market appeal.

The focus of design in the present volume will outline the process to follow while selecting a style that suits the individual reader. Midwesterners are in large part a pragmatic bunch and maybe that is what partially defines a midwestern landscape design. I have watched how grasses have gradually become accepted in most landscapes and a more natural style has transpired from their use. People also are moving less often so there is more of a movement towards long-term landscape choices rather than instant gratification and resale value. Ultimately, a design, whatever its style, has to be meaningful to its homeowner, creator, or user. That search for meaning is why certain gardeners embrace a natural landscape style.

The Design Process

I earned bachelor's and master's degrees in landscape architecture and these three steps in the planting design process are forever ingrained in my head: inventory, analysis, and scheme.

INVENTORY

To create a successful planting design that utilizes native plants to their optimum, you must begin by doing an inventory of the site you are planning. Foremost is to identify the

Lauritzen Gardens in Omaha, Nebraska, showcases the use of native plants in a variety of landscape styles.

soils on the site. Are they wet, moist, or dry? Comprised of sand, loam, clay, or gravel? Is there bedrock and, if so, what kind of stone is it?

It's ridiculous to amend a site so significantly to change it for your desired plants. Instead, pick plants that will thrive under the existing soil conditions. When in doubt, check your local soil survey and/or get a soil test.

Next it's important to note the directions from your site. Think of R.E.M.'s song "Stand": "Stand in the place where you live / Now face north." It's critical to know north from south, east from west. Here in the Midwest, in the Northern Hemisphere, it is critical to know this because of the angle of the sun and where it casts cooler shadows, where it beats down most intensely (from the southwest), and how the sun changes through the seasons. The sun rises directly in the east on the vernal and autumnal equinoxes, but rises in the northeast in summer, southeast in winter.

The midwestern winds are also a factor, almost universally southwesterly in summer and from the northwest in winter. You even have to think about the aspect of the land. Is it flat? Is it sloping, and if so which direction? Note these variations if you have topography. Those who garden on the east sides of bluffs never see the sunset and have a much more sheltered site while those on the west sides of a hill may never see the sunrise and may get baked by the afternoon sun.

The inventory also should note all existing plants. This is a great clue in interpreting the site's conditions and an indicator of what plants will do well in there. Gardeners with a bare, blank slate can look nearby for such plant clues. Trees are your best indicator because they have been there a long time.

A good inventory must also map all utilities including overhead and underground wires, gas lines, water and sewer lines, and septic fields so that plantings are compatible with and allow access to such infrastructure, thus preventing future headaches. You also want to call 811 and get underground utilities flagged, a service provided in all regions to prevent homeowners and contractors from accidentally digging into them.

ANALYSIS

Once you've identified existing plants on or near your site, look into the conditions under which they thrive. What are you doing with that research? You are analyzing the situation. Do the same for the soils: analyze the conditions and note any unique anomalies. It is often best not to fix them but to work with them. Again, analyze which plants thrive in your soils. Don't forget compass directions and the sun and wind changes through each season. Think of where sunny and shady sites are, how to screen the southwestern sun from your home or outdoor seating space in summer. Note to block the cold northwesterly winds in winter.

Here's an example based on my own, most recent, experience of buying a home. Since it's in a rural setting, I looked at the county's soil survey and learned that I had a uniform droughty clay soil, shallow to limestone bedrock in places. I inventoried where there were exposed rocks. The ridge-top property sloped eastward, draining into a ravine to the southeast. I noted the directions and existing large trees to the southwest that cool the house in afternoon and shingle oaks with marcescent leaves (holding rather than shedding lifeless leaves in winter) to the northwest that make a nice windbreak. The north and east sides of the house reflect a more cool and moist microclimate, the west side of my house hotter and drier.

The existing trees were all second-growth species (shingle oak, honeylocust, black cherry, elms, hackberry) that are known to be very drought tolerant. Under the trees grew woodland delphinium (*Delphinium tricorne*), mayapple (*Podophyllum peltatum*), and round-leaved groundsel (*Packera obovata*) so I knew I would be able to grow woodland wildflowers that thrived in dry, upland woods. An open meadow area to the south had a few prairie plant remnants including low green and blunt-leaved milkweeds (*Asclepias viridiflora*, *A. amplexicaulis*)—indicators of well-drained, dry soils. That demonstrated that upland, dry prairie species would do well in that site. There were no trees or wildflowers that indicated moist or wet conditions.

The location of the site southeast of Kansas City places me in the Osage Plains ecoregion where plants from the Lower Midwest will thrive in winter hardiness zone 6 (USDA) and summer heat zone 7 (AHS). My droughty upland soils indicated that I should choose from a palette of plants from the oak-hickory woodlands and dry prairies to fit the horticultural restrictions of the site without extra water or soil amendments.

SCHEME

Putting together a basic scheme is the next step to the final design. As this book is about native plants, schemes of readers probably include things such as creating a landscape that celebrates their spirit of place; is ecologically sound and sustainable; has edible and medicinal plants; attracts birds and other wildlife; is insect-friendly, attracting bees, pollinators, butterflies, moths, so on; or is easily cared for yet beautiful and a sound investment.

With any scheme chosen, you have to think about the style that you are comfortable with and that suits your tastes. Do you embrace traditional landscapes of order, or do you like natural landscapes? If you want a more natural look, be sure to research the regulations or landscape ordinances of your neighborhood or community first. Be prepared to get a variance and discuss what you are doing with your neighbors. Think about what you are capable of maintaining or how you plan to maintain your landscape. Landscape maintenance is a critical part of your scheme, so review the landscape use recommendations in the plant profiles that describe a plant's behavior and suitability for a formal or natural landscape.

Design with native plants is meant to be a practical guide to all styles. Yes, there are native plants that can be used traditionally; already most trees and some shrubs are embraced and readily utilized in landscaping. Native evergreen shrubs, vines, perennials, groundcovers, bulbs, and annuals are much less understood and cultivated. There are many great examples from all plant types and some are quite popular now including our native wisteria as a vine, purple coneflowers (*Echinacea* spp.) and prairie dropseed (*Sporobolus heterolepis*) as perennials. In the plant descriptions, a plant's behavior in a garden setting is described under landscape use.

Traditional styles where plants are grown in an orderly fashion show the hand of humans over nature. Plants usually must stay put and be segregated, only groundcovers are allowed to spread and then only uniformly. Plants must be well behaved and under control. This is the typical style of suburbia where shade trees, select ornamental trees, foundation plantings, hedges, and a few perennials prevail, adorning a lush, turf grass lawn.

Natural styles embrace the hands of Mother Nature and allow plants to naturalize freely and behave as they would in the wild. This allows for complex relationships and mixes of species that look unkempt, untidy, or out of control to many tastes. To the trained ecologist's eye, there is pattern here best described as "disordered hyperuniformity." I read that term as describing the light receptors of a chicken's retina (order at large distances, disorder at shorter distances) and it immediately reminded me of how plants are arranged in a native prairie.

There are in-between styles, the average perennial border being a good example where relatively well-behaved plants are planted in groupings that create compositions that are usually synchronized for bloom through the seasons and in such a way so that the ornamental characters of the plant create artistic compositions based on color and texture. Piet Oudolf is a master of this design, and one can be inspired by his work at Chicago's Lurie Garden in Millennium Park.

Here's how I schemed my landscape: I read the property restrictions of my subdivision and, luckily, leaving woodland and natural landscapes is allowed. I wanted a more orderly look immediately around the house and a more natural setting filled with wildlife beyond—that was why I moved to this semirural site. I retained a sweep of lawn around the house for access to maintain the house, for access to utilities, and as a place to walk and observe wildlife and the garden's plants. Though the Midwest is not in a major wildfire zone, I did keep that in mind so the sweep of lawn also acted as a firebreak. I remember recent droughts and how areas of dried native vegetation can be explosive if on fire.

My scheme also set parameters of function and aesthetics for the plants I chose. I already mentioned shade for the house during the hottest part of the day and winter windbreak. I didn't have any particular visual screens I needed to make—be sure and look out all your windows, especially in winter and note what you see. I just worked to create beautiful edges to the existing woodland surrounding my home. I also knew my house created some more sheltered moist and shady microclimates on its north and east sides where I could grow some plant treasures I enjoy without wasting resources having to constantly baby them to thrive.

When it came to aesthetics, I played with color and season of color. I chose a palette of primarily orange-yellow gold flowers through the seasons for my west-facing front door space—a place I often sit on the front steps and watch the sunset, so why not choose those colors that play off that? Bright red and vibrant purple colors for the hot sunny, south side of the house are colors that invigorate. On the east side of my house, where I unwind and relax on the deck after work, calming green foliage and flowers of pale shades and white that show up at night are the scheme; it's also where I have my best spring-flowering plants.

Follow your own choice of aesthetic schemes to make your landscape fit your needs and help give it some parameters so that it is not all hodgepodge. Gardeners simply restoring a native landscape usually work aesthetically well too as the color palettes of our native plants through the season for different habitat types have their own splendor. White, golden yellow, and lavender purple are recurring color schemes of midwestern prairie plants.

Clean, defined edges to natural landscaping along with signage are helpful for neighbors accepting of only a formal landscape. I certified my landscape with the National Wildlife Federation, North American Butterfly Association, and Monarch Watch and have posted the signs they provided. Other local and national groups will do the same, including Wild Ones and the Xerces Society. You also can just post your own signs, but communicate what you are doing and what your natural landscape reflects.

As this book is also about gardening sustainability, the above process of design will ensure that you select plants workable for your site without added inputs. When the right plant is put in the right place, its need for fertilizer, watering, and pesticides is greatly reduced. A happy plant is more vigorous and resistant to weather calamities and pests.

Always embrace problematic site conditions. If you can't beat 'em, join 'em. A wet spot can become a wetland garden; a dry locale should embrace what does well in that habitat. Dense shade can usually support moss and ferns or other woodland wildflowers. Sandy sites have a whole suite of plants that thrive under those conditions. Make peace with what you are given, and a better, more sustainable design unique to you will ensue. Capture your spirit of place and bloom where you're planted. There are native plants that thrive in every niche.

Simplicity and Repetition

I find the most beautiful, impactful, and memorable designs repeat plants that do well on the site. I recall the wooded home landscape of friends Dan and Barbara Williams near Rockford, Illinois, where they utilize wild ginger (*Asarum canadense*) extensively as groundcover, Virginia creeper (*Parthenocissus quinquefolia*) as both vine and groundcover,

A row of eastern redbud (*Cercis canadensis*) trees creates a line of color along a street.

and pagoda dogwood (*Cornus alternifolia*) as an understory tree. These three selections tie the space together, woven with a fairly diverse ground layer of native wildflowers and a canopy of native trees from bur oak to hackberry and black cherry.

Sometimes it takes time to figure out just what really thrives on your particular site, and then you can divide these plants and transplant their seedlings or saplings to create quite a splash. In my garden, various violets (*Viola sororia, V. striata*), roundleaf groundsel, woodland phlox (*Phlox divaricata*), and redbud (*Cercis canadensis*) were behaving in a way that told me to repeat them in my young garden.

Popular Plant Uses and Choices

Well over 95 percent of homeowners do not want a natural style. Because of that I'll review typical plant use and selection criteria and make suggestions on how to choose a native plant for these situations: shade trees, evergreens, select ornamental trees, foundation plantings, hedges, flower border, containers, and lawn alternatives.

HOW TO PICK A SHADE TREE

Native shade trees are readily available at most nurseries. They are the largest and most long-lived plant in your design, so they should be carefully selected and located in the landscape. Make sure to follow local ordinances in the selection and placement or get a variance. Be mindful of overhead and underground utilities.

Common mistakes are planting trees in the middle of the lawn. Would you put a tall lamp in the middle of your living room? The best location for trees is where they will shade your home or outdoor living space from the hot afternoon sun. Plant them to frame your outdoor spaces and views. Street trees really help cool a community and their biomass helps mitigate the environmental impact of street pavement.

On very small properties, only one shade tree may be necessary or use small trees that fit the scale better. It is well documented that trees improve the value of a home both economically and ecologically.

HOW TO PLACE WINDBREAKS AND EVERGREEN TREES

In winter, fierce winds blow in the Midwest from the Northwest and Arctic. A windbreak that buffers a home from these winds can reduce the home's energy costs by 20–40 percent. The most important criteria are the height of the windbreak species and the density of its branches and/or evergreen foliage (55–85 percent is ideal).

A 20-foot-tall dense conifer will slow the wind speed down by around 75 percent within 100 feet, which is five times its height. Adding depth and layers to the planting, including a couple of rows of evergreens along with a row of dense shrubs, increases effectiveness.

Windbreaks work best on both northern and western boundaries around an open farmstead where they create a triangle of protection. They should be at least 75 feet beyond the home and extend past any structure on both their west and north boundaries in snowy regions. Former practices of woodlots that provided tree resources worked well, but with the high cost of prime farmland and commodity prices, windbreaks are reduced to a minimum anymore. Search your local extension service for ideal plans in your region; Iowa State University Extension, Pm-1716 is a great resource.

Windbreaks work on a smaller scale too, even in suburbia where space is limited. The use of evergreens and dense plantings to the north and west sides of homes makes a significant difference. Increased use of windbreak plantings offers the greatest reduction in energy use in the Midwest, a situation that is not getting any press in these times when energy conservation is paramount.

Using evergreens for ornamental and screening purposes works well within the framework of a windbreak, but why are they not recommended for planting on the south side of a home or structure? It's because there they cast winter shadows, blocking sunlight and passive solar heating. In such settings, ice can remain in the shadows cast by evergreens and indoor space will be gloomy even on a sunny winter day. Evergreens also simply look like dark shadows when backlit and the details of their needle texture and color as well as their cones are lost. Use smaller evergreens and evergreen shrubs for appropriate sized screens or seasonal ornamentals to the south.

HOW TO SITE ORNAMENTAL TREES

Many homeowners go to a nursery to buy a small tree with colorful spring flowers, foliage, bark, or fruit to stand out in their landscape. Well-placed ornamentals are an important design element to every landscape. Think how these trees will perform in all seasons beyond the flowers or whatever character for which you selected them. Those sited where you can enjoy them from indoors also adds to their appeal. Remember they are often focal points as your eye is drawn to their uniqueness.

A great site is often off to the side of the front door to create a comfortable entrance giving "human scale." They also work that way around outdoor living spaces and are used off corners of buildings to soften hard edges. You also can use them in place of shade trees where overhead wires would interfere with a larger tree.

WHAT TO USE FOR FOUNDATION PLANTINGS

Say the words *foundation plant* and gardeners usually envision needle evergreen junipers (*Juniperus* spp.) and yews (*Taxus* spp.) along with broadleaf evergreen boxwood (*Buxus* spp.) and holly (*Ilex* spp.). We feel our homes are naked without this ring of bushes around them: low or squat under windows, taller upright types at the corners or to frame a doorway. Most standard choices ultimately become large plants and require annual shearing to keep them in bounds. That can create some unusual shapes over time.

Do we really need foundation plantings? It's a matter of personal choice but do keep in mind future home maintenance and the ultimate size of the plants you choose. It's often wise to set them back from the foundation for maintenance reasons.

Consider dwarf selections of native conifers, which usually grow much slower than standard evergreens. They may cost a bit more up front but will save with lower maintenance in the long run. Deciduous shrubs, sturdy perennials, or groundcovers also should be considered. Be cautious when using native prairie grasses that can be highly flammable when they're dormant.

WHAT MAKES A GOOD HEDGE

Conventional hedge plantings are generally chosen for three attributes. First of all, they must be woody. Second, they need to take shearing "buzz cut" pruning. Lastly, they should create a dense mat of foliage or twigs. Exotic plants are the most prevalent choices including various privets (*Ligustrum* spp.), alpine currant (*Ribes alpinum*), and boxwood.

Think outside the box since who says hedges must be sheared and why can't they be herbaceous plants that hold up well all winter? (Jens Jensen hated manipulative pruning on plants.) Herbaceous switchgrass and its cultivars, shining bluestar (*Amsonia illustris*), and aromatic aster (*Symphyotrichum oblongifolium*) make an outstanding hedge that needs to be maintained only once a year: cut down in early spring. Shrubs including gray and rough-leaf dogwood, and arrowwood viburnums (*Viburnum dentatum*) and related species are also good native unsheared hedge candidates.

If you need that sheared look, consider a native plant. I have seen hedge trials at the Minnesota Landscape Arboretum with American arborvitae (*Thuja occidentalis*), eastern white pine (*Pinus strobus*), and even American fringetree (*Chionanthus virginicus*) looking beautiful. I discovered two I never thought of for such by accident thanks to deer browse shearing them: lanceleaf buckthorn (*Rhamnus lanceolata*) as a shrub and New England aster (*Symphyotrichum novaeangliae*) as a perennial. Lastly, uniform hedges may be your style but why not create a tapestry hedge that alternates or repeats various plant patterns?

HOW TO CREATE FLOWER AND MIXED FLOWER-SHRUB BORDERS

Create a successful flower border by following these basic selection criteria. First, it's a good plan if all the plants you choose can grow under the conditions of the bed, especially its soil, moisture, and sunlight conditions. Secondly, know the plants' ultimate size and lay them out accordingly, shorter species show better in the front with taller species, including shrubs, working well in the back. It's okay to have a few taller species forward as focal points to create some drama. Lastly, select plants that will fill a sequence of bloom or other ornamental attributes through all the seasons. Many spring flowers go dormant by summer so pair them with compatible plants that will fill in when they are absent. Actually listing on paper a matrix of the plants you choose with their seasonal interest helps visualize weaknesses. The plants are the rows and the season of bloom or ornamental interest are the columns—review that you chose plants of seasonal ornamental attributes from spring through winter and that these are laid out in a balanced manner across your planting design.

HOW TO USE CONTAINERS

Many of us live in apartments, condominiums, patio homes, or other places where our gardening is limited to containers. The use of native plants in containers is a perfectly great way to include them in any landscape.

Typical principles of container design can be followed, namely, including plants that thrill, fill, and spill. Choice of thriller plants varies with the size of the container. Maidenhair fern (*Adiantum pedatum*) is a thriller in a small container in shade while magnificent interrupted fern (*Osmunda claytoniana*) would work as such in a larger container where maidenhair would be stunning filler. Prairie grasses make excellent thrillers for sun, prairie dropseed a fine thriller in a small container, but switchgrass a knockout choice for a large container where dropseed would be good filler. Rose verbena makes the finest spiller for sun, as do wild strawberries (*Fragaria* spp.) in sun or shade.

Water plants can be shown to a tee in a container that is sealed or watertight. Place containers of water plants at the appropriate depth, putting a brick or block beneath them to adjust their placement. Make sure to utilize mosquito dunks, an organic, biological control for mosquitoes. If the container is large and if it is situated so it won't overheat, you can add tiny mosquito fish or guppies to it. I store my water garden container in the basement for the winter, simply removing most of the water and keeping the plants just alive until I put them back outdoors in spring.

Most native plants, even when grown in a container, are perfectly hardy left outdoors year-round. Verify that the container is made of such material that it will survive the freezes and thaws of winter.

HOW TO SELECT LAWN ALTERNATIVES

The status symbol of lawn has gotten out of control as probably the greatest waste of land and a significant threat to our watersheds because of pesticide misuse and excessive water runoff. Turf is our most important groundcover but should be limited to where it is utilized as for golf and other games and for light-use walking paths, where it is the best groundcover for such purposes.

Consider abandoning perfect monocultures of grass in favor of communities of plants that thrive under the mower. Where I grew up, virtually no one used lawn chemicals and we enjoyed playing outside on turf filled with dandelions, clovers, veronicas, and chickweeds. Today such lawns are flagged as "unmaintained" but they are much more healthy and I think they actually enhance play. Opt for lower maintenance turfs wherever available, and one day our native sedges will hopefully usurp turf grasses as the best lawns, especially the ones no one ever walks or plays on.

Choose other groundcovers or shrub masses as more sound alternatives to turf lawn. It is well documented that turf beneath trees is not good for them, so that's a great place to start reducing the lawn. I reduced my lawn by well over 50 percent. I created turf-free beds linking all the masses of trees while maintaining a sweep of lawn around the house. Doing this left me with a much smaller front yard lawn panel that I readily utilize for observing nature. I also retained lawn closer to the road to define the natural-style beds and better link with the neighbors. I use all the lawn I have as a place to walk without worrying about ticks and chiggers. I have virtually no lawn in my backyard, which is a heavily wooded and essentially woodland with paths maintained for access.

CONSTRUCTING RAIN GARDENS

Rain gardens are a smart way to mitigate your footprint on the landscape. Rooftops, drives, other hardscape, and even turf are usually impervious or nearly so, thus sending rainfall off-site and contributing to overloaded storm sewers, downstream flooding, and erosion. Precious water that

otherwise could be allowed to replenish groundwater and help quench your site is wasted.

Rain gardens are simply depressions that collect runoff and allow it to percolate into the soil; they are rarely wet more than a couple of days. They are relatively inexpensive to install and their size should be based on the size of rooftop or driveway from which they collect water. The "Rainscaping Guide" on the Missouri Botanical Garden website (missouribotanicalgarden.org/sustainability/sustainability/sustainable-living/at-home/rainscaping-guide.aspx) offers great guidance on how to construct a rain garden and gives other, wise landscaping options for utilizing rainfall.

Design Guidance for Specialty Use of Native Plants

Several garden styles lend themselves to incorporating native plants. Here we will look at prairie, woodland, water, and rock gardens, as well as edible landscapes and gardens that attract birds, butterflies, and other wildlife.

PRAIRIE GARDENS

No other planting scheme provides more midwestern appropriate, beautiful, and productive biomass to a landscape than planting a prairie. I use the terms *prairie reconstruction* when starting from scratch and *prairie restoration* when adding to remnant degraded prairies, meadow, or grassland with existing natives.

Prairie plantings provide exceptional food and cover for beneficial insects and wildlife. Virtually all prairie reconstructions comprise midsummer- to fall-blooming wildflowers, as the spring-blooming species are most costly to plant and challenging to establish. Prairie plantings also usually favor the larger warm-season grasses that are easier to establish. Smaller grasses like prairie dropseed, and cool-season grasses like Junegrass (*Koeleria cristata*) and Kalm's brome (*Bromus kalmii*) get ignored, along with the many wonderful sedges.

A prairie planting is the epitome of a natural landscape so must be planned and well thought out regarding how it works with local landscape ordinances. Its long-term maintenance must be carefully considered along with its flammability. In urban contexts, clean edges and signage are usually a necessity along with a landscape variance. Burning is often not allowed, so an annual or occasional cut will be necessary. A flail mower is an ideal tool for cutting a prairie and chopping it into lovely mulch.

At Linda Hall Library in Kansas City, Missouri, turf under shade trees has been replaced with low-maintenance Pennsylvania sedge (*Carex pennsylvanica*)

There are many ways to plant a prairie but almost all begin with a clean, weed-free plot in full sun. Prairies are most economical when started from seed, but using plant plugs can speed up the project. Select species for the planting mix that fit the site's soils whether sand, clay, or loam and wet, mesic or dry. Be careful with certain exuberant species: some may be integral but better added after more conservative species are established so there is competition to keep them in check.

There are many books and articles on prairie restoration and your local native plant nursery will have recommended methods best suited to particular regions. I strongly suggest a mix of plugs and seed for smaller projects starting with plugs or plants of some of the neglected spring wildflowers, smaller grasses, and sedges. Keeping weeds out through mowing high the first season or two also will help while the long-lived prairie plants develop their roots.

Sow seed of black-eyed Susan (*Rudbeckia hirta*), partridge pea (*Chamaecrista fasciculata*), common evening primrose (*Oenothera biennis*), and other short-lived, disturbance dependent annual or biennial plants in restorations. They provide early color and suppress weeds and will fade out as long-lived species become established. Leave ubiquitous daisy fleabane (*Erigeron annuus*) and frost aster (*Symphyotrichum pilosum*)—both usually emerge on their own—as they also provide early color and weed suppression. These plants bloom quickly, proving to the uniformed that you didn't just plant a patch of weeds.

I find the most meaningful prairie reconstructions include only native plants found in the local area rather than from a region-wide seed mix. Regarding restorations, botanist Mark Leoschke reminded me that he does not support adding species to native prairies (or forest), even if they are somewhat disturbed. To him, it makes the site less natural, as opposed to managing a site (removing invasive and exotic plants, burning) to enhance the existing native plants. Mark wrote, "There are so few natural areas left in states like Iowa so why alter them by adding species?"

I concur as registered native remnant prairies at Powell Gardens are managed by brush removal, exotic species

The Winnebago County Forest Preserve District's headquarters landscape in Rockford, Illinois, contains a prairie reconstruction as its core.

control, and burning, but no plant may be added as per Powell Gardens' collection policy. It's part of the historic legacy of the 970-acre site. In my personal 1/3-acre meadow with remnant prairie species, I do add species because it was so degraded and its context in a subdivision makes it essentially a garden.

WOODLAND GARDENS

Our beloved spring ephemeral wildflowers, spring-flowering trees, and plants of dazzling fall colors define most gardener's love for woodland gardens. They are a no-brainer for wooded landscapes. Just be sure to pick plants compatible with the types of trees and the environment you have.

Some gardeners in new, treeless landscapes want woodland gardens and that is not a problem either: start on the shaded north or east side of your home and plant trees that will eventually allow you to expand your woodland plantings. Most trees grow surprisingly fast when well sited and cared for, so refrain from choosing short-term, quick trees. I always recommend oaks and hickories.

Oaks are more readily available at nurseries, but hickories can be grown from nuts that you collect. Hickories grow well in the shade, taking their time and being quite beautiful from seedling to sapling stage with attractive buds and tropicalesque foliage. Some highly desirable small woodland trees such as redbud and pagoda dogwood grow really fast and jumpstart your woodland garden.

One of the beauties of a woodland garden is that every type of plant should be used to create the most complete picture: canopy trees, understory trees, woodland shrubs, perennials, and so on. Vines must be carefully matched with young trees so they don't smother their host and are best added to existing, more mature trees or against buildings or structures. Use fallen leaves for natural mulch, but don't let it get too thick until perennials are established. Surprisingly, most woodland wildflowers need bare earth to germinate so you can rake away select spots or let the turkeys scratch through the leaf litter to open up new seedbeds.

Woodland gardens always have a focus on spring, but the other seasons must be included as well. Ferns make striking foliage that replaces spring ephemerals after they go dormant, but there are some wonderful summer-blooming wildflowers from biennial American bellflower (*Campanulastrum americana*) to perennial Culver's root and starry campion (*Silene stellata*).

Fall in the woodland garden is often exuberant with various shade-loving asters, goldenrods, and bonesets that create an outstandingly colorful display with abundant nectar,

This woodland garden thrived in the shade of my former home, which was newly built and had no trees.

pollen, and seeds for winter songbirds. Many woodland plants have exquisite berries in late summer and autumn too, namely, blue berries of blue cohosh (*Caulophyllum thalictroides*) and Solomon's seal (*Polygonatum* spp.), red and white baneberries (*Actaea* spp.), and red fruit on false Solomon's seal (*Maianthemum racemosum*) and Jack-in-the-pulpit (*Arisaema triphyllum*).

Winter in the woodland garden should highlight some of the fine evergreens including the stunningly patterned leaves of hepatica (*Anemone acutiloba* and *A. americana*), Christmas fern (*Polystichum acrostichoides*), groundsels (*Packera* spp.), heucheras (*Heuchera* spp.), and foamflowers (*Tiarella cordifolia*).

WATER GARDENS

Artist Claude Monet popularized the water garden for eternity and rightfully so. Water plants are the showiest midwestern wildflowers. The American lotus (*Nelumbo lutea*) is by far our most spectacular flower, but it's too invasive for a garden. Many other lovely grasses and related plants along with gorgeous wildflowers require wet feet. The sound of water, its exquisite reflective properties, its cooling nature, and the unique creatures it attracts make it a very popular garden type.

Water gardens are not rain gardens and must be sited out of major rainwater drainage or they will quickly fill in with silt and debris. They obviously must be lined with soils that hold water or with a waterproof liner to be effective. Circulating water is needed for aeration and filters are required to maintain water clarity, so gardens do not become cesspools. New construction styles pump water from below the pool so that it filters through gravel for purification and does not get clogged with leaves or other debris.

Design a water garden with various water depths for all sorts of plants from those that like wet feet to those that grow in shallow water to those like waterlilies (*Nymphaea* spp.) that require at least 18-inch deep water. Water plants can be containerized and moved about, overwintered in unheated pools in the deepest parts where the water doesn't freeze.

Water gardens with waterfalls, bubblers, and heaters that provide fresh water to wildlife and birds through all seasons are a naturalist's delight. There is no finer way to observe many of our colorful migrant songbirds, especially warblers which relish drinking and bathing in such features.

Water is also a requirement for the life cycle of garden frogs and toads, whose peeps and trills are so welcome after a long winter. Pools without fish or with shelves or nooks that fish cannot get to are best for these amphibians. Dragonflies and damselflies are another group of dynamic insects gaining in popularity that require water for their eggs and juvenile life stages while they become wonderful predators of nuisance insects. Adult dragonflies and damselflies eat many nuisance insects in the air.

ROCK GARDENS

Rock gardens are ideal places to cultivate drought-tolerant plants that require well-drained soils and grow wild on rock outcrops, cliffs, glades, talus slopes, and glacial deposits. They were a garden fad in the 1920s and are now seeing resurgence in popularity in the new millennium.

Any site with natural rock outcroppings is a place to start, but rock gardens are a good choice where steep changes of grade occur and can actually be living walls where stones are dry stacked with a soil mix between them. Drainage is essential for all rock gardens, though some species thrive in wet scree or gravel.

When constructing a rock garden it is most appropriate to use local stones set in such a manner as they would appear naturally. Soil mixes between the rocks should be equal parts local topsoil, gravel, and compost. A gravel mulch can be applied to give a clean and neat look and to keep plants from rotting in wet weather.

Rock gardens that also function as retaining walls must have a sturdy footing and good drainage (be backed by coarse gravel and drainage tile) to drain excess moisture and to not frost heave in winter. Each layer of stone on a living wall should be stepped back about ½ inch for buttressed support. The same soil mix should be placed between stone layers as recommended for any rock garden. An engineer should approve any wall taller than 4 feet to ensure that it is stable in the long term. Planting a living wall should be done in spring so the plants have a chance to root in before summer's heat and to be well rooted so they aren't frost-heaved out during the winter. A mix of clay and sphagnum moss can be pressed around new plants to hold them in place as they establish.

EDIBLE LANDSCAPING

Rosalind Creasy made popular the term *edible landscape*, and it has undergone resurgence with the sustainability and foodie movements. Edible landscaping simply means utilizing food plants just like one would ornamentals in a functional and aesthetically pleasing way, but the benefits are the delicious treats produced by the plantings. Gardeners and nongardeners alike are embracing edible plants as a trend that has regained popularity and it's translating into edible landscaping. Yes, there are native plants of all types from shade trees to groundcovers that are edible and should be valued for that resource.

Pecan (*Carya illinoinensis*), blueberries (*Vaccinium* spp.), and native blackberry hybrids (*Rubus* spp.) are commercial successes. Other edible species are just as delicious but don't meet shipping or shelf life criteria so are rarely found beyond local growers' and farmers' markets.

Incorporating native edibles in the landscape definitely adds to a sustainable garden that allows one to celebrate and savor the bounty of the seasons. In my 3-acre garden, I have edible shade trees from black walnut (*Juglans nigra*) to black cherries (*Prunus serotina*) and persimmon; small trees of red mulberry, serviceberry (*Amelanchier* spp.), and pawpaw; shrubs including clove currant (*Ribes odoratum*), Missouri gooseberry (*Ribes missouriense*), and many brambles; vines of wild grapes (*Vitis* spp.), and greenbrier spears; and perennials that include wild strawberries. Do I enjoy the food off these plants? You bet.

A collection of native food-yielding plants grown in a natural woodland manner is termed a *food forest*. Such a garden is both productive and ecologically sound. It's a good way to use many of the native edible plants that don't conform well to recommended planting and care for prime production.

Wild black raspberry plants, for instance, are too disease prone to be cultivated in a formal, trained bed. When allowed to run through open woods or woodland edge, some plants may suffer from native rusts and other maladies, but there are always some that produce fruit to enjoy. In the plant profile sections of this book, edible landscaping information is provided in the landscape use paragraph.

GARDENING FOR BIRDS

My home landscape's bird list is nearing 180 species, I love having birds around and their song and presence are a visceral component of a garden. But what does it take to really have a bird-rich landscape? Food, water and cover.

Bird food comes in a bag right? Yes, I feed the birds all year, from squirrel-proof feeders hung in ways the raccoons can't get to them. But what does it really take to raise most songbirds? Watch any nest and look at what the parents are bringing their young. It's almost exclusively protein-rich invertebrates, especially spiders, and caterpillars of butterflies and in particular our underappreciated and abundant moths.

Marketers tout plants with berries for the birds, but plants that host the most caterpillars are the best choices for a bird-rich garden. Berries and seeds are utilized after nesting season for sustenance. Doug Tallamy has spread this word well with his nationwide lectures and his book *Bringing Nature Home*. His website (bringingnaturehome. net) provides a list of plants and the number of species of Lepidoptera that they host which is a good place to start. Native trees excel in this category and now we know why there are no chickadees in the urban deserts of Norway maples (*Acer platanoides*), London plane (*Platanus* ×*acerifolia*), Japanese zelkovas (*Zelkova serrata*), ginkgoes (*Ginkgo biloba*), Callery "Bradford" pears (*Pyrus calleryana*), and other plants that host nigh a single caterpillar for young birds to eat.

The mast and fruit of plants are still important after a brood of birds is raised so this cannot entirely be abandoned. Acorns are at the top of the list along with native cherries, dogwood berries, red cedar berrylike cones, poison-ivy berries, and the seeds in pinecones and birch fruits. Provide a diversity of plants with fruit that ripens through the season from early summer through fall, along with those that persist into winter, and you will provide the most bird-friendly place.

GARDENING FOR BUTTERFLIES

Where do butterflies go in winter? It's surprising how few gardeners think about that and it's the cornerstone of sound gardening for Lepidoptera.

The International Crane Foundation in Sauk County, Wisconsin, may be the ultimate bird garden with restoration of the native plants there as a metaphor for the work they do to preserve and protect the world's 15 species of cranes. Whooping cranes once nested in wetlands across the Upper Midwest.

A cryptic, wild cherry sphinx caterpillar feeds on a wild plum and only feeds on wild cherries and plums (*Prunus* spp.).

Grapes (*Vitis* spp.) are a host for pandorus sphinx caterpillar, which in turn provides protein-rich food for young birds.

Each species has a particular way of surviving winter beginning with several that overwinter as adult butterflies. Yes, even in the frigid north, several species survive sheltered in cracks and crevasses of wood, in an outbuilding but probably not in your butterfly house purchased for garden décor. Many species survive the winter as a chrysalis hanging from a stalk, branch, fence, or just about any obscure support or as with the skippers, their pupae snuggled in a nest of dead vegetation. Still many survive the winter as a caterpillar; two species create little sleeping bags of remnant leaves and tether the leaf stalk (petiole) to the plant with their silk while more just overwinter in the presence of fallen leaves. Finally, some hairstreaks in particular overwinter as eggs, poised to emerge and feed on young growth. Surprisingly, almost a third of midwestern butterfly species are colonists and migrants and do not survive the winter here.

What this all means is that garden maintenance underlies the most butterfly friendly landscape. Rake up all the leaves, cart off all the dead stems, clean up every brush pile, and you will inadvertently kill a lot of butterflies. It also points out that butterflies have a complete metamorphosis and all three other stages are needed for the butterfly. You may apply pesticides thinking you are not hitting a butterfly, but are you impacting its unseen eggs, caterpillar, or chrysalis?

Most butterflies need a specific set of related hosts or a single host plant for their caterpillars to eat and grow and make more butterflies, so provide the host plant and the butterflies will find you and colonize. Adult butterflies need nectar or other sustenance to survive, so planting nectar-rich flowers or providing a butterfly feeder filled with spoiled fruit does the trick. I've had well over 70 species of butterflies in my yard, and there are times when almost 1000 are present at one time. If you plant and maintain properly, butterflies will come and certainly enrich the experience of your landscape.

Moths are even more important to a healthy garden's web of life, but they don't get the respect they deserve compared to butterflies. Moths require the same gardening plant selection, care, and maintenance considerations as butterflies. Moths also have a complete metamorphosis from egg to adult though they form a pupa, often protected by a silken cocoon. There are more than 20 times the species of moths as there are for butterflies in the Midwest.

NUISANCE WILDLIFE

Several creatures have names that are four-letter words and create challenges to gardeners, namely, deer, vole, and mole.

Squirrels, rabbits, raccoons, woodchucks (groundhogs), various gophers and chipmunks are often problematic wildlife too and there is no easy solution to these creatures. Healthy gardens are simply habitat for native wildlife and each owner will have to assess their own threshold of tolerance to creatures that damage prized plants.

I do not include a list of deer-resistant plants in this book, for example, because a hungry deer will eat almost anything. I refuse to make plant choices based on these omnipresent and often overpopulated though beautiful creatures. I have fenced them out of part of my landscape and routinely use deterrent sprays elsewhere. I always support professional wildlife managers' decisions on their local control.

Historical records refer to the abundance of deer in the Midwest, and Lewis and Clark's journey supports that. I often hear people say we have built into their (deer) home, but that is not the real problem as deer have adapted well to our suburban landscapes. These new "living lawn ornament" deer are not fit from fear of being food for a wolf or cougar so they are not on the move. They don't just browse a bit here and there and move on, they continually browse the same location to the point that their favorite floral salads are exterminated.

The overpopulation of deer has now become a top threat to our native plants and the creatures that rely on them—overtaking the bulldozer since the Great Recession, but overtaken by the recent expansion of agriculture into conservation reserve areas. I have witnessed this firsthand on my own acreage. When I bought the property, it was a deer park, young oak trees all trimmed, not one redbud seedling, no violets, every aster eaten off before bloom. Now, sapling oaks have a chance to grow, violets have returned along with the fritillary butterflies, asters provide nectar to native pollinators in late summer and fall. My lone huge native redbud now has seedlings. This landscape rebirth is a result of fencing and deterring deer, though they are present nearly every day.

Maintenance Tips: Weeding and Mulching

The most-time consuming weeds in most gardens are tree seedlings. Squirrels plant oaks, hickories, and walnuts. Nature often produces an abundance of elms, maples, and ash to the point their seedlings can germinate almost in a carpet. These activities demonstrate the conundrum of planting the right tree in the right place (or any plant for that matter).

With ash and some maples and other dioecious trees, a solution has been to plant all males. It is something to consider on a site specific, local basis, but this strategy has shown it will haunt us when all males are widely utilized as street trees: we end up with copious pollen allergens. Removing tree seedlings (and other weeds) is just a garden chore that promotes physical fitness. I want my trees to be productive as their fecundity is what creates such a rich web of life around me.

Most of our native perennials also readily self-sow where they are grown in ideal conditions. It's what nature intended. That is why many native landscapes are a lot of work to maintain if you don't want them to look natural. One option is to "deadhead" the plants and remove the seed heads, but that often ruins a plant's beauty, its winter interest, and its value to wildlife. Another option is plant perennials that don't produce copious seedlings. Gardens like Chicago's Lurie Garden use only select native plants. It's a pragmatic approach for such a garden where a controlled look is needed to carry out the landscape art of the designer and cater to the demands of the public in such a public space. Towards that end, I have included a list of the top native plants for formal landscapes in each plant category of the plant profile section of this book.

Bare ground and light are what it takes to germinate most seeds, so a thick cover of plants with their natural duff and leaf litter really helps deter weeds including many tree seedlings. I recommend mulch on new plantings where it helps hold moisture and deter weeds in bare soil until plantings are established. I am a member of what I tongue-in-cheek call "mulch gardeners anonymous" and prefer to show plants, not mulch. Many homeowners still prefer plants tidily set in bare mulch as a very popular style, especially in traditional landscapes.

Mow your lawn with a mulching mower to efficiently recycle clippings. Mowers can be used to make passes over fallen leaves and grind them up in place rather than bagging them up and sending them to the landfill. Mulching and flail mowers also can be used on herbaceous plant borders; consult Roy Diblik's book *The Know Maintenance Landscape* for further tips.

I cart nothing off my own landscape, utilizing the mulching mower on excess leaves, chopping perennials in place, breaking up sticks. When disease- and storm-damaged trees had to be removed around my home, I had them chipped on site by the arborists to utilize the material for paths or mulch. Note that certain wood infected with verticillium wilt, for example, cannot be used that way or you threaten to inoculate and infect new territory. EAB infested ash can be safely chipped and utilized.

Guide to the Plant Profiles

The plant profiles are arranged in groups based on type of plant because a gardener or designer picks plants by how they use those plants in the landscape. Plant type—whether it is a large shade tree, evergreen, small tree, shrub, vine, perennial, groundcover, and so forth––and its use in the landscape are most important. The most beautiful and successful landscapes utilizing a full range of plants from shade trees to annuals.

The plant type groups are organized into what has been called the planting pyramid: shade trees are on top and annuals at the bottom. It's a great hierarchy to work from because the most critical choice, shade trees, is on top. Shade trees are your biggest investment and take the most time to grow into their function in the landscape. You also should need the fewest of them because of their large size. They are the biggest commitment that impacts all other choices, so are the wisest investments and should be carefully chosen. Annuals are at the bottom because they are relatively cheap and changeable with a short-term impact on your landscape. Within each group, plants are organized alphabetically by scientific name.

SHADE TREES are deciduous trees that grow at least 50 feet tall at maturity and are the largest, dominant plants in the Midwest. It is critical to separate them from evergreen trees because they are used so differently in the landscape and each group plays a significant role. A shade tree's role in cooling a large space is vital in a sustainable landscape. The tree's shading power greatly reduces cooling costs when it blocks the summer sun from a house, but overall reduces summertime temperatures beneath its lofty crown, creating human-comfortable landscapes. In the winter, the tree's bare twigs allow warming rays of sun and translate into passive solar heating. Oaks are the foremost shade trees in the Midwest, followed by hickories and maples.

EVERGREEN TREES follow shade trees in the region. Unlike in the Pacific Northwest or other areas where evergreen trees are dominant, in the Midwest, they play a secondary role but are nonetheless important. Evergreen trees grow at least 15 feet tall (tree-sized) and hold their foliage through all seasons. They play a critical role as windbreaks in the Midwest and also provide screening to unsightly views in all seasons. Their year-round foliage is a pleasure to look at through the bare winter months when they provide important shelter for many species of wildlife. Red cedar is the only universally native evergreen tree in the Midwest and thus the most prominent in the region, but other native species are found in various localities.

SMALL TREES AND LARGE SHRUBS are the third group and also play an important role in the landscape as they include the "ornamental" and "flowering trees." These also are deciduous plants and are simply smaller versions of shade trees. They are often the trees and large shrubs found in the understory of woodlands, while many naturally grow on the edges of woodlands and along fence lines. These plants create "human scale" to a landscape—more in size with people, creating a very comfortable landscape similar to the walls and ceiling of a home. They grow at least 12 feet tall but some can reach 30 feet or more at maturity. These plants are often multitrunked, blurring the line between tree and shrub. In the Midwest, the rose family is the most important group of small trees and large shrubs.

EVERGREEN SHRUBS are next with a very different role from other shrubs. They block views year-round, create windbreaks on a smaller scale, and add greenery in all seasons for continuity in a landscape. Their favorite use in the Midwest and nationwide is as foundation plantings. They range from dwarf bushes to large multitrunked plants, but do not reach tree size. Native evergreen shrubs are very rare in the Midwest yet important to a well-designed landscape. The plant profiles in the evergreen shrub chapter include dwarf conifers, that is, clones of naturally dwarf forms, sports, or brooms regularly found in native evergreen trees. Apart from these clones, there are few native choices. Besides, dwarf conifers are a much wiser choice than most typical foundation plants found at a nursery because they will need little pruning and not the annual shearing required of most yews, junipers, boxwoods, and hollies.

DECIDUOUS SHRUBS follow as woody perennials that grow 1–8 feet tall. Because they do not die back to the ground each year, they are integral to provide year-round low baffles and hedges and as backdrops to herbaceous plants. Shrubs should be selected for seasonal appeal beginning with what they look like as bare stems in winter. Summer foliage is their next best attribute, while seasonal spring and fall foliage colors, along with bloom and fruiting, make them stand out at various times of the year as an important, engaging landscape element.

VINES are next as they are the chameleons, so to speak, of the plant world. What other plant type can make a blank, human-made vertical wall come alive? Vines can fit into narrow spaces or cover structures with greenery, seasonal flowers, and/or fruit. It is important to consider how vines climb. Do they simply sprawl or twine around objects, or do they have tendrils or leaf stems that wrap around any support? Some vines have rootlets that grow from the stem to help them stick to vertical surfaces while others even have adhesive disks. Be sure that how a vine climbs is compatible with whatever structure you plan for it to grow on. Vines also can be paired with other larger plants, but confirm that the vine will not overwhelm and smother its host.

PERENNIALS are herbaceous plants that return in subsequent growing seasons; in other words, they do not have woody stems so die back to the ground each winter but return with new stems for several to many years. The lifespan of perennials varies from species to species, some being rather short-lived while others may live for as long as a human or longer. They are usually dynamic and important plantings in a landscape because many have very showy flowers or dramatic seasonal foliage. What they look like, as a dormant plant with dead winter stems and seed heads, always should be considered.

Orchestrating perennials for beauty of flower, foliage, or fruit from spring through fall is the challenge of designing with perennials. Most do not bloom for more than a couple of weeks and each has its own bloom time. Spring ephemerals are beautiful in early spring but go dormant by midsummer, while other species don't bloom until fall. Many have exquisite winter stalks and seed heads that should be left for winter interest in the landscape. Perennials are by far the largest plant type group in the region, and the descriptions have been sorted into three chapters based on the main habitat origins: prairie, forest, and wetland.

GROUNDCOVERS are usually low shrubs, perennials, or vines that readily spread via rhizomes, stolons, or prostrate stems to cover the ground over a relatively large area. The most desired groundcovers spread quickly and to such a degree that they deter other plants from seeding in and growing among them, though larger plants and appropriate bulbs can be planted among groundcovers. None of our native groundcovers are as smothering and invasive as some popular exotic plants such as English ivy (*Hedera helix*) and wintercreeper (*Euonymus fortunei*), both evergreen vines; periwinkle (*Vinca minor*) and Japanese pachysandra (*Pachysandra terminalis*), both evergreen subshrubs; lilyturfs (*Liriope muscari*, *L. spicata*), evergreen perennials; or bigleaf periwinkle (*Vinca major*), a deciduous perennial. Most native groundcovers are best utilized as a living tapestry of several compatible species for best effect as few are the invasive thugs like the aforementioned exotic groundcovers readily available.

BULBS are perennial plants with underground swollen stems or roots that are usually bought and transplanted bare root. I use the term *bulb* beyond the botanical definition to include true bulbs, corms, and rhizomes based on how a plant is selected and installed in the landscape. Typical exotic bulbs include tulips (*Tulipa* spp.), daffodils (*Narcissus* spp.), and hyacinths (*Hyacinthus* spp.), while native bulbs are not produced in such commercial quantities, are far more expensive, and not as readily available. They tuck nicely in compatible groundcovers or between perennials and shrubs and thus are almost at the bottom of the planting hierarchy. Many are spring ephemeral plants that emerge early, bloom in springtime then quickly set seed and go dormant (wither and disappear) by summer. Because of that landscape behavior, they are best planted in the fall as companion to other plants that will fill the void during the growing season when a particular bulb is dormant.

ANNUALS are plants that grow, flower, set seed, and then die in one growing season. Biennials are similar plants that do not have flowers for their first growing season (usually just a rosette of basal leaves), then bloom, set seed, and die the next. Together annuals and biennials are colorful plants, many with longer bloom times than perennials. Most require disturbed soil to germinate, so they can sometimes be challenging to have as a sustaining planting in a landscape without continual replanting. They can provide instant gratification and are relatively inexpensive so are at the bottom of the planting pyramid.

of a plant in the region is given, but keep in mind there are many variables from genetics to soils that affect a plant's ultimate size.

Lastly the third paragraph, "Ornamental Attributes," describes the beauty of the plant's flowers, foliage, fragrance, or other aesthetic characteristics that make it a beautiful ornamental in a garden or landscape.

I include a story about my experience with each plant and how the plant is important in our lives. I have personally grown or widely observed all the plants in this book. I always think a story about experiences with a plant help bring it to life and make for a more enjoyable read. I know that my experiences will not agree with everyone, but my agenda is simply to educate gardeners and landscapers about our native plants and to inspire their use to create a landscape more relevant to our place.

The Most Garden- and Landscape-Worthy Midwestern Native Plants

The plant profile chapters present descriptions and photographs of the most garden- and landscape-worthy midwestern native plants. More than 6000 plants are native to the Midwest, and this reference is not a field guide for their identification but a compilation of the best native plants for gardening and landscaping. The selection features plants that are fairly readily available at nurseries and have proven performance in a cultivated landscape. Lists of the best dozen or so plants for a formal landscape are provided for each plant type, and there are notes for each of the three main regions of the Midwest (Upper, Lower, and Eastern). Some of the described species are often inherited wild on a property though shunned in gardens or are generally not cultivated by nurseries. I also mention some "garden thugs" that play a critical role in the midwestern landscape as a way to think about how to utilize them wisely.

The plant profiles describe the main horticultural, functional, and aesthetic characteristics of each species. The first paragraph, "How to Grow," covers the horticultural aspects of a plant: its habitat, range, cold and heat hardiness, adaptability to Midwest gardens, and propagation notes. For more info on seed germination, see Prairie Moon Nursery's website: prairiemoon.com/How-to-Germinate-Native-Seeds.html.

The second paragraph, "Landscape Use," suggests where a plant should be utilized in a garden so that it performs its best function in the landscape. Garden styles for which the plant is most suited are mentioned. The general size

Challenging Qualities of Native Plants

Plant profiles include how to (and not to) use a garden thug (every plant has a place!). I find it hypocritical that our favorite exotic groundcovers are accepted for their invasiveness but when it comes to a true native, the same behavior gives it the label of "garden thug." We simply don't utilize our native thugs to our advantage. Common sense says to use these plants where their "take over" spirit can be harnessed to advantage: filling in a problematic area, stopping erosion or slowing runoff, making a quick and dramatic mass, and so on. I have relegated most of these plants to the groundcover section (many are also vines) but will always make note of a plant's behavior under "How to Grow" and "Landscape Use." Some species, when planted in less-than-ideal situations, are controlled by their own struggles in a harsh environment.

How does one embrace the usually robust size and fecundity of our native plants? The tidy, dwarf, evergreen, sterile look is in. I've heard the terms *evergreen, ever-dwarf, ever-fruitless,* or *ever-dull* used to describe the traditional landscape. The Midwest has the richest soils on earth while our climate is quite benevolent and plant friendly between the extremes. We're not in a desert thirsting for water or in an alpine region with a two-month growing season so that everything blooms at once and all the plants grow low for warmth and protection. The tidy, compact plant looks are "in" and desired. Jens Jensen commented about this issue as his clients or constituents often remarked that native plants were coarse. He responded by saying how humiliating it was

Widely planted in times past for its abundant shade, cucumbertree (*Magnolia acumintata*) is strong enough to resist damage by ice and snow.

to speak that way of the plants with which the Great Master decorated this land.

I embrace the bounty of my native plants because I garden for wildlife. I will admit to doing added maintenance to alter some of my favorite plants in formal settings, namely, rattlesnake master (*Eryngium yuccifolium*) and silphium sunflower (*Helianthus silphioides*). I love both those plants for their elegant and striking forms and they are in my front door landscape for that reason. I allow them to grow and bloom; the rattlesnake masters have had such clouds of summer azures (a butterfly) nectaring on them that even a repairman took notice. But when seeds are ripe, I cut the plants to their basal foliage. Do I throw them in the compost? Not a chance, but I do tuck them in my native meadow and woodland edges where I don't mind seedlings and actually want more. They are too seedy for growing by my front door. Before I started cutting them back after flowering, I had hundreds of seedlings.

I will note a plant's seeding behavior in the plant profile, but I must give the caveat that soils and local conditions make all the difference in the world. Cup plant silphium (*Silphium perfoliatum*) can be a self-sowing thug in rich soil,

but in my droughty ridge-top garden I am happy if the plant even flowers every year. And talk about birdseed, silphiums trump sunflowers. Every time you plant a sterile hybrid remember that you are robbing the bounty of our land from other creatures that rely on small seeds or fruit.

Rare Natives with Great Garden Attributes

In keeping with the celebration of our spirit of place, the plant descriptions cover mainly native plants widespread and iconic to the region. Several plants rare in the Midwest (whether just naturally isolated or peripheral) also are included because when they are cultivated, they grow so extraordinarily well in our region. Northern catalpa (*Catalpa speciosa*), yellowwood (*Cladrastis kentuckea*), and cucumbertree (*Magnolia acuminata*) are great examples. All become extraordinary trees across nearly the entire Midwest and reaching their most magnificent size and stature. Midwestern icons indeed, these trees fit the whole region's web of life and are utilized across the board by local insects from bees to beetles. The catalpa hosts its special moth, the catalpa sphinx, which follows the plant wherever it's planted. The caterpillars of eastern tiger swallowtail feast on the cucumbertree's foliage, while a yellowwood draped in its white bloom is abuzz with bumblebees.

Cultivars, Selections, Strains, and Ecotypes

Everyone reading this book has a different goal in how and why they want to grow a native plant. Some will be more focused on restoring what once was, while others may be more concerned with aesthetics. If your goal is to restore a native habitat, then use locally sourced plants from your ecoregion that will have the genetics closest to what once was in your particular site. Native plants absent from your region of the Midwest should not be included.

Others may want a beautiful garden, utilizing a certain selection of a native plant for a particular reason. That choice may be a variety that is more adapted to their site's cultivation, has showier flowers or other unique ornamental/edible attribute, or simply adds diversity to their landscape. In my garden, there is a combination: more formal and aesthetic selections around my home, but local strains in the woodland and prairie restorations.

Whenever a native plant has an important midwestern-sourced variation, it is noted in the plant's profile. Most native species are so adaptable that they do well over a wide range. Pawpaw is a good example whereas fruiting cultivars 'Mango' from Georgia, 'Pennsylvania Golden' from Pennsylvania, or 'Sunflower' from Kansas do equally well and I can't tell the plants apart. Other plants are much more specific and it is critical to use local strains. There are well-documented examples involving far-ranging little bluestem, for example. When little bluestem plants sourced from the southern Ozarks or southern Illinois are grown in the Upper Midwest, they are usually killed by frost before they even flower and set seed because the region's growing season is neither long enough nor hot enough for them.

Flowering dogwood exemplifies a native comprised of regionally adapted wild strains. Dogwoods from the western Ozarks possess tolerance to the drier winter winds that will damage the flower buds on trees sourced from farther east, while southern-sourced trees won't have enough heat if grown in Michigan even though flowering dogwood is native there too. Make sure to grow proven cultivars or local strains of flowering dogwood for your particular area

of the Midwest. Flowering dogwood is also simply not winter hardy in the Upper Midwest west of Lake Michigan, though that may change by finding a yet-undiscovered hardier strain.

Some northern plant strains won't take the heat of the Lower Midwest. I have failed with northern-sourced rose-shell azalea (*Rhododendron prinophyllum*) as it languishes in the heat of my Lower Midwest garden, but an Ozark-sourced strain has finally survived. There are always exceptions, however, one being pagoda dogwood that is fickle to grow in western Missouri. The strain we've had the best success with at Powell Gardens is from my hometown in the Upper Midwest. Whenever I have experienced an issue regarding a plant's source, I make note of it in the plant descriptions.

Cultivars of a native species (sometimes called "nativars") are included in the plant profiles if they're proven in the Midwest and still perform all the ecological functions of the wild plant. 'Sprite' winterberry (*Ilex verticillata*) is a great example. It's a female cultivar with large and abundant fruit on a shrub that is far more compact than the typical species. The birds still relish the fall and winter fruit, the bees and other pollinators still readily visit its flowers, and insects like the pawpaw sphinx's (a moth) caterpillars readily feed on its leaves.

Garden phlox (*Phlox paniculata*) is a prime example where the myriads of cultivars are not included because they simply do not have the valuable nectar resource that the wild strain has. I have tried many cultivars in my own garden and all the butterflies and other pollinators never or almost never visit the cultivars while they swarm the wild-sourced plants.

A variety of nativars and other unique variations of selected native plants are shown on pages 74–77:

Aster novaeangliae "Powell Pink" (New England aster)

Monarda fistulosa "golden-leaved" (wild bergamot)

Rudbeckia subtomentosa "semidouble" (sweet black-eyed Susan)

Rudbeckia subtomentosa 'Henry Eilers' (sweet black-eyed Susan)

Liatris scariosa forma *alba* (savanna blazingstar)

Boltonia asteroides 'Pink Beauty' (boltonia)

Coreopsis tripteris "golden-leaved seedling" (tall coreopsis)

Prunus nigra 'Princess Kay' (Canada plum)

Ratibida pinnata "double" (gray-headed coneflower)

Geranium maculatum f. *alba* (wild geranium)

Polemonium reptans 'Stairway to Heaven' (spring polemonium)

Lobelia cardinalis pale seedling (cardinal flower)

Replacing Invasive Exotics

Prolific nonnative plants that invade the Midwest's wildlands and usurp native plants have become nothing less than an ecological disaster. The plants are often called "invasive exotics." Shrub types, for example, choke many understories in our woodlands, leafing out earlier than natives and thus robbing the spring ephemeral wildflowers of the early sun they need to survive.

These exotics stop most of the native trees from reproducing, preventing seed germination and robbing light from seedlings, as their dense foliage holds late in the year as well. Some produce natural chemicals that prevent the growth of other plants (these are known as allelopaths). It's even been documented that some invasive shrubs can contribute to the decline of nesting songbirds because their early leaf-out acts as bait, while their stiff structure allows predators to climb and access nests more easily.

Exotic plants may have nectar-rich flowers beneficial to pollinators or abundant fruit to feed songbirds, but most have foliage that is not utilized by native insects as a host. Their abundance contributes to a black hole in the web of life associated with our native lands.

Invasive exotics come in all forms from annuals to shade trees. If you utilize the plants in this book, you will not contribute to their spread. The plant profiles note good native alternatives to the worst and most widespread of these exotic plants. Also noted are a few of our native plants that are just as invasive, especially if grown outside of their native haunts.

My experience has shown firsthand how removal of invasive exotics has a positive impact on the surrounding environment. My acreage's woodland was choked with multiflora rose (*Rosa multiflora*) and contained autumnolive (*Elaeagnus umbellata*). The meadow/remnant prairie was and still is choked with tall fescue (*Festuca arundinacea*), smooth brome (*Bromus inermis*), and cheat grasses (*B. tectorum*). I thought I would never remove all the roses, but eventually I got them all (sometimes looking like I had been herding cats). After removing much of the fescue and brome from the meadow, I was amazed at what wildflowers came back even without seeding them, but the fescue and brome continue to be problematic. The removal of the shrub invaders has allowed native shrubs to fill back in and the diversity of plants and animals to improve.

High Consequence Exotic Diseases and Pests

I grew up on a street shaded by tall vase-shaped American elms. What midwesterner my age or older didn't? Also in the neighborhood and local woodlands were grand butternut (*Juglans cinerea*) with silvered bark and sticky husked nuts. There were two isolated American chestnuts (*Castanea dentata*) in the county. Green ash (*Fraxinus pennsylvanica*) is ubiquitous to local floodplain forest while black ash (*F. nigra*) and white ash (*F. americana*) are more localized. Eastern hemlock (*Tsuga canadensis*) graces a few of the old historic homes and cemeteries. Black walnut (*Juglans nigra*) thrives in rich soil forests and reaches impressive size when spared for its valuable timber.

Why would I mention my local trees? Because I watched the American elms die *en masse* of alien Dutch elm disease, their carcasses cut up and hauled to the landfill. Young elms are left but only a handful of any size. The butternuts got a canker disease (*Sirococcus clavigignenti-juglandacearum*), leaving a few struggling saplings—sick as patients in intensive care. The American chestnuts prevail, though their nuts are always empty, as they are miles from another chestnut to pollinate. Their isolation spared them chestnut blight (*Cryphonectria parasitica*) that has reduced other trees to stump sprouts that repeatedly die before flowering in their native haunts farther east. My home county is under quarantine when it comes to ash trees, the dreaded Asian emerald ash borer (EAB) found, probably brought closer and faster by fools moving infested firewood. Hemlocks farther east are having the life sucked out of them by an imported pest, the hemlock woolly adelgid (*Adelges tsugae*). Black walnuts live on, but I know walnut twig beetle (*Pityophthorus juglandis*) and thousand cankers disease (TCD, *Geosmithia morbida*) loom to the west.

In other parts of the Midwest, escaped exotic beech bark disease threatens American beech forests, dogwood anthracnose (*Discula destructiva*) is maiming our beloved flowering dogwoods, and newly imported alien redbay ambrosia beetle (*Xyleborus glabratus*) is spreading laurel wilt (*Raffaelea lauricola*) and has the potential to kill our sassafras and spicebush.

What's a native plant gardener to do? I read about alternatives: Asian kousa dogwood (*Cornus kousa*) instead of flowering dogwood, for example. Is this wise? Unfortunately, I find the replacement recommenders ill-advised on the ecological impact. Migrating birds find our dogwood's fat-rich fruit important fuel; many insects rely on its foliage, in turn feeding hungry baby birds. The Asian replacement has sweet edible fruit meant for snow macaques; our native insects don't utilize its foliage. Replacement? I think not.

Because of these catastrophic diseases, urban foresters adopted a 10–20–30 formula on the makeup of any given community's street trees: no more than 10 percent should be any single species, no more than 20 percent a given genus, and no more than 30 percent a given family of plants. For example, a community should have a maximum of 10 percent bur oak in its tree population, 20 percent other oaks, and no more than 30 percent of trees in the beech family (oaks, chestnuts, beech).

New formulas have gotten even more strict because of EAB. A new 5-percent rule has been proposed—no more than 5 percent of any community's trees be one genus. These are well intentioned but the ecological impact to the Nature of a community is never considered with these arbitrary numbers that don't take cloned cultivars into consideration. The forests of the Midwest were nowhere anything close to that makeup and our wildlife will certainly suffer by it. I hope some sort of compromise comes to light.

I am disheartened by the lack of research and information to tackle these destructive exotic pests and diseases. There is some success with a handful of these species. Kudos to organizations such as the U.S. Forest Service testing American elms and the American Chestnut Foundation for its work to find and breed tolerant or resistant strains of American chestnut, including almost pure hybrids that possess the disease-resistant gene.

Help the fight by doing whatever it takes to save your remaining threatened trees and support community efforts at managing these pests. Plant resistant/tolerant strains when they fit your needs, and you are helping native creatures that rely on these native hosts. Be wise with alternatives and carefully consider the ecological impact to the web of life so tied to the Midwest's original native trees.

Shade Trees

Oaks and hickories reach their greatest importance in the Midwest and are critical to a sound landscape here. Maples are the most popular planted shade trees and are another critical component of the Midwest.

Trees grown to shade a house should be wind-, ice-, and snow-resistant. I witnessed an unprecedented catastrophic ice storm in 2002, and a handful of trees passed with flying colors while others were ripped limb from limb, most losing the top third of their canopy. Bur and post oaks, sugar maple, Kentucky coffeetree, honeylocust, shagbark and shellbark hickories, and persimmon were largely unscathed. Strong trees must be used if they are to enhance your largest investment.

American beech, sugar maple, basswood, and most of our mesic forest trees have dense surface roots that, along with dense shade, limit what will grow beneath them. These trees also need aerated soil, so adding hardscape or even turf beneath them is hard on their health, and turf languishes under their deep shade. Midwestern oaks and hickories found in upland woods have deep coarse roots and cast a lighter shade, thus allowing more diverse plantings beneath them. They like well-aerated upland soils and do not like root disturbance from construction.

Trees native to floodplain settings are the most forgiving of our disturbed soils and construction projects. Floodplain trees are subject to occasional saturated soils caused by flooding that dumps sediment over their roots. This tolerance is why these trees make the best street trees where paving and concrete cover much of their roots. American elm, green ash (both no longer recommended because of disease), silver and red maples, pin and swamp white oaks, and Kentucky coffeetree and honeylocust are adapted to root disturbance and construction projects.

Successional trees are the first trees to colonize open land. They appear in the landscape, unplanted by humans, and provide the shade needed for other forest dwellers to thrive. In the Midwest, the predominate successional trees are paper birch, persimmon, Osage orange, black cherry, and sassafras.

There is something spectacular about a mature open-grown shade tree. Many communities even have ordinances on how close together one can plant trees. Gardeners should consider planting trees in clumps and groves, especially in woodland situations. Most of our trees grow in clumps including our mighty oaks. I can think of wonderful compositions of shade tree groups throughout the Midwest but

A mature shagbark hickory, encased in ice, stands tall without a broken branch.

A swamp white oak thrives in a parking lot at the Missouri Botanical Garden.

SHADE TREES FOR USE AS STREET TREES
Acer ×freemanii (Freeman maple)
Acer rubrum (red maple)
Acer saccharinum 'Laciniatum' (thread-leaf silver maple)
Carya illinoinensis (pecan)
Celtis occidentalis (hackberry)
Gleditsia triacanthos var. *inermis* (honeylocust)
Gymnocladus dioicus (Kentucky coffeetree)
Quercus bicolor (swamp white oak)
Quercus imbricaria (shingle oak)
Quercus macrocarpa (bur oak)
Quercus muehlenbergii (chinkapin oak)
Quercus shumardii (Shumard oak)

SHADE TREES FOR TRADITIONAL LANDSCAPES
Acer nigrum (black maple)
Acer saccharum (sugar maple)
Betula spp. (birches)
Carya ovata (shagbark hickory)
Cladrastis kentuckea (yellowwood)
Fagus grandifolia (American beech)
Liriodendron tulipifera (tulip tree)
Magnolia acuminata (cucumbertree)
Nyssa sylvatica (blackgum)
Quercus alba (white oak)
Quercus rubra (northern red oak)
Quercus stellata (post oak)
Tilia americana (basswood)

An open-grown white oak spreads its spectacular limbs wider than tall.

usually Mother Nature was the designer. Why are we so afraid to plant shade trees close together? It seems we are hypocrites when it comes to birch and aspen found in a nursery. Wild sassafras and persimmon usually grow naturally in multitrunked groves. What about a clump of black cherries, bur oaks, or chinkapin oaks? Be daring and pair up compatible trees to intertwine and stand out in various seasons.

Plant an eastern white pine with a northern catalpa, sycamore with a Kentucky coffeetree, hackberry with a sugarberry, sugar maple with basswood. I dare you. As long as you make sure the trees are compatible with one another and the site conditions, you will create some memorable living sculptures that will last for generations.

Acer rubrum

Red maple

Sapindaceae (soapberry family)
Landscape groups: mesic forest trees and floodplain/
bottomland trees

Because of its many cultivars, red maple is probably the
most widely cultivated tree in the Midwest. It is well loved
for its brilliant red fall color but not all trees display that
characteristic and thus seedling trees are almost never
available at nurseries—just clones of trees with notable fall
color. *Acer rubrum* Red Sunset ('Franksred') was Iowa tree
of the year in 2000. Trees are usually male or female; occasionally trees produce both sexed flowers.

HOW TO GROW The tree grows wild in wet to mesic forests,
usually in areas of lower pH and is increasing and spreading
in abundance and range by seeding into closed woods. Red
maple is found across the Eastern Midwest but its range
westward diverges north and south around the Central and
Western Corn Belt. It grows fast and is easy to transplant
but suffers in the western Midwest where not native, often
becoming chlorotic yellow, with scorched leaves in summer
and severe sunscald on the trunks. Few mature trees occur
west of its native range even though it's widely planted.

LANDSCAPE USE A fine shade tree for appropriate soils and
conditions. Its use should be carefully considered in regions
where it constitutes more than 10 percent of street trees. A
Kansas City area nurseryman told me he tries to promote
other long-term trees, but homeowners usually choose this
tree for its quick, short-term use and gorgeous fall color. It
makes a fine tree for a rain garden, moist woodland garden,
or edge of a water garden. Red maples grow 50–80 feet tall.

ORNAMENTAL ATTRIBUTES The fall color is spectacular red on
some trees though varies from dull yellow through orange
to reds and burgundy. The underside of the leaf is a lighter
color that enhances the leaf's appearance. Flowers are tiny,
yellow to ruddy red (varies on individual trees but female
flowers are usually red) in early spring before the tree leafs
out and are a good source of nectar and pollen (male flowers) for early pollinators. The helicopter seeds on female
trees are a pale salmon color to a showy red, varying region
to region.

NOTES Be sure and select a tree that is appropriate for your
region. Do this in the fall so you can pick the fall color
you want.

RELATED PLANTS Many cultivars have consistent red fall
color including Scarlet Jewel ('Bailcraig') from Minnesota
and Burgundy Belle ('Magnificent Magenta'), selected from
a cultivated tree in western Missouri.

Acer rubrum Red Sunset ('Franksred') is known for its brilliant fall color.

The early spring bloom of *Acer saccharinum* (silver maple) is a
much-anticipated flower after a long winter.

Acer saccharinum
Silver maple

Sapindaceae (soapberry family)
Landscape group: floodplain/bottomland trees

Silver maple grows lickety-split and if improperly pruned, often develops poor branching structure then literally splits apart in high winds. The tree matures in just 50 years and rarely lives longer than 125 years, with soft wood earning it the title "soft maple."

HOW TO GROW A native floodplain tree found throughout all but the northwestern Midwest, this maple tolerates all soil types. I planted a seed of this tree in 1969 and it's already a massive tree with a trunk more than 3 feet in diameter. Fast growing is an understatement and because of that it is widely planted as a quick shade tree. The tree does become a behemoth with aggressive water-seeking roots, too. Most plants produce copious amounts of "helicopter" winged samaras in mid to late spring, which germinate immediately.

LANDSCAPE USE A massive, wide-spreading shade tree best for wide-open spaces or riparian restorations. A phenomenal silver maple with low limbs spanning 118 feet once graced the central campus lawn of Iowa State University and it was replaced by another silver maple. The maple lane at the Edsel Ford estate near Detroit, Michigan, is comprised of silver maples too.

ORNAMENTAL ATTRIBUTES The leaves are beautiful, deeply lobed with a silvery white underside and usually turn a good yellow in fall. The springtime flowers are always a delight to see, often starting in February in the Lower Midwest but waiting until early April in the north—providing fresh pollen and nectar to early emerging insects. Flowers are either male or female, usually on the same tree but some trees are solely male: the seedless cultivar 'Silver Queen' is a male plant.

RELATED PLANTS Several cultivars have been selected for deeply cut "thread" leaves, including 'Skinneri' from Kansas. The thread-leaf cultivars are quite handsome and grow much more moderately, actually making good street trees.

Freeman (red-silver) maple (*Acer* ×*freemanii*) is a natural hybrid, occasionally found growing wild where its parent trees grow near each other. It's intermediate between the two parent species, with the drawbacks of each—yet is easy to grow and transplant. Several cultivars with red fall color have been selected and are now widely planted in landscapes as a quick shade tree. The most common of these are Autumn Blaze ('Jeffersred') and Sienna Glen ('Sienna'), an "improvement" with better branching resistant to breakage. In the Upper Midwest, Matador ('Bailston') and Firefall ('AF#1') are Minnesota selections more adapted to a shorter growing season.

This *Acer saccharinum* (silver maple), is at its peak of fall color.

Two trees of *Acer* ×*freemanii* Autumn Blaze ('Jeffersred') provide fiery fall foliage.

Acer saccharum
Sugar maple

Sapindaceae (soapberry family)
Landscape group: mesic forest trees

My scholastic introduction to this tree was an elementary field trip to the local sugar bush—a wild stand of sugar maples tapped for their sweet sap—utilized for making maple syrup as well as candies. The tree grew along streets in my neighborhood and I collected its colorful yellow, orange, and red fallen leaves each fall. The locals knew it as "hard" maple. This is the maple leaf on Canada's flag but is also an iconic tree where native across the northeastern quarter of the United States. It is the state tree of Wisconsin.

HOW TO GROW Sugar maple grows in mesic forests across the entire Eastern Midwest and westward to central Minnesota, eastern Iowa, throughout the Ozarks, and up the Missouri River to Kansas. Sugar maple is, along with beech, our most shade tolerant of trees so is fast increasing in the understory, as woodlands have become closed forests. Sugar maple is displacing oaks. It can be grown from seed, or select varieties can be grafted on seedling rootstock. It's a highly variable plant and local strains are a must, especially in the western part of its range where northern-sourced plants languish and die prematurely during hot, dry summers.

LANDSCAPE USE A fine shade tree perfect for a woodland garden or other moist, upland soil sites. It casts dense shade and has dense surface roots, so limits turf beneath it when mature. It can be planted in food forests but a tree should be 50 years old before tapping for sap. Sugar maples grow 50–80 feet tall.

ORNAMENTAL ATTRIBUTES Some trees display spectacular fall color from brilliant reds to orange and yellow. The spring bloom is underappreciated as the tree is cloaked in pendant chartreuse flowers before it leafs out—what a complement above a carpet of spring wildflowers. Trees on the southwestern edge of the range have smaller leaves with a thicker, lustrous leaf surface.

RELATED PLANTS Many cultivars of sugar maple exist including new super sugar strains for syrup production. Crescendo ('Morton'), Fall Fiesta ('Bailsta'), 'Legacy' and 'Oregon Trail' (from northeast Kansas) are good midwestern cultivars.

Black maple (*Acer nigrum*) is a close relative of sugar maple. I was able to identify it after acquiring the Golden Guide's *Trees of North America*, which depicted it. Many of my neighborhood trees fit the description with their drooping lobed, cupped leaves that were fuzzier underneath. When I attended Iowa State University, black maple was the only locally native hard maple; its unique leaves make the plant more heat and drought tolerant so it ranges past eastern Iowa with relict stands in western Iowa. It's also often found in mesic floodplain forest where sugar maple won't survive. Look for black maple across the Eastern Midwest

Acer saccharum (sugar maple), in peak fall color, is one of the most beloved shade trees.

The convex foliage of *Acer nigrum* (black maple) is mainly three-lobed.

This sugar maple fits the description of *Acer saccharum* "Illinois sugar maple."

Betula alleghaniensis (yellow birch) has subtle but sublimely beautiful bark.

spent time with the late George Ware, renowned tree expert from the Morton Arboretum, at the Klehm Forest Preserve in Rockford, Illinois. Both times he showed me what he called "Illinois" maple—a naturally occurring "hard" maple intermediate between sugar and black maples with three lobes but a smooth underside. Every time I see trees like it, I recall our walks through what's now the Klehm Arboretum, but I never see any literature verifying a description of Illinois maple.

Betula alleghaniensis
Yellow birch

> Syn. *Betula lutea*
> Betulaceae (birch family)
> Landscape group: mesic forest trees

Yellow birch is found from the Maritime Provinces of Canada southwestward down the Appalachian Mountains and westward across the Great Lakes region to Minnesota. Its midwestern range includes the North Central Hardwood Forests, Driftless Region, Southern Michigan, and Northern Indiana Drift Plains and the western Lake Erie Drift Plains. It grows southward in relict stands to the Iowa River in central Iowa, Rock River in northern Illinois, Indiana Dunes, the rugged western edge of the Allegheny Plateau in Ohio, and even a few sites in the Shawnee Hills of southern Indiana and Kentucky. It's always found in rich, moist sites that are well drained.

HOW TO GROW Yellow birch is easy to grow from seed and is easy to transplant as are most birches. It is tolerant of more shade than the other birches but does best in full sun.

LANDSCAPE USE Yellow birch is rarely cultivated but makes a fine shade tree in a moist sheltered site. Plant it on the sheltered north or east side of a home to mimic a ravine's or cooler microclimate habitat. It is a fine woodland garden tree and it displays nicely when planted in clumps or masses. The tree should be included in food forests since it contains oil of wintergreen, used for flavorings including brewing root beer.

ORNAMENTAL ATTRIBUTES The trunks and limbs of this tree are subtly beautiful with silvery to coppery gray exfoliating bark. The crushed buds and stems give off a strong scent of wintergreen and I can't resist sampling them every time I meet the tree. Fall color shows rich yellows.

NOTES Yellow birch often germinates on stumps, logs, or exposed rocks and shows an interesting root system (with bark like the trunk) around such features, creating interesting leglike roots after wood has rotted away.

westward in parts of the Ozarks and up the Missouri River to Kansas City, northward to central Iowa and southeast Minnesota. Oddly, sugar maple grows farther west in Minnesota and Missouri. Black maples in the landscape are pretty much all hand-me-downs—trees salvaged from local stands many years ago. It grows even slower than a sugar maple, not conforming to efficient nursery production like hickories and many oaks as it grows its drought-tolerant roots before growing above ground. The fall color is good but the thicker, less translucent leaves create less vibrant color and usually turn yellow and orange. The convex foliage is mainly three-lobed.

Acer saccharum "Illinois sugar maple" is a form I have seen but never found in the literature. On two occasions, I

Betula nigra
River birch

Betulaceae (birch family)
Landscape group: floodplain/bottomland trees

River birch has striking exfoliating bark that is an adaptation to deter vines from growing on it, as it's a tree almost exclusively of floodplain forests where vines are abundant. In the Midwest, it's found along sandy or igneous-based streams and is absent from limestone-based streams. Look for river birch along the Kankakee, Fox, Wisconsin, Black, Chippewa, Wapsipinicon, and Minnesota Rivers well into the Upper Midwest. Though mainly a Southern tree, it does grow more widely northward to southern Ohio in the east and southeastern Kansas in the west. It's fully cold hardy through USDA zone 4 when grown from regional sources.

HOW TO GROW River birch is easy to grow from seed and cuttings. It grows fast, is easy to transplant, and is found in almost every midwestern nursery. Most cultivated river birches are cultivars selected for exceptional bark characteristics; Heritage ('Cully'), selected by renowned plantsman Earl Cully of Jacksonville, Illinois, is the most widely grown cultivar. River birch can suffer iron chlorosis (yellowing) and even die in soils with a high pH. Thus it is not recommended for street tree use, where alkaline concrete and rubble are in the soil.

LANDSCAPE USE It is commonly cultivated in clump form as a focal ornamental tree. It shows best planted in groves; a marvelous planting in downtown St. Louis reflects that. It does become a large shade tree over time, and bark on older trunks becomes dark charcoal gray. It has water-seeking surface roots that compete well with turf but can water-starve other herbaceous perennials beneath. It will often lose many of its leaves during summer dry spells in non-irrigated sites. Too often the tree is planted just for short-term appeal in tight spaces where it must be prematurely removed.

ORNAMENTAL ATTRIBUTES The bark on young trunks and limbs is gorgeous, exfoliating outwardly into curly papery sheets and revealing cream to salmon and cinnamon shades beneath. The winter tracery of twigs tipped by catkins is subtly beautiful and become the underappreciated golden pendant bloom before the leaves emerge in spring. The whole crown of mature trees often cascades all around the edges with lovely weeping branches. Fall color can be a good yellow but not consistently so.

Betula papyrifera
Paper birch

Betulaceae (birch family)
Landscape group: successional trees

No tree is more beloved for its stunning white bark than the paper birch, namesake of Silvercrest Golf course in the area where I grew up, but disappearing there as in much of its native midwestern range as it is a successional tree that invades the edge of woods and prairies and rarely lives to be 100 years old. Once it was a very popular ornamental,

Betula nigra (river birch) creates a woodland at Citygarden in downtown St. Louis.

This mass of *Betula papyrifera* (paper birch) highlights the wildflower garden at the Minnesota Landscape Arboretum.

and homeowners planted a clump of this birch in their front yards across much of the Midwest. Bronze birch borer, a native pest, eventually infested these trees and killed most of them prematurely. Borer-resistant river birch became the replacement, while borer-resistant cultivars are now available from Wisconsin and North Dakota, and a few majestic clumps of this tree defy the borer as far south as Kansas City. Paper birch is native from New England westward to the Upper Midwest and northwestward to Alaska. It prefers the cooler and more moist Upper Midwest, but is often found on the hottest, rocky outcrops adjacent to hill prairies in the Driftless Area. Relict populations occur southward to oak openings near Toledo, Ohio, the Iowa River "greenbelt" in central Iowa, and along the Niobrara River in Nebraska.

HOW TO GROW Paper birch is easy to grow from seed, which naturally drops through winter and germinates on bare mineral soils in spring. The tree is easy to transplant whether container grown or even bare root. Paper birch requires full sun or very high light and naturally is shaded out by the next wave of trees it provided shelter. It requires well-drained soil though it grows on sand lenses adjacent to wetlands—growing along the edges of Cowles Bog in the Indiana Dunes National Lakeshore.

LANDSCAPE USE Paper birch makes an exquisite mass planting along the edges of woodlands where it is native. Though it becomes a shade tree, it is often grown by nurseries in clumps to show off its beautiful bark and often grows that way in the wild. It does make a nice focal planting, and is a fine tree to plant near an outdoor seating area where night lighting will reflect off its trunks. The tree has very high wildlife value as host to many moths and seeds that are relished by finches in winter, especially the redpolls that winter in the Upper Midwest from the Arctic.

ORNAMENTAL ATTRIBUTES The white bark is like that of no other native tree and peels off into smooth plates with contrasting black horizontal lenticel scars. A tree in spring is draped in the golden catkins, and fall color is consistently a rich yellow.

Carya spp.
Hickories
Juglandaceae (walnut family)
Landscape group: oak-hickory forest

Hickories are an amazing group of trees that signify eastern North America as they are not found, other than as fossils, in Europe, and only a single species (*Carya cathayensis*) exists in eastern Asia. Each species has a slightly different habitat preference, but they all can, though rarely, hybridize and make identification challenging. They are not standard landscape trees that conform to economical nursery production as they grow an amazing taproot before they grow much above ground, which makes them challenging to transplant. A few nurseries grow them in deep, open-bottomed containers to force them to make a more fibrous root system. A tree spade can best move smaller trees. Virtually all trees are inherited as relict native trees in a landscape and since they all dislike soil compaction and salt, they can suffer after construction practices and development so are not seen in commercial landscapes. They are simply beautiful, long-lived trees worth planting and protecting for future generations.

Carya cordiformis
Bitternut hickory
Juglandaceae (walnut family)
Landscape group: oak-hickory forest and mesic forest trees

Bitternut hickory is the most widespread and hardy species of hickory and is becoming even more common and widespread as it thrives in undisturbed, mesic forests. It's reproducing just fine while white and northern red oaks are not and so may eventually displace them. Its nuts are bitter and unpalatable (very rarely sweet) with a thin skin and nutshell.

Mature bark of *Carya cordiformis* (bitternut hickory) is the most finely textured gray bark of any hickory.

HOW TO GROW Bitternut hickory is not available in the nursery trade but is easily grown from nuts. It's hardy throughout the Midwest in all but wet or compacted soils.

LANDSCAPE USE Virtually all trees in the landscape are remnants of native woodlands. They make fine shade trees 60–80 feet tall, but also produce abundant crops of nuts most years.

ORNAMENTAL ATTRIBUTES Bitternut hickory has beautiful butter yellow fall color, more pure yellow than any species of hickory. The bark is beechlike being smoother and bluish gray but does become rough gray with age. Leaf buds are golden yellow (tree's nickname is yellow-bud hickory) all winter and in spring emerge a stunning aqua to turquoise upon close inspection. Add observing them to your bucket list of trees to look at in spring.

Carya glabra / Carya ovalis
Pignut hickory / Red hickory

Juglandaceae (walnut family)
Landscape group: oak-hickory forest

The pignut–red hickory complex sometimes defies classification with variable foliage: veins and petioles range from fuzzy (tomentose) to smooth (glabrous). Plants between these two extremes intergrade just to keep botanists on their toes. "Pignut" refers to the somewhat smaller nuts, fit only as pig fodder.

HOW TO GROW These species are found across the Eastern Midwest westward across almost all of Illinois (absent in the northwest), and throughout most of the Ozark Highlands as far northwest as the Kansas City region. True pignut hickory is more apt to be found in richer floodplain soils, while red hickory is usually found on dry upland sites.

LANDSCAPE USE Pignut and red hickories provide good shade in the landscape, and their leaves turn an attractive yellow in fall. Because they produce nuts in abundance, they are suitable for wildlife gardens, but may be too messy for pristine sites.

ORNAMENTAL ATTRIBUTES True pignut hickories have smooth leaves and the husks of its nuts split from end only partway to the base, while red hickories have fuzzier leaves and the husks of its nuts split from end to base. Intermediate forms are common.

RELATED PLANTS Black hickory (*Carya texana*) can be described as "tough as nails." It is the smallest of the hickories and is named for its near charcoal black–patterned trunk. The tree grows on some of the hottest, driest, rocky soils in the Ozarks and adjacent regions, eastward to southwest Indiana and southward into Texas. It's often a small

The bark of *Carya glabra* (pignut "red" hickory) matures from rough to smooth.

The charcoal-colored bark, darkest of any hickory, identifies *Carya texana* (black hickory).

Hickories have rich yellow fall color. This tree happens to be *Carya texana* (black hickory) in Tulsa, Oklahoma.

Large leaves and rough, dark gray bark are distinctive of *Carya tomentosa* (mockernut hickory).

With its huge leaves, this seedling of *Carya tomentosa* (mockernut hickory) contrasts with the finer foliage of other woodland plants.

tree (it grows very slowly) but does reach well over 50 feet tall at maturity. It has the smallest leaves and buds of any of our hickories and sweet, delicious nuts too. Fall color is often golden yellow.

Mockernut hickory (*Carya tomentosa*) get its common name from the large but very strong nutshells that make a mockery of anyone who tries to crack one. This hickory is found in drier upland woods across the Lower Midwest and northward into southeastern Iowa, along the Wabash River into northern Indiana, and throughout eastern Ohio. The leaves are large and spectacular with a noticeably fuzzy underside that turns golden-yellow tones in autumn. This long-lived tree grows with a sturdy upright trunk of rough, gray bark.

Carya illinoinensis

Pecan

> Juglandaceae (walnut family)
> Landscape group: floodplain/bottomland tree

Pecan is a species of hickory. Think pecan and one thinks of the American South but the tree's botanical name "of Illinois" is correct in that the tree is a widespread native across

Carya illinoinensis (pecan) creates a luxuriant tree at Chicago Botanic Garden.

The shaggy bark of *Carya ovata* (shagbark hickory) is without equal.

the Lower Midwest. These midwestern trees are marketed as "northern" pecans with smaller, thicker shelled, more oil-rich, and flavorful nuts. Wild pecans thrive in rich alluvial soils of rivers and are abundant through the Osage Plains, Lower Missouri, Mississippi, and Ohio Rivers and up the Illinois River to Peoria. The tree is actually very cold hardy through USDA zone 4, but the growing season is not long enough in the Upper Midwest for the tree to ripen nuts. In 10 years of living in Rockford, I never saw large local trees with fully ripened nuts. There are also trees at the Minnesota Landscape Arboretum.

HOW TO GROW Easy to grow from nuts but many local cultivars with better nut quality and production are grafted for planting. Pecans, along with the entire walnut family, have large, deep taproots so are challenging to dig and transplant unless grown in deep or open-bottomed containers that root-prune the plant. These trees have amazing heat, drought, and flood tolerance. Late freezes or extreme cold can damage flower buds, also inhibiting nut production.

LANDSCAPE USE Pecan becomes a premier and magnificent shade tree 80–100 feet tall and tolerant of a wide range of soils. The nuts are deemed messy and a liability, so the tree is not considered suitable for street planting, though it is perfectly adapted for such conditions except for its intolerance to salt. It's a perfect choice for natural

landscapes, edible landscapes, nut orchards, and food forests (demand for northern pecans exceeds current production capabilities).

ORNAMENTAL ATTRIBUTES The lush green foliage through the hottest and driest seasons and through record floods is the beauty of a pecan tree. The ashy, light gray bark with intricate shagginess upon close inspection is also beautiful. After the nuts have fallen, I also find the blackened husks, which linger on a bare tree into winter and are quite ornamental.

NOTES Specialty nurseries sell many cultivars of pecan selected for nut production in a particular region. The trees also require cross-pollination with an appropriate cultivar.

RELATED PLANTS Pecan's natural hybrid with the shellbark hickory is known as hican (*Carya ×nussbaumeri*).It is one amazing tree with huge blimp-shaped nuts.

Carya ovata
Shagbark hickory

Juglandaceae (walnut family)
Landscape group: oak-hickory forest

Here's an easy-to-identify shade tree because of its very unique bark that splits off in flat gray strips that bend outwards from both their upper and lower ends. (Seedling and

The emerging buds of *Carya ovata* (shagbark hickory) look almost like red-petaled flowers.

The midsummer bloom of *Castanea dentata* (American chestnut) is spectacular.

young trees have smooth medium gray bark and are much more challenging to identify from other hickory species.) Its wood is used to produce hickory-smoked meats.

HOW TO GROW The tree grows wild over all of the Eastern and Lower Midwest and is found in the eastern Upper Midwest as far west as southeastern Minnesota and southeastern Nebraska. It's cold hardy throughout in USDA zone 4. Almost no nursery carries shagbark hickory, but state nurseries often sell bare-root seedlings, as do a few specialty mail-order nurseries. Bare-root seedlings do not transplant well (expect low survival rates). Shagbark hickory is extremely drought tolerant, surviving in almost any well-drained soil, and is easy to grow from nuts.

LANDSCAPE USE Shagbark hickory is occasionally planted in native plant arboreta and edible landscapes and makes a fine food forest tree. Young trees are tolerant of all but dense shade but grow best in full sun. The nuts are delicious and nutritious but not easy to crack or extract the nutmeat. Shagbark hickory usually grows 60–90 feet tall.

ORNAMENTAL ATTRIBUTES The bark on mature trees is stunningly beautiful, but my favorite ornamental attribute is the emerging buds with scarlet scales, which unfurl like petals of a flower surrounding the splendid form of the emerging compound leaves. Leaves of young trees are comprised of five leaflets and tropical looking. Fall color is yellow to golden brown.

RELATED PLANTS Shellbark hickory (*Carya laciniosa*) is a closely related species found across the Eastern and Lower Midwest and northward in the southern Iowa–northern Missouri Drift Plains, but is also cold hardy through USDA zone 4. It is found in deep, rich soils of mesic floodplains. Its bark is not as shaggy as that of shagbark hickory. It also differs from shagbark hickory by having seven even larger leaflets and much larger nuts (often called kingnut hickory).

Young trees in a woodland understory produce simply spectacular foliage, looking like they belong in a rainforest.

NOTES In restoring my oak-hickory woodland, I planted shellbark hickory nuts and seedlings along the lower draw and shagbark hickory nuts and seedlings elsewhere. The young trees are strikingly stunning with their large leaves contrasting nicely with other vegetation. Why would anyone want a Japanese maple instead? I know I will never see the mature size of the hickories, but one day they will reward future owners with beautiful bark and delicious nuts.

Castanea dentata
American chestnut

> Fagaceae (beech family)
> Landscape group: oak-hickory forest

When the tree canopy is at its midsummer peak of verdant growth, this tree explodes like the fireworks of the season, with showy sparklerlike creamy flowers. Historically the tree produced a heavy mast by autumn, relished by people and wildlife: the classic chestnut to roast on an open fire. The trees grew huge until 1902 when this magnificent icon was brought to its knees by a fungus, chestnut blight, unknowingly introduced on logs of Japanese chestnut imported to New Jersey. Within a few short decades, the entire native population of trees was infected, killing the aboveground portions of the tree while the roots lived on, sending up doomed stump sprouts, some still present today. American chestnut once dominated Appalachian forests but ranged into the Midwest through almost all of eastern Ohio, northwestward to southeast Michigan and southwestward locally to southern Illinois (once "abundant" in

Pulaski County, Illinois). The tree never grew wild west of the Mississippi River but was sparingly planted beyond its native range across all of Lower Michigan and across lower Wisconsin as far westward as it would survive in southern Minnesota, central Iowa, and southeastern Nebraska. Cultivated trees survived the blight because of their isolation. As a young man, I learned there were two such trees in my home county and a row of them at a farmstead in adjacent Houston County, Minnesota.

HOW TO GROW This tree once reached monstrous proportions in mesic, acidic to slightly acid humus-rich soils. It requires good drainage and tolerates neutral soils but is not drought tolerant in lean soils in the western Midwest. It's cold hardy through USDA zone 4. Only plant disease-resistant hybrid trees, or plant trees in isolation.

LANDSCAPE USE Best as a food forest tree as its nuts are encased in sharply spined husks that drop in autumn and are not compatible with a lawn or suitable for a compost pile. Two different trees are needed for cross-pollination and nut production, but a lone tree will produce empty nuts encased in the spiny husk. Groundcover or woodland plantings are a must under the tree where husks can naturally compost themselves.

ORNAMENTAL ATTRIBUTES The midsummer flowers add color when few other trees are in bloom. The monoecious flowers are aromatic but the scent is not agreeable to all though rich in pollen for insects. Fall color is lackluster bronze to tan and mature trees have a beautifully rugged, wide crown with deeply furrowed bark.

NOTES Plant scientists formed the American Chestnut Foundation (TACF) to protect the tree from extinction. Founding members Philip Rutter, David French, and Charles Burnham developed backcross breeding with disease-resistant Asian species to make almost pure American chestnut with resistance to chestnut blight fungus. The Midwest's Dawe's Arboretum, east of Columbus, Ohio, with TACF, installed an American chestnut research plot of 458 seedlings in 2014 (dawesarb.org/chestnutproject). Kudos to all those committed to conservation of this magnificent tree species. May it one-day repopulate its former native haunts and our gardens across the Midwest.

RELATED PLANTS If there ever was an aptly named oak, chestnut oak (*Quercus montana*) would be it. Its rugged, furrowed bark looks like that of a true chestnut (*Castanea*) and its native Midwestern range mimicked the chestnut almost to a tee. Chestnut oak is in the white oak group with distinctive large, elongated acorns. The leaves look somewhat like chinkapin oak leaves, but you can pick out the tree by its distinctive bark. In the wild the tree grows on thin soils over bedrock and also was native only east of the Mississippi

A grove of *Quercus montana* (chestnut oak) clings to a bluff on the edge of the Appalachian Plateau in Ohio.

though there are trees on the southern Illinois bluffs where you can see Missouri across the river. The tree is heat and drought tolerant, grows well westward in USDA zone 5 across the Midwest, and has naturalized in southeastern Michigan (Waterloo Recreation Area near Ann Arbor) and as far west as M. L. Thompson Nature Sanctuary near Kansas City.

Catalpa speciosa
Northern catalpa

Bignoniaceae (bignonia family)
Landscape group: floodplain/bottomland trees

A northern catalpa tree decked out in bloom in late spring (early summer in the Upper Midwest) is reminiscent of a tropical tree with large leaves and bright, showy flowers. It is related to the colorful *Tabebuia* trees from the American tropics grown widely in frost-free areas. Northern catalpa was originally native to the magnificent floodplain forests in a small area from the lower Wabash and Ohio

Naturalized *Catalpa speciosa* (northern catalpa) in full bloom catches the evening light along the Rock River in Rockford, Illinois.

Here's an outstanding specimen of *Celtis occidentalis* (hackberry) with a perfect vase shape reminiscent of an American elm.

Rivers in southwest Indiana and western Kentucky downstream to where the Ohio meets the Mississippi River and southwest to northeastern Arkansas. This region grows some of the largest deciduous trees on the continent and catalpa reaches magnificent proportions when mature. One can easily observe native trees at phenomenal Beall Woods State Park on the Wabash River in Illinois. The tree is so widely planted across the entire Midwest and naturalized near plantings so that the native range is a mere remnant of where the tree grows today.

HOW TO GROW Northern catalpa is easy to grow from seed, grows very fast as a small tree, and blooms at an early age. It requires full sun and thrives in rich soil. It is extremely susceptible to verticillium wilt, so its use should be cautioned in areas where that disease is prevalent and it should never be mulched with wood that could be infected with the fungus.

LANDSCAPE USE Northern catalpa is tolerant of disturbed soils and actually suited to urban use. The tree was widely planted on estates, parks, and home landscapes in Victorian times, then fell out of favor but is enjoying a renaissance of planting. The massive size of a mature tree, combined with its big leaves, flowers, and beanlike fruit that all drop in their season, place it in the messy tree category. Reiman Gardens in Ames, Iowa, uses the tree beautifully in a terrace

area surfaced in fine gravel. It's a good tree for beekeepers and for anglers, who utilize the catalpa sphinx caterpillar off the tree as popular bait. The caterpillar can only feed on trees in this genus.

ORNAMENTAL ATTRIBUTES Northern catalpa has flouncy white flowers with colorful purple streaks into the throat of a flower with a yellow spot above—that act as pollinator guides. Native bees relish the flowers and people think they look like orchids. Pollinated flowers produce long, string-bean podlike capsules that hang through winter and drop by spring. The capsules are quite interesting and ornamental. The tree occasionally produces yellow or yellow-green fall color but can freeze black on the tree.

Celtis occidentalis
Hackberry
> Ulmaceae (elm family)
> Landscape group: floodplain/bottomland trees

The woodland surrounding my home is full of hackberry trees, and every other year when they are laden with fruit, I look forward to a birdie winter. Flocks and exceptional numbers of robins, bluebirds, waxwings, flickers, sapsuckers, and other fruit-eating birds visit. In summer there are two flights (early and late) of the most common butterfly in the region—the hackberry emperor—which remind me of the wildebeest of the butterfly world as I can see hundreds in flight at a time. They overwinter as caterpillars rolled up in the fallen leaves so natural beds beneath the trees provide safe haven for them—lawn mowing and raking up leaves and debris spells their demise.

HOW TO GROW Hackberry is native throughout the Midwest and is incredibly tolerant of soils and weather extremes. It does well in full sunlight to light shade. Originally native mainly in floodplain woods, it has colonized all disturbed woodlands and is now found in all but the shadiest mesic forests. It requires a warm summer and is heat tolerant and survives extreme cold and extreme drought. It's easy to grow from seed. The wood is prone to rotting from storm damage or limb removal/excessive pruning. Prune only branches or small limbs so they can quickly heal.

LANDSCAPE USE Some strains of hackberry make fine shade trees with a vase shape similar to its classic relative, American elm. Hackberry is a premier bird and butterfly tree: the edible fruit are a feast for many birds and the foliage hosts five species of butterflies.

ORNAMENTAL ATTRIBUTES The bark on young to middle-aged trees is furrowed with stunning corky warts, becoming shaggier with age. The foliage almost always has nipple gall, a purely cosmetic issue harboring the hackberry psyllid bug.

In some areas, the trees also readily have witches' brooms (clusters of twigs also considered unsightly but of no harmful consequence to the plant). Fall color is yellow.

RELATED PLANTS Sugarberry (*Celtis laevigata*) is the American Southeast's relative to the hackberry and is native across the Lower Midwest and cold hardy to USDA zone 5b. Its use is interchangeable with the hackberry, hosting the same birds and butterflies though its fruit is usually red. The trunk looks very beechlike.

Cladrastis kentuckea
Yellowwood

> Fabaceae (legume family)
> Landscape group: mesic forest trees

A yellowwood dripping with its pendant clusters of white flowers is a sight not forgotten. This tree is rare in the wild, found sparingly and sporadically west of the Appalachians in moist, limestone coves as far west as the Ozarks. The tree shows its best in gardens across the Midwest, seemingly brought back to life by gardeners.

HOW TO GROW Yellowwood thrives in moist, well-drained soil but tolerates dryness in summer. When planted in a harsh site with compacted, wet or acidic soil, the tree becomes stressed and susceptible to cankers that kill it. The

Celtis laevigata (sugarberry) has a bluish gray, smooth trunk with just a smattering of corky warts.

Cladrastis kentuckea (yellowwood) in full bloom is a memorable experience.

Fruit of *Diospyros virginiana* (persimmon) must be fully ripe to sweet and tasty.

tree will grow in full sun as well as light or high shade. It does best where it is sheltered from summer's hot, drying winds in the western Midwest but is cold hardy throughout almost the entire region—to USDA zone 4. Prune yellowwood after it leafs out; otherwise you will be amazed how much it bleeds sap.

LANDSCAPE USE A premier shade or ornamental tree with deep coarse roots that make it easy to cultivate other plants beneath the tree. The nectar-rich flowers attract droves of medium and large bumblebee species.

ORNAMENTAL ATTRIBUTES The showy white flowers draped through the tree in late spring are sublime. This show is not necessarily an annual event as many trees flower well only on alternate years. The flowers produce light brown seedpods (legumes) that hang on the tree through winter. The pinnately compound foliage is medium textured and light green, turning clear yellow in the fall. Yellowwood trunks are sumptuously smooth and bluish gray and splay upward and outward into a rounded vase shape. The branching can be weak structurally and subject to splitting, sometimes cracking but remaining sound for decades.

Diospyros virginiana
Persimmon

Ebenaceae (ebony family)
Landscape group: successional trees

Trees are usually smaller than shade tree size though readily grow over 50 feet tall at maturity. They usually sucker into groves, but I have a male tree that is single trunk with never a sucker. The wood is very strong.

HOW TO GROW Persimmon is native across the entire Lower Midwest but is cold hardy northward into USDA zone 5. It's easy to grow from seed but has a deep, coarse root system making it hard to transplant. Select fruiting plants must

be grafted as they are challenging to divide off groves. The tree is best in full sun, tolerant of many soils from periodically wet to dry, and is amazingly heat and drought tolerant once established.

LANDSCAPE USE Persimmons are either male or female and when in bloom they are abuzz with bees. Female trees produce delicious edible fruit, sweet when ripe; they'll give you a memorable puckered mouth if eaten too early. Persimmon is thus a great food forest tree and select female clones with larger, tastier fruit are available from specialty nurseries. Female trees hold their fruit into winter, dropping and creating a messy situation for tidy landscapes but a boon to wildlife so they make a welcome addition to a natural garden. The tree is usually inherited with development of woodlands.

ORNAMENTAL ATTRIBUTES The stunning blocky bark, like an alligator's hide, shows best in winter. The early summer flowers are small white urns little noticed among the foliage. Fruit on female trees can be quite beautiful when they are peachy yellow and ripe right after leaf drop.

NOTES Do not plant persimmons near native prairie remnants, as they are nigh impossible to remove after invading a prairie. I've been chided for promoting red cedar as it's a prairie invader but it's easy to remove, killed by fire or cutting it off. Persimmon is a challenging invader with tenacious roots that resprout prolifically.

Fagus grandifolia
American beech

Fagaceae (beech family)
Landscape group: mesic forest trees

American beech, along with sugar maple, is the ultimate climax tree found in mesic forests. It is in the beech family along with oaks and chestnuts. Possibly the finest remnant forest of American beech is found in the Midwest at Warren Woods State Park in southwest Michigan. This area was destined to become charcoal when an area philanthropist, Edward Kirk Warren, purchased the property and saved the trees. American beech is found in eastern North America as far west as the Upper Peninsula of Michigan, southward along the western shore of Lake Michigan in Wisconsin, almost perfectly east of the Illinois–Indiana state line, but ranging westward in southern Illinois into southeast Missouri and southwestward to eastern Texas and points east to the Atlantic. With appropriate genetic stock, it is cold hardy through USDA zone 4 but intolerant of dry conditions. Its native range, along with tulip tree, defines the more humid, moderated climate of the Eastern Midwest.

Mature trees of *Fagus grandifolia* (American beech) at Cave Hill Cemetery in Louisville, Kentucky, are truly magnificent.

HOW TO GROW American beech is easy to germinate from its periodically abundant mast of edible nuts but is challenging to transplant. Beech seedlings may carpet the ground in moist conditions and are gaining ground over the oaks and hickories of more disturbed and fire-adapted landscapes. American beech is intolerant of extreme drought and requires moist, well-drained soils. It will survive in moist, sheltered sites when cultivated west of its current native range. Fine trees occur at Minnesota Landscape Arboretum, Iowa State University campus, and Linda Hall Library in Kansas City.

LANDSCAPE USE American beech is a magnificent shade tree for parks, cemeteries, and woodland gardens where its dense shade and surface roots can be accommodated beneath the tree's canopy. It reaches 60–100 feet tall. Outstanding, mature trees at historic Cave Hill Cemetery in Louisville are worth the visit.

ORNAMENTAL ATTRIBUTES The smooth, voluptuous, bluish gray trunks almost beg humans to carve symbols and language on them so larger trees often display graffiti. A trio of wild, mature American beech trees at the internationally renowned Frederik Meijer Gardens and Sculpture Park in Grand Rapids, Michigan, grabbed my attention as living sculptures—as fine as any by the famous sculptors' work on display. The fall color is rich golden to orange-brown. Every time I visit a beech, I have to touch its incredibly thin and smooth leaves that are elegantly arranged horizontally to capture all available light.

NOTES Unfortunately, beech bark disease, an invasive pathogen complex comprised of an exotic soft-bodied scale insect (*Cryptococcus fagisuga*) attacks the bark, creating an entry point for two types of invasive exotic fungi (*Neonectria faginata*, *N. ditissima*), which cause cankers and ultimately kill the tree. There are resistant trees and the U.S. Forest Service initiated a disease resistance–breeding program with the Michigan Department of Natural Resources. Holden Arboretum outside Cleveland is a partner. Learn more at nrs.fs.fed.us/disturbance/invasive_species/bbd_resistant_beech.

Fraxinus spp.
Ashes

Oleaceae (olive family)
Landscape group: floodplain/bottomland trees

Ash trees were never favorite trees of mine, but I always enjoyed them in their native haunts for their beautiful yellow to purple fall colors. Fast-forward to 2014 in the midst of the plague of a foreign beetle, the emerald ash borer (EAB): as I was walking the perceived pristine Indiana Dunes Nature Preserve, the towering giant black ash in the swamps and white ash in the rich mesic forests stood dead, and most of their offspring, too. It suddenly hit me: this is it; future naturalists and tree enthusiasts will never know the splendor of this tree just like I will never know a mature stand of wild American chestnut. Like too many experiences in life, you don't fully appreciate what you have until it's gone.

If you own a magnificent ash tree, you should try and save it through pesticide treatment. The value of the live tree exceeds its removal. Sadly, many insect species that are exclusively linked to ash trees as their only source of food have it much worse and are little discussed as to how they are poisoned by the pesticide. I am a butterfly and moth enthusiast, so it pains me that four large moths—great ash, ash, waved, and fawn sphinx—all of which fly like hummingbirds, may be lost as with many untold smaller species of insects. Ashes have few midwestern relatives.

The Asian native insects that specifically control EAB on its home turf have been introduced and released in our wilds on a limited basis. The jury is still out on their effectiveness and we all worry about unintended consequences (though EAB control insects were thoroughly tested showing no harm to other native species). Gypsy moth control parasites ended up destroying the marvelous saturniid moths in the Northeast, though nature in its marvelous way has shown one of our native hyperparasites (parasite

Fraxinus americana (white ash) is the first tree to change color in the fall.

Leaves of *Fraxinus nigra* (black ash) look more like a hickory but are opposite on the stem.

This magnificent specimen of *Fraxinus pennsylvanica* (green ash) in a historic cemetery is worth treating against emerald ash borer.

of a parasite) is controlling the alien parasite. Yes, a handful of ash trees are showing resistance near ground zero in Michigan and time will tell how these trees will fare in the long run. I'll admit I hate harming any ash tree seedling in my woods, just in case it may be "the one" resistant to EAB. Learn more about emerald ash borers at emeraldashborer. info.

HOW TO GROW Although ashes are no longer recommended for planting, they are valuable native trees and worth saving when in good condition. See each species for specific details.

LANDSCAPE USE Ashes excel as shade trees and are often used as street trees. Female trees produce clusters of flattened fruits, like the samaras of maples, and can therefore be somewhat messy.

ORNAMENTAL ATTRIBUTES Fall color and year-round textured bark.

Fraxinus americana (white ash) is widespread across eastern North America but absent from the western Upper Midwest. It is at its best in the Ohio River valley where its wood makes the finest baseball bats (Louisville sluggers). This upland species is found in more mesic woods westward and is usually the first forest tree to turn in early fall, becoming purplish. Fall color is usually burgundy to purple though some trees can have yellow fall color, especially in shaded situations. White ash is very locally adapted, and regional cultivars and varieties have to be grown for hardiness issues.

Fraxinus nigra (black ash) is an Upper Midwest ash native from the Northwoods southward to northeast Iowa, northern Illinois, and southward sparingly to the lower Wabash River in Indiana and across the northern two-thirds of Ohio. This tree is often found in swamps or mesic forests and was also a fine street tree in northern parts of the Midwest. The name black ash comes from the very dark brown buds.

Fraxinus pennsylvanica (green ash) was the most widespread midwestern ash species found across the entire temperate eastern half of North America as well. It is a floodplain and disturbed ground species with two distinct forms; the fuzzy twig and leaf form is named red ash. Green ash was highly adapted to street tree conditions and widely planted after DED destroyed the elms. Now we are looking for another substitute. Fall color is often a beautiful clear yellow.

Fraxinus quadrangulata (blue ash) is a tree of thin soils over bedrock and has a midwestern range from the Ozark Highlands into southeast Iowa, up the Rock River in Illinois barely into Wisconsin, eastward to eastern Ohio and Kentucky's Bluegrass Region. It's magnificent in the Louisville

Fraxinus quadrangulata (blue ash) literally has the ashiest gray-colored bark of any ash species.

Pendant fruit adorns the canopy of *Gleditsia triacanthos* (honeylocust).

to Cincinnati region. This tree has distinctive square twigs often with four corky ridges. The name blue comes from the inner bark that was used as a blue dye. Blue ash is showing more resistance to EAB than the other ash species.

Gleditsia triacanthos

Honeylocust

Fabaceae (legume family)

Landscape group: successional trees and floodplain/bottomland trees

Spay and disarmed forms of this tree were planted in almost every urban and suburban landscape across the Midwest (and wherever hardy in the North America). I hate the word *overplanted* but it does suit the "thornless-seedless" honeylocust. Now I live in woodland where the native version of the tree is all around, so my attitude toward the tree has changed and for the better.

Wild honeylocust is armed with fierce branched spines up and down its trunk and along its stems. Its flowers are little noticed, golden green, and abuzz with pollinators.

Female flowers are in separate, looser clusters than male flowers that are usually on separate trees. The female flowers produce beautiful pea-pods (actually legumes): long and pendant in a corkscrew pattern. In midsummer my neighbor asked me what tree was flowering in my front yard. I looked at her perplexed, then realized she was noticing the golden young honeylocust pods catching the evening light. The pods turn almost black when mature and rain down the lawn through winter.

The mature wild tree in my front yard no longer has spines. Some trees quit producing them at maturity, though there are true thornless trees known as *Gleditsia triacanthos* var. *inermis*. I had to ask why wild honeylocust trees had such incredible armor, and paleobotanists provided the answer: it was probably protection against extinct giant ground sloths and mastodons that once roamed its range. Honeylocust is sometimes separated into the senna family (Caesalpiniaceae).

HOW TO GROW At one time, honeylocust was native mainly to floodplain areas. With settlement, it spread into the prairies and became a common second-growth tree as cattle spread its seeds into areas of pasture, feasting on its nutritious pods, scarifying, and fertilizing its seeds. The tree's range has expanded because of this, but it was originally native between the Appalachians and the Great Plains from all of Ohio to southernmost Michigan westward around the Illinois prairies and up major rivers like the Mississippi into southeastern Minnesota and up the Missouri River into southern South Dakota and southward to the Gulf of Mexico. Honeylocusts are easy to transplant, tolerant of urban soils, and grow in tree wells set in concrete. No nursery sells the wild form (though occasionally thornless but seeded varieties are available). The species is easily grown from seed while all cultivars are grafted onto understock. The tree's hardiness varies from north to south so make sure to plant locally hardy forms of the tree. Northern strains are fully cold hardy through USDA zone 4. The tree grows 50 to almost 100 feet tall with a massive trunk at maturity.

LANDSCAPE USE The thornless-seedless cultivars are premier shade trees because they have deep coarse roots and fine leaflets and rachises that allow light to filter through so that turf grass can grow well beneath the tree. Seedless trees produce only male flowers, which are still visited by pollinators. This no-need-to-rake, no-mess attribute along with the heavy and strong wood make it the ultimate urban street tree. Seeded trees have value in food forests as a premier permaculture tree. The pods are actually edible and make great fodder for chickens and livestock while the flowers produce nectar. Honey-locust indeed. Wild trees also are treasured for wildlife gardens as they are the sole host for a plethora of phenomenal moths from two species of honeylocust silk moths to moon-lined moth, Magdalen underwing, and my favorite, the orange-wing—a moth that readily lands on people to imbibe sweat. There is beauty in the beast (the wild form) and one can plant groundcovers and shrubs beneath wild trees to provide places for the pods to drop, along with cover for the overwintering moths. Thorns can be wicked, piercing through soles of shoes and tires but providing protection for nesting songbirds. Give wild honeylocust a chance if inherited in a landscape but use common sense where you plant one.

ORNAMENTAL ATTRIBUTES The finely textured leaves are beautiful and can turn pure yellow in the fall. The leaves are comprised of pinnately or doubly pinnately compound tiny leaflets off fine rachises (leaflet stems). Mature trees show rugged trunks of curled-edged, stuccolike plated

Female trees of *Gymnocladus dioicus* (Kentucky coffeetree) produce purplish brown pods that hang on the tree into winter.

gray bark. The tree's form is a skyline like the acacias of the Serengeti—a fitting sight for the savanna of the Midwest. Urban trees seldom mature to this beautiful stage.

NOTES Unfortunately, the thornless-seedless strains are now mired by mimosa webworms across the Lower Midwest. Oddly, that pest leaves the wild forms alone.

RELATED PLANTS Thornless-seedless cultivars include 'Sunburst', with bright golden new leaves, and 'Rubylace', with purplish red new leaves.

Gymnocladus dioicus
Kentucky coffeetree

Fabaceae (legume family)
Landscape group: floodplain/bottomland trees

Coffeetree is the coarser cousin of the honeylocust. It has much larger leaflets on doubly compound rachises so that the main central rachis is 1–2 feet long and substantial enough to hold a compound leaf up to 3 feet long on the stout branchlike twigs of the tree. The whole thing drops in autumn so one has a network of rachises under the tree even though the leaflets curl up and are relatively no mess. It makes small trees look ridiculous in winter because all its "twigs" drop. Don't be fooled by this ugly duckling; it becomes a swan as it matures.

The coffee part of the common name comes from the short but stout pods (actually legumes) that have large, hard seeds in them. Allegedly, a coffeelike brew was made from them but I wouldn't try that at home and the pods are listed as toxic.

Coffeetree is a heart-of-the-continent species native sporadically between the Appalachians and the Great Plains usually in floodplain forests but oftentimes near where Native Americans found shelter. It was found wild farther north than honeylocust well into Lower Michigan, Wisconsin, and Minnesota at settlement and absent from most of the South. It's common in the Osage Plains and Flint Hills too. In Kentucky, it's found along the Ohio River and Bluegrass Region but not statewide. It is fully hardy through winter hardiness zone 4. Coffeetree grows 60–100 feet tall. This species is sometimes separated into the senna family (Caesalpiniaceae).

HOW TO GROW Kentucky coffeetree is grown from seed; the male cultivars are grafted on seedling stock. The seed coat is amazingly thick but, if broken, allows the seed to germinate readily. I put the seed coat in a vice grip and use a hacksaw to cut its hard coating. The tree tolerates a wide range of calcareous soils as long as they are not wet for long periods. Plants are somewhat shade tolerant. The tree develops a carrotlike taproot, making it a bit challenging to transplant, and young trees remain a stick because the large compound leaves (up to a 3 feet long) comprise its deciduous twigs.

LANDSCAPE USE A fine shade and street tree with attributes similar to honeylocust. The light shade and deep coarse roots allow for cultivation of turf to most shade plants beneath. The pods on female trees are considered messy, as are all the leaf rachises that drop in fall. The tree is amazingly wind and ice firm and because it loses its "twigs" in winter, it's also a premier home-shading tree letting more passive solar sun through in winter. Some trees sucker into clumps and thickets, usually male trees but occasionally a female tree will while most trees never send up a sprout unless the trunk is removed.

ORNAMENTAL ATTRIBUTES The rugged bark is stunning as it splits into plates that curl up on the edges—sharp enough to deter any would-be human climber and probably a protective armor for long extinct ground sloths. The pale green flowers have a unique aroma and are pollinated by a few native bees. The pods become dark polished brown with purplish highlights and are quite beautiful but must be raked into groundcover beds or composted. The foliage emerges coppery to purplish with silvered edges so is beautiful in springtime but gorgeous when mature because of the mesmerizing herringbone pattern of leaflets it creates. Fall color can be rich yellow but is usually greenish yellow.

Juglans nigra
Black walnut

> Juglandaceae (walnut family)
> Landscape group: widely adapted(?)

Walnuts are in the same family as hickories. Mention black walnut and gardeners cringe since you can't grow prized tomatoes under or near the tree because it produces a compound called juglone that inhibits the growth of some species of plants. Most native understory trees and wildflowers are immune to this.

Black walnut is tenacious. During the droughts of 1988 and 2012 when many plants were scorched, established walnuts remained verdant green. Now, the tree is threatened by thousand cankers disease (TCD) carried by an insect pest from the American Southwest: the walnut twig beetle (*Pityophthorus juglandis*). I hope this will not play out as another tragedy; walnut is the most valuable timber tree across much of the Midwest. Black walnut's foliage supports many insects; maybe not as many species as oaks but the biomass of its associated fauna may be surprising. Black walnut's nutmeat is edible and oil rich with a very

unique flavor utilized in many confections and desserts.

HOW TO GROW Black walnut is native across the Midwest absent only from the northwestern edge where its range is expanding from cultivated plants. It grows in almost any rich, well-drained soil from mesic floodplains to dry upland woods and is hardy everywhere in the Midwest. It's easy to grow from its nuts, though select cultivars for better nut quality must be grafted. Seedling trees have a deep taproot so the tree can be challenging to transplant.

LANDSCAPE USE Black walnut is a premier shade tree for large parks and gardens, usually inherited from original wild plantings in a home or commercial landscape. Its drawbacks are the large nuts, which drop in autumn, not a pleasant sound on a rooftop or car. The nuts' husks stain concrete or other hardscape surfaces. It's a premier food forest tree but also welcome in a woodland garden or other nature-friendly landscape. TCD is a probable concern—anyone who moves walnut wood from west of the Great Plains or the few infested areas is an accomplice to a crime against humanity and nature.

ORNAMENTAL ATTRIBUTES The pinnately compound foliage is quite beautiful though it is slow to emerge in spring, often first to drop in fall, and frequently mired by early fall by a plethora of harmless insects from fall webworms to masses of its namesake walnut datana moth (these hairy caterpillars are important food for yellow and black-billed cuckoos). It's the favorite midwestern host plant to the magnificent green and tailed Luna moth—you must have unraked

Juglans nigra (black walnut) creates a spectacular shade tree of exceptionally beautiful foliage.

groundcover beneath a walnut if you want this moth to overwinter as its cocoon falls with the leaves and needs insulation with fallen leaves and contact with the ground to survive winter. Fall color can be a true yellow some years.

RELATED PLANTS Butternut or white walnut (*Juglans cinerea*) is native across all but the westernmost Midwest—west to central Minnesota and southward through most of the Ozarks. It was a magnificent tree native in mesic forests with silvery, striated bark and a distinctive, widely spreading crown when grown in the open. A canker disease (presumed alien/imported) has tragically destroyed virtually all butternuts. The canker was first discovered in Wisconsin, but has now spread throughout the butternut's range. I remember magnificent trees as a teenager and handling the sticky, football-shaped nuts and tasting their unique "buttery" nutmeats. The tree now thrives only where planted away from infected trees, almost exclusively outside its native range. Let's hope black walnut doesn't suffer a similar fate from TCD.

Canker disfigures the silvery bark of *Juglans cinerea* (butternut) and will eventually kill the tree.

Liriodendron tulipifera
Tulip tree

Magnoliaceae (magnolia family)
Landscape group: mesic forest trees

Tulip tree is the state tree of Indiana and Kentucky and was the second tallest tree (after eastern white pine) in eastern North America. Trees of this species invade a disturbed site, grow like a rocket skyward, and then live on for centuries to become some of our largest native trees. It is native only in the Eastern Midwest but has been widely planted westward in USDA zone 5 and warmer. A few stunted trees survive in sheltered sites in USDA zone 4, but the species isn't adapted to areas that cold. It has naturalized near plantings as far west as the Kansas City area. Almost monumental wild trees can be seen at Beall Woods State Park in southeastern Illinois.

HOW TO GROW Seed has low viability, but fertile seed readily germinates. Seedlings grow quickly provided they have sun. Tulip tree requires rich, moist soils that are never wet for any period of time. The tree is easy to transplant.

LANDSCAPE USE This monstrous shade tree reaches 80–120 feet tall or more and is suitable for parks and large gardens. It's very beautiful planted in mass where its straight trunks quickly create cathedral-like columns with a shady space for woodland garden plants beneath.

ORNAMENTAL ATTRIBUTES Young greenish trunks are striped and speckled with white, described as looking like the belly of a pike (fish). The leaves and their unique shape are

Flowers of *Liriodendron tulipifera* (tulip tree) are uniquely colored but rarely noticed high in the leafy canopy.

always memorable and turn a rich yellow in the fall. Flowers are also special, being light green with a flame of orange and yellow at the base of each tepal (they are neither petals or sepals) surrounding a central golden cone. Flowers are often unseen as they bloom after the leaves emerge in late spring and are often high in the tree but sometimes found clipped by squirrels beneath the tree. The blond, conelike fruits adorn the tree in winter.

RELATED PLANTS More columnar and smaller cultivars are sometimes available.

Female trees of *Maclura pomifera* (Osage-orange) produce softball-sized fruit.

Maclura pomifera
Osage-orange, hedge apple, bois d'arc

Moraceae (mulberry family)
Landscape group: successional trees

I debated whether to include this plant as a midwestern native as its relict range at settlement was very small in the southern Great Plains though Native Americans cultivated it into Missouri where Lewis and Clark "discovered" it. Before barbed wire was invented, settlers planted thorny and tenacious Osage-orange as a living fence across the prairies. Osage-orange is related to mulberries and in the same family.

HOW TO GROW Easy to grow from seed, it thrives in almost any soil that is not wet. The tree requires full sun, grows fast, and yet has extremely strong wood. It's naturalized across the Lower Midwest and northward to southeast Nebraska, central Iowa, northern Illinois, southern Michigan, and all of Ohio. It's usually listed as cold hardy to USDA zone 4, but having grown up in that zone, I can attest that is not the case—it's hardy through most of USDA zone 5.

LANDSCAPE USE A historic tree for landscape restorations. Thornless male trees make fine street trees while female trees have such large fruit that they must be located where they won't damage anything beneath them when the fruits fall in autumn. Old hedgerows of Osage-orange persist, often spared when land is developed.

ORNAMENTAL ATTRIBUTES The fruit are large (about softball size), lime green, and beautiful in autumn. They are used

The green-yellow flowers of *Magnolia acuminata* (cucumbertree) are little noticed among the spring foliage.

Red foliage and blue fruit make *Nyssa sylvatica* (blackgum) an astounding specimen tree for the fall landscape.

as decor and as a deterrent for spiders. Mature trees have lovely brownish orange, furrowed bark.

NOTES The tree invaded abandoned pastures and prairies to the point of being considered a pest in some areas. It's a favorite browse of deer to the point it no longer survives within their reach—no longer reproducing. It has a namesake sphinx moth also called Hagen's sphinx that has followed the tree into its current range and whose caterpillar feeds on no other plant.

RELATED PLANTS 'White Shield' is a male, thornless tree while 'Cannonball' is a massive fruited female cultivar for those who may want a conversation piece.

Magnolia acuminata
Cucumbertree

Magnoliaceae (magnolia family)
Landscape group: mesic forest trees

Cucumbertree is a true magnolia and, along with tulip tree, the only two genera in the magnolia family (Magnoliaceae). I grew up near the former national champion cucumbertree that resided in Waukon, Iowa (the tree is no longer there). We would always stop by and visit the inspiring giant tree on family trips to and from the Mississippi River. I've since witnessed wild cucumbertrees, none remotely so large. They're scattered in a rich ravine leading towards the Mississippi River in Trail of Tears State Park near Cape Girardeau, Missouri. Were they witnesses to the Cherokee's forced march?

HOW TO GROW Cucumbertree is mainly an Appalachian tree, hardy in USDA zone 4, but ranges northward into Ontario's Carolinian Forest and westward to the Ozark Highlands and Ouachita Mountains, barely crossing Arkansas's western state line into Oklahoma. It's easily grown from seed and readily transplants balled and burlapped or from containers but is rarely in the nursery trade anymore. It's native to mesic forests of mixed species and nowhere common.

LANDSCAPE USE A phenomenal shade tree 60–80 feet tall that was widely planted across the Midwest in cemeteries, estates, and arboreta in Victorian times. It makes a huge, sturdy tree (ice and snow resistant) best planted in a lawn or woodland garden and is not tolerant of street tree planting or similar disturbed urban soils.

ORNAMENTAL ATTRIBUTES The rather large, coarse leaves that make the tree stand out from other trees in the landscape. Young trees show perfectly pyramidal crowns but become wide-spreading (as wide as tall) with age. It's the only magnolia with yellow-pigmented flowers though they bloom after the foliage emerges and are a blend of yellow and green. Some midwestern trees have glaucous, bluish-tinged flower buds that are exquisite. The young fruit look like little cucumbers but become pregnant with vermillion fruit that burst out of the follicle when ripe. The fall color is consistent golden brown.

Nyssa sylvatica
Blackgum, tupelo

Cornaceae (dogwood family)
Landscape group: mesic forest trees

The fiery, blaze-red fall color of many blackgums is the first thing that comes to mind when I think of this tree. Blackgum is in the dogwood family (Cornaceae) but with little-noticed, un-showy, dioecious flowers. Female trees (with a male nearby) produce rounded bluish black fruit that are relished by fruit-eating birds and that looking attractive in the fall, rarely lasting long after leaf fall as they are high in fat and stripped by migrating birds. This tree is decidedly pyramidal with pagodaesque branching when young but develops a squat, rounded crown with age. Blackgum grows in moist, acidic soils across the entire Eastern Midwest and westward through most of the Ozarks. Outlying trees were once found in southeastern Wisconsin.

HOW TO GROW Blackgum often grows on well-drained, sandy lenses of soil on or adjacent to wetlands but is not a swamp tree. It requires slightly to strongly acidic soils. Depending on its source, a tree on rich soil may show extreme drought tolerance but it can be lost to dryness. I have yet to have a tree survive on my dry ridge. Blackgum is cold hardy to USDA zone 5 with select northernmost strains surviving as far north as Minneapolis.

LANDSCAPE USE Blackgum makes a stunning fall-focal shade tree 50–60 feet tall but needs a woodland or edge of wetland site. It's a great choice for a garden focusing on fall songbirds or for a honeybee garden.

ORNAMENTAL ATTRIBUTES Fall color is usually astounding, always grabbing attention from a great distance because few trees flame so vibrantly red. The form of the tree, with its horizontal tiered branching, accentuates the flat ground plain of the Midwest while the trunks of mature trees are always memorable with their blocky bark reminiscent of the pattern on a reticulated giraffe.

RELATED PLANTS Many new selections of this species are available in the horticultural trade, including weeping, variegated, and spring red-leaved forms. Ask nearly any nursery if a potential tree is a male or female and you usually will get a blank stare; you will have to buy the tree in flower or fruit to be sure.

Platanus occidentalis
American sycamore

Platanaceae (planetree family)
Landscape group: floodplain/bottomland trees

American sycamore was the largest tree in eastern North America, our giant sequoia so to speak. In the lower Ohio River valley, it grew to massive proportions and its nature of hollowing out with age made it an ideal shelter, used for livestock in the absence of barns. The hollow tree was also the original home of chimney swifts and it makes a fine tree for all sorts of cavity-nesting and roosting wildlife. An American sycamore's hollow nature rarely compromises the tree's strength and its shock-resistant wood was used for butcher blocks.

HOW TO GROW American sycamore is found in floodplain forests across the entire Eastern and Lower Midwest and into parts of the Upper Midwest—northward up the Missouri River to Omaha, from the Des Moines River to central Iowa, and sparingly along the Lower Wisconsin River. It is easy to grow from seed and grows as fast as any tree: 3–4 feet in a single season is commonplace. A 25-year-old tree can reach 100 feet tall when growing in rich alluvial soils. American sycamores are easy to transplant, even bare root. The tree is heat resistant and drought tolerant.

LANDSCAPE USE American sycamore is ideally suited to urban use but is slighted by urban foresters because of its huge mature size and susceptibility of anthracnose that blackens new foliage during cool, wet springs. For this reason, its hybrid with the European planetree (*P. orientalis*), the London planetree (*P. ×acerifolia*) is utilized instead. When I walk down streets where both are planted, I fail to see London planetree's advantage. American sycamore is a fine tree for riparian restorations and large gardens. Its continual exfoliating bark, large leaves, and round fruit that disintegrate make the tree too messy for tidy landscapes.

ORNAMENTAL ATTRIBUTES It's all about the spectacular bark and the massive, awe-inspiring size and form of the tree. American sycamore's adaptation to floodplain forests filled with vines is to regularly slough off its bark and competitive vines. This reveals fresh, snowy white bark underneath and an ever-changing marvelous pattern from white to grays on older patches of bark. American sycamore's massive gnarly white trunks and limbs delineate riverways and are marvelous sights through winter's changing light. Any anthracnose simply gives one more time to look at a tree's marvelous bark, but can cause stem dieback in severe cases. Leaves are large and coarse with a distinctive pleasant aroma. Fall color is golden brown. Many songbirds including

A massive tree of *Platanus occidentalis* (American sycamore), with characteristic white trunk, towers above Kansas City's renowned Ward Parkway.

American goldfinch and purple finch relish the seeds (actually achenes) found in American sycamore's round fruit heads (which has just a single fruit head per stem while London planetree will have multiple fruit heads per stem). Read Nancy Ross Hugo's "American Sycamore" in her book *Seeing Trees* where she states: "This tree has more visual attractions than an art gallery" and "There are few trees more dazzling."

NOTES In the Upper Midwest, it is important to plant local or proven strains of American sycamore as it's not fully hardy and can suffer from frost cracking and dieback from both cold and severe anthracnose enhanced by the cooler springs. There are a few great trees northward into USDA zone 4b; a marvelous tree in Wold Park in Decorah, Iowa, has thrived through amazing cold and defies all the issues normally attributed to the tree in the Upper Midwest. I know of other anthracnose-resistant American sycamores too, and wish they would be tested and utilized rather than the hybrid. A true American sycamore shines with the spirit of our place.

Populus deltoides
Eastern cottonwood

Salicaceae (willow family)
Landscape group: floodplain/bottomland trees

Eastern cottonwood, the state tree of Kansas and Nebraska, is among the largest of trees in the Midwest and can readily grow 120 feet tall and wide with a massive trunk 10 feet or more in diameter. Historic trees are becoming fewer and farther between, but they grow quickly and rarely live more than 150 years. In the presence of stately cottonwoods, I always recall renowned conservationist John "Jack" White's lecture "Why Save Nature" where he begins with his experience as a boy in central Illinois to witness such a historic tree dynamited, bulldozed, and burned. To this day it's hard for him to speak about what he witnessed, as it would be with all plant lovers. Listen to his story at the beginning of his lecture "The Illinois Prairie and Its People" at youtube.com/watch?v=fjLEWAVplnI&feature=share.

I grew up just a few blocks from a massive cottonwood still standing on Luther College's campus. Jens Jensen planted this cottonwood in 1911 as a symbol of the American plains. I'm always glad to see that beauty when I'm home and I mention this because the tree has a reputation of being weak-wooded, short-lived, and a poor choice for landscapes. Then why does it still thrive and reach gargantuan proportions through all the wild weather the Midwest can throw at it? Survivor is an understatement. Everywhere in the Midwest, there are spectacular cottonwoods, leaves

dancing and sparkling in the summer breezes. The tree inspired Walt Disney's characters from his rural north Missouri farm.

HOW TO GROW Cottonwood is easy to grow from fresh seed, which immediately germinates in wet, exposed ground. In nature that meant quickly colonizing the aftermath of a flood or an old buffalo wallow, but it'll do the same where bare soil is exposed. It can also be rooted from cuttings taken in the dormant season and is easy to transplant bare root. Cottonwood is heat and drought tolerant and thrives in soils from wet to dry. It does require full sun.

LANDSCAPE USE Cottonwood is an iconic midwestern tree for open parklands, riparian restorations, or wherever there is space for such a massive tree. Its roots seek water so it is not a choice where there are sewer and septic lines. Female trees produce the capsules that burst, when ripe, with tiny seeds attached to its namesake cotton that give it the ability to go airborne. Mature female trees produce copious amounts of cotton for a very brief period and this "mess" can clog window screens, heating and cooling units, and even stick to clothes drying on a line. Siberian elm was promoted as a replacement for this reason and we know how well that mistake went. I always love a cottonwood's summer snowfall (seed release), which should be witnessed by all midwesterners as the cotton captures the intense, near summer solstice sun.

ORNAMENTAL ATTRIBUTES Mature trees have deeply furrowed bark on older limbs and trunks that are a real standout in winter. Fall color can be resplendent yellow to gold in drier autumns. The tree's summer foliage is exquisite with a shiny surface that sparkles in sunlight and the tree's thin, flat petiole allows the leaf to wobble in the wind, creating a lovely pitter-patter in the slightest breeze. Pendant catkins of reddish flowers adorn the tree in spring before the leaves emerge.

RELATED PLANTS Cottonless cottonwoods are simply male trees. A hybrid sold under that name grows quickly and is subject to canker diseases that usually kill the tree by the time it reaches 20 years.

Populus deltoides (Eastern cottonwood) stands taller than Luther College's seven-story Main building in Decorah, Iowa.

Populus tremuloides
Quaking aspen

Salicaceae (willow family)
Landscape group: successional trees

Quaking aspen is closely related to cottonwood and in the willow family. The rounded leaves of this tree wobbling in the slightest breeze on their thin, flattened petioles create a sparkling look and a pleasant sound giving the plant

Bark of *Populus tremuloides* (quaking aspen) is reminiscent of paper birch but does not exfoliate.

its name. The bark is reminiscent of paper birch but does not exfoliate.

HOW TO GROW Quaking aspen is found across the Upper Midwest with a few isolated locations farther south. It's a classic successional tree that invades open ground and spreads into groves by underground rhizomes—individual trunks are rather short-lived. It can be grown from seed or by dividing off suckering plants. It requires full sun and does best in areas with cooler summers. It is hardy in AHS heat zone 6 and northward. It will grow southward in moist, sheltered locations.

LANDSCAPE USE A memorable ornamental shade tree, especially when allowed to grow into a naturalistic clump or grove; it makes a fine summer screen, especially good for buffering noise.

ORNAMENTAL ATTRIBUTES Quaking aspen is best known for its pure yellow fall color but locally adapted strains are needed. Trees are narrowly upright, usually growing in tight groves. The trunks are very showy grayish white to white with black horizontal markings where branches have abscised.

RELATED PLANTS A selection 'Prairie Gold' from Nebraska possesses good heat and drought tolerance and is a good choice for western and southern midwestern plantings.

Bigtooth aspen (*Populus grandidentata*) is a bigger cousin of quaking aspen that grows quite similarly as a successional species that suckers into groves. It's found from central Minnesota to the northeast corner of Missouri and eastward to almost all of Ohio, ranging southward sporadically into the southern Eastern Midwest. It is rarely available from any nursery but does move well with a tree spade. The leaves have a strongly larger toothed edge than quaking aspen and the trunk is more whitish brown.

Prunus serotina
Black cherry

> Rosaceae (rose family)
> Landscape group: successional trees

Doug Tallamy, with his lectures and books, may have finally swung the nursery industry around to consider growing black cherry and given homeowners reason to plant black cherry. Its leaves host such a wide variety of marvelous insects; it really is the epitome of a great bird plant beyond its fruit, which they also devour. It's in the rose family along with other cherries, plums, crabapples, and hawthorns. Robert Dyas at Iowa State University long praised black cherry's virtues, but for the most part the tree is relegated to weed status by arborists and horticulturists. The tree is common in upland woods and successional areas like fencerows throughout almost the entire Midwest, absent only as a native plant from the Dakotas, western Minnesota, and the northwest corner of Iowa. Black cherry can be an understory tree for many years but is long-lived. Mature trees readily grow 80–100 feet tall or more with a trunk diameter greater than 3 feet.

HOW TO GROW Black cherry is easy to grow from seed, readily reseeds, and grows rapidly as a young tree. It is oddly difficult to transplant, so it is best purchased as a container-grown plant. Seedlings are available from some state nurseries. The tree grows best in full sun but will grow in light or open shade. Well-drained soil is a must. It is very heat and drought tolerant once established.

LANDSCAPE USE Black cherry is a premier timber and food forest tree. Rockford, Illinois's former furniture industry developed from its fine local stands and most consumers

The flaky bark of *Prunus serotina* (black cherry) has earned it the name "potato chip tree."

Oaks are the official national tree of the United States, the state tree of Iowa, and the dominant trees of the Midwest in numbers, coverage, and importance to the web of life. No group of trees even comes close to their value to wildlife as both forage and mast. In much of the Midwest, oaks are not regenerating and are not being planted with species that reflect their historic makeup. The dense, closed forests of today do not allow for oak seedlings to survive and the trees themselves are more susceptible to diseases and environmental stress. Most oaks take time to grow and are not economical for nurseries competing with big box store prices nor in tune with our desire for instant gratification.

I look at most mature oak woodlands and see elm, hackberry, bitternut hickory, sugar maple, and beech thriving, filling the gaps where veteran oaks blow down or die. Foresters and land managers are working with timber cuts and fire to improve oak regeneration but outside of areas with lean soils, there is rarely success. What everyone can do is plant appropriate oaks for your area and site conditions. Do it for untold generations of wildlife and future generations of people. A person can hardly leave any finer living legacy.

Oaks are comprised of two groups: white oaks and red oaks. White oaks are found across the entire Northern Hemisphere, are more disease resistant, longer lived, and produce leaves that have rounded lobes. They produce acorns in one season that are "sweeter" with less tannin. Their wood holds water so is used for barrels. In the Midwest, the white oak group includes bur, chestnut, chinkapin, post, swamp white, and white oaks.

Red oaks are found only in North America, take two seasons to produce their acorns, and have leaves with pointed or bristle-tipped lobes. Their wood does not hold water. Midwestern red oaks include black, blackjack, northern pin, northern red, pin, scarlet, and Shumard oaks.

cherish cherry lumber for its beautiful color. The flavor from its wood on smoked meats is also becoming prominent (much sweeter than hickory). The fruit is edible and makes a fine juice more boldly flavored than cultivars of sweet cherry (*Prunus avium*) currently labeled "black cherry." The tree should be considered for planting in clumps and masses like river birch for a premier wildlife friendly natural garden.

ORNAMENTAL ATTRIBUTES The upright, open crown of a mature tree is not highly rated by most gardeners (I've done informal polls during many tree tours). The flaky, charcoal gray bark can be beautiful, and has earned black cherry the name "potato chip tree." A tree at the International Crane Foundation in Baraboo, Wisconsin, had the shaggiest bark I have ever seen on one. The tree would rate higher if grown in a clump to show the bark and create a more fully formed plant. The flowers are lovelier than given credit, in bottlebrush-like clusters, and remind me of white lace adorning the plant in midspring as the leaves emerge fully. Fall color can be some of the best of any tree in the Midwest with colors ranging from golden yellow to apricot orange and scarlet red, all often on the same tree.

PLANTING AN OAK

EASIEST: swamp white, pin, and shingle oaks
EASIER: northern red, Shumard, northern pin, chinkapin, and chestnut oaks
CHALLENGING: white, bur, post, and scarlet oaks
DIFFICULT: black and blackjack oaks

Quercus alba
White oak

Fagaceae (beech family)
Landscape group: oak-hickory forest

If pressed to pick a favorite tree beyond the one in front of me, it would have to be the white oak, the state tree of Illinois. It may be the first tree drawn to my attention for its beauty by my parents, who admired a perfectly formed tree near one of the golf course tee offs. It's a survivor too, readily reaching 200 years and occasionally 400 years of age. I think of it as the "Queen of Trees" as it is more feminine with finer branching and unmistakable exquisitely curved, deeply lobed leaves.

HOW TO GROW White oak is characteristic in dry mesic to dry woodlands across most of the Midwest as far west as central Minnesota, southeastern Nebraska and easternmost Kansas. It's easy to grow in containers from acorns that sprout in the fall. It is more challenging to transplant than most oaks but that is mainly because it abhors construction-damaged, compacted, or otherwise destroyed soils. Plant it in well-drained soils that are moist to summer dry and have not been graded or filled. In appropriate soils, it actually grows surprisingly fast with two flushes of growth per summer season. Next to the northern red oak, it's also the most shade tolerant so I have planted these two as the future forest of my young woods. One day, long after the existing trees and I are gone, white oak will rule these woods.

LANDSCAPE USE This premier oak makes a fine, long-term shade tree choice. It grows 50–100 feet tall or more with a much wider crown than height.

ORNAMENTAL ATTRIBUTES White oak is known for its beautiful and uniform branching, stout (wider-than-tall) crown, and consistently stunning fall color, by far the most colorful of the white oak group. Young leaves are silvery, while young trees often hold their foliage into winter.

Quercus bicolor
Swamp white oak

Fagaceae (beech family)
Landscape group: floodplain/bottomland trees

Swamp white oak, currently the darling of city foresters, is easy to grow in urban-damaged soils. About 400 of these trees grace the World Trade Center Memorial in New York City.

HOW TO GROW Native to select river floodplains and wetlands that are more sandy-based, this species is distributed across the Midwest from southeast Minnesota southward to Missouri north of the Ozarks. It's the dominant upland

Quercus alba (white oak) leaves paint the fall with a range of warm colors.

The undersides of *Quercus bicolor* (swamp white oak) leaves are strikingly whitened.

Quercus coccinea (scarlet oak) has the most vibrant red fall color of any oak.

Quercus coccinea / *Quercus ellipsoidalis*
Scarlet oak / Northern pin oak

Fagaceae (beech family)
Landscape group: oak-hickory forest

113

There is something almost otherworldly about a scarlet oak in peak fall color. Its name says a lot but the translucency of the leaves capturing the low sunlight of autumn combined with the intense saturated color is simply stunning. Scarlet oak is found wild from the eastern Ozark Highlands to the Shawnee Hills, Appalachian Highlands, and southward through the Piedmont of the Southeast and northeastward to southern New England. It grows in sandy, igneous-based, or other well-drained acidic soils. Its midwestern range is scattered northward into northern Indiana and Lower Michigan, but I think no botanist will agree where it transitions into the northern pin oak (*Quercus ellipsoidalis*), an extremely similar oak native across almost the entire Upper Midwest. Northern pin oak thrives in clay and more alkaline soils. As a young botanist, I would key perfectly trees near my hometown to scarlet oak and some to northern pin oak. *Flora of the Chicago Region* lumps them as one species, but it's good to know a plant's origin to match the appropriate tree to your local conditions.

HOW TO GROW Lower Midwest scarlet oaks require well-drained acidic soils and are fickle to transplant into other soils; they also suffer from chlorosis in alkaline conditions. Upper Midwest northern pin oaks are more tolerant of a wide range of soils as long as they are not wet.

LANDSCAPE USE These make fine shade trees 50–60 feet tall, but on a limited basis. They are highly susceptible to oak wilt so are best isolated and not planted in groves as the wilt is transferred by natural root grafts. Both species work well for a natural landscape. They produce abundant smaller acorns relished by wildlife.

ORNAMENTAL ATTRIBUTES These oaks have some of the finest fall color of any shade tree in glowing scarlet red.

RELATED PLANTS Majestic Skies ('Bailskies') northern pin oak from Minnesota is a fine choice for the Upper Midwest and areas like Powell Gardens near Kansas City where typical scarlet oak is more challenging to grow.

Black oak (*Quercus velutina*) is the red oak of poor, dry soils, often on pure sand or gravel but occasionally on rich upland soils. It's a tree of many sand savannas and dry oak-hickory woodlands. The species ranges across the Eastern and Lower Midwest and much of the Upper Midwest northward to central Wisconsin, southeast Minnesota, central Iowa, and southeastern Nebraska. Although it is not in the commercial nursery industry, it is available from some

tree at Powell Gardens (its range ends just 8 miles west), where clay soils create wet conditions in spring, even on high ground. Such habitats with clay hardpans are often called flatwoods. Along with pin oak, swamp white oak is the most easily transplanted oak, field dug or grown in containers. It can easily be grown from acorns that sprout in the fall. It prefers neutral to acidic soils and is absent from limestone-based rivers and suffers chlorosis in areas where soils are limey and alkaline. New trees in my childhood neighborhood have died because of the high pH soils, so the use of this tree in such areas should be cautioned.

LANDSCAPE USE A premier shade tree in appropriate soils, the swamp white oak also literally shines when night-lit as the underside of the leaves is whitened and reflects light. It usually grows 60–80 feet tall and wide.

ORNAMENTAL ATTRIBUTES This tree has an upright oval crown when young with a skirt of drooping branches, but as it matures, its shape becomes broad spreading. An open-grown swamp white oak is a site to behold with magnificent horizontal limbs. The two-toned leaves sparkle as their undersides show in a breeze.

RELATED PLANTS Most homeowners loathe acorns, but for wildlife lovers the selection 'Bucks' Unlimited' is a very vigorous swamp white oak with extraordinary acorn production providing a favored fall feast for deer and other wildlife.

Quercus velutina (black oak) in bloom is an underappreciated ornamental shade tree.

Leaves of *Quercus marilandica* (blackjack oak) are mainly three-pointed and turn shades of red in the fall.

state nurseries. Black oak is challenging to transplant but easy to grow from acorns. It is usually inherited in a landscape and does make a fine shade tree but is completely intolerant of soil compaction or disturbance. The foliage is usually glossy dark green and turns a chartreuse-golden-olive fall color that is quite distinctive and differentiates it from northern pin and scarlet oaks. Botanists consider many trees in the Upper Midwest hybrids with northern pin oak. The new leaves are rich maroon to pinkish red, always a beautiful contrast with the golden pendant male flowers. The tree usually has a strong-buttressed base and the mature, somewhat pebbled trunk is almost black, giving this oak its name. 'OakRidge Walker' is a cut-leaf cultivar from Illinois.

Blackjack oak (*Quercus marilandica*) is another Lower Midwest oak that requires hot summers. It often grows in the same habitat as black oak and they do hybridize. It is a smaller-statured oak, often called a scrub oak as it grows slowly on some of the worst soils but does eventually reach shade tree size. It also grows below ground and little above so is not in the nursery industry. Like black oak, it hates soil compaction or disturbance. An established tree shrugs off heat, drought, and ice storms. The tree is usually inherited in a landscape and is quite beautiful with a sturdy pyramidal crown and lovely lustrous leaves that can turn vibrant shades of red in the fall. The mainly three-pointed leaves are exquisite when backlit. Mature trees have black, somewhat warty trunks.

Quercus imbricaria
Shingle oak

> Fagaceae (beech family)
> Landscape group: successional trees

With its easy-to-identify simple (unlobed) leaf, this oak is the first oak to colonize open ground and truly a successional species. Its marcescent foliage is iconic in the winter landscape wherever native.

HOW TO GROW Shingle oak is found from the Osage Plains across the southern Iowa–northern Missouri Drift Plains, eastern Ozarks, through the central and eastern Corn Belt Plains and into the Bluegrass and Shawnee Hills. It is hardy almost everywhere in the Midwest but northern strains must be grown in USDA zone 4. Shingle oak grows very fast, tolerates disturbed soils, and readily transplants—making it easily available in the nursery industry. It is one of our shortest-lived oaks, fully mature and declining by age 100. This oak is highly susceptible to gaudy oak galls that create woody ball-like growths that can weaken a tree under severe conditions. The tree is also susceptible to oak wilt.

LANDSCAPE USE Shingle oak holds its leaves through winter for up to 50 years, making it a wonderful windbreak and screen in soils or situations where evergreens won't grow. The tree starts out with pyramidal growth somewhat like a pin oak but with less-pendant branches and matures with a large rounded crown nearly 100 feet tall and just as wide

Quercus imbricaria (shingle oak) trees decked in marcescent winter foliage create beauty in the winter landscape.

This monumental specimen of *Quercus macrocarpa* (bur oak), with a trunk diameter of 10 feet, grows near Columbia, Missouri.

Quercus macrocarpa (bur oak) is named after its fringe-cupped acorns.

when found in good soils. It makes a fine street and shade tree with abundant tiny acorns relished by wildlife.

ORNAMENTAL ATTRIBUTES Fall color may be red on saplings but small to large trees simply turn brown late in fall. The winter leaves are a wonderful composition with companions such as red cedar, broomsedge, and purpletop grass in abandoned fields.

RELATED PLANTS Oaks are wind-pollinated and species occasionally hybridize where they grow in close proximity. You will find a hybrid in almost any locale. Powell Gardens, for example, has two documented wild hybrids on its grounds: northern red crossed with shingle oak (*Quercus* ×*runcinata*) and swamp white crossed with post oak (*Q.* ×*substellata*). Hybrids usually look intermediate between both parents. A few oak hybrids make excellent shade trees and are available from specialty nurseries, including white oak crossed with bur oak (*Q.* ×*bebbiana*) and its cultivar 'Taco', bur oak crossed with chinkapin oak (*Q.* ×*deamii*) and its acorn-free cultivar 'Champion Seedless', and bur oak crossed with swamp white oak (*Q.* ×*schuettei*) and its selection 'Kimberley'.

Quercus macrocarpa
Bur oak

Fagaceae (beech family)
Landscape group: oak-hickory forest

Bur oak is one of the, if not the iconic tree of the Midwest. It is found across the entire region and its range is centered here. There are two quite distinct forms of the tree: an Upper Midwest version with smaller acorns, corky twigs, artistic growth, and weirdly fickle to transplant. The Lower Midwest version has huge acorns reaching the size of golf balls, a straighter trunk, and is more easily cultivated. They gradually change from north to south so are not given distinct varietal names.

Groves of bur oaks grace landscapes all across the Midwest and are the dominant tree across major portions of the Upper Midwest from Chicagoland to Nebraska and northward through all of Iowa, southern and western Minnesota and southern Wisconsin. They're found in savanna areas eastward, more commonly called oak openings in western Ohio and southern Michigan. Very few of these trees are being planted or replaced. On a drive across northern Iowa, I saw bur oak trees anchoring many a farm or home but did not see any recently planted one. It will be a sad day if they disappear, replaced by cheaper "ornamental" trees that will never match their ruggedness and the spirit of the place. We are replanting the tree with swamp white oak simply because its cheaper and easier, regardless of site conditions, historic, and botanical integrity.

HOW TO GROW Bur oak is challenging to transplant, especially the Upper Midwest form which has a deep taproot when young. It is easily grown from acorns but saplings are few around parent trees. Young trees grow moderately fast but growth slows considerably as trees mature. Bur oak tolerates a wide variety of soils as long as they are not continually saturated. It is a prairie invader tree that requires full sun to thrive. Bur oak is a major component of upland savannas and drier woodlands in the Upper Midwest, while more restricted to deeper soils in the unglaciated Lower Midwest where it reaches its most massive size in mesic floodplain forests.

LANDSCAPE USE: This prime large shade tree grows 60–100 feet tall and wide and is resistant to wind, ice, and heavy snow, with a longevity of at least two centuries and often much longer. It is tolerant of urban conditions and disturbed soils, but the southern form is shunned because of its large acorns and the northern form is rarely planted because it is challenging to produce by the nursery industry. Bur oak has tremendous wildlife value, hosting a wealth of insects and providing acorns for other creatures.

ORNAMENTAL ATTRIBUTES A magnificent tree with a sculptural character that may exceed all other native shade trees, bur oak often displays strong horizontal limbs that can sweep to the ground. Young northern trees can have exquisitely corked twigs, an adaptation to protect against fire and browsing. Foliage is decidedly lighter underneath so is quite beautiful in the wind. Fall color is dull olive-golden-tan. Flowers appear as the leaves emerge in spring. Male flowers are pendant, gold pollen-laden catkins while tiny female flowers produce distinctive fringe-cupped acorns relished by wildlife.

RELATED PLANTS 'Big John' is an upright, acorn-less cultivar. Post oak (*Quercus stellata*) is reminiscent of upland bur oaks of the Upper Midwest but found only in the Lower Midwest. Here it grows in marvelous groves of dome-shaped trees of rugged character as far north as northeastern Kansas to southeastern Iowa and eastward to southern Ohio. It's most common in the Osage Plains and Ozarks. Post oak is incredibly heat and drought tolerant with lustrous leaf surfaces and wood strength of iron, immune to ice and all but tornado winds. Its fall color is much more colorful than that of bur oak, often turning rich burnt reds. Post oak is native into USDA zone 5b but requires summer heat so is best planted only where native in the Midwest and is not hardy in most of the Upper Midwest.

Quercus muehlenbergii
Chinkapin oak

Fagaceae (beech family)
Landscape group: oak-hickory forest

Chinkapin oaks crowned the bluff overlooking the spot where the Kansas River meets the Missouri River—where downtown Kansas City now stands. There stands a monument to Lewis and Clark, who, at the time of their historic discovery, described the scene, which included the first herds of bison they encountered, elk, Carolina parakeets, and chinkapin oaks. Do you think there's a chinkapin oak there today? Sadly, no. These iconic trees are limited to limestone outcrops on bluffs and hills throughout much of the Midwest including all of eastern Kansas to southeastern Nebraska, to southeast Minnesota (no extant trees there now), southern Wisconsin and across Lower Michigan into Ontario and in most of Ohio except for its eastern flanks. The Midwest is the core of this oak's range but there are outliers eastward to Lake Champlain and the Hudson Valley and southward to the Texas Hill Country. Trees near the northern edge of the range in northeastern Iowa along bluffs of the lower Upper Iowa River are barely shade tree size.

HOW TO GROW Chinkapin oak prefers neutral to alkaline soils that are well drained though it can be found in some mesic floodplain forests. Its native haunts translate into adaptability to droughty conditions with urban concrete and rubble. It's easy to grow from acorns and readily

Quercus stellata (post oak) produces fabulous burnt red fall color.

Quercus muehlenbergii (chinkapin oak) in bloom shows its emerging leaves, which are easily identified by their characteristic saw blade-like teeth.

Quercus palustris (pin oak) in peak fall color is a sight to behold.

transplants. It also grows fast where happy, usually putting out a second flush of growth in summer.

LANDSCAPE USE This marvelous long-lived shade tree is adapted to most street tree locations. It has deep coarse roots that are great for gardening beneath and aren't a menace to underground utilities. Chinkapin oak grows 50–80 feet tall and wide.

ORNAMENTAL ATTRIBUTES The tree's unique growth pattern translates into a gnarly crown of branching character that really stands out all winter. The light ashy gray bark is also quite noticeable. Fall color can be marvelous from burnt orange into red tones.

NOTES Chinkapin oak is a bit challenging to find but many native plant nurseries do carry it.

Quercus palustris

Pin oak

Fagaceae (beech family)
Landscape group: floodplain/bottomland trees

Pin oak dominates the urban forest of many communities across the Lower Midwest and it is planted wherever cold hardy across the entire Midwest (USDA zone 4). Because of its common use and problems with chlorosis in many soils, this tree is currently demonized though it's a far superior choice than current nonnative trees that are widely

planted. It is native to floodplain forests and to upland flatwoods where a clay hardpan creates more poorly drained soil. Pin oak ranges from abundant in the Osage Plains, areas around the Ozark Highlands (it's absent from the core of the Ozarks), across the Lower Mississippi River, almost the entire Ohio River basin, Lake Erie Lake Plains, and eastward to the Mid-Atlantic states.

HOW TO GROW Pin oak is easy to grow from acorns, grows as fast as any tree, and is the easiest oak to transplant. It requires rich clay to loam soils that are neutral to acidic; it turns an anemic yellow and suffers from chlorosis (lack of iron) in alkaline soils. Regionally native sources, adapted to your local soils, are best choices. Prune out any multiple leader growth on young trees to maintain a strong central leader; multiple leaders are often subject to storm damage on this species. Pin oak is occasionally subject to oak wilt disease (*Ceratocystis fagacearum*).

LANDSCAPE USE Pin oak is a premier shade tree for appropriate soils and in locations with room for its skirt of lower branches. If its lower limbs are allowed to sweep to the ground, it creates a premier children's play space or fort. Young trees hold their leaves through winter, creating a good screen and windbreak for a couple of decades or more. Pin oak is often used to line entrance drives and creates a grand entrance where branches over the drive are pruned up while outer branches are allowed to cascade. It also makes a good street tree in areas with limited use of concrete and where the natural soil profile is intact (it's not happy in areas where the earth was heavily sculpted with rubble fill for a development). It was heavily planted as a street tree after Dutch elm disease, so in some areas its use should be tempered to allow for tree diversity. The wildlife value of pin oak is extremely high as it supports a plethora of insects and produces a heavy mast of small acorns relished by many songbirds. Pin oak grows 60–100 feet tall.

ORNAMENTAL ATTRIBUTES Young trees create a very special pyramidal crown with a strong central leader, upper branches ascending upward, middle branches growing outward, and lower branches pendant towards the ground. With age, the tree loses its lower branches and forms a massive, rounded crown. Fall color can be lackluster in areas with warm autumns, but is brilliant burnt red to scarlet in cooler areas or in cooler autumns.

Quercus rubra
Northern red oak

> Fagaceae (beech family)
> Landscape group: mesic forest trees

Northern red oak is the granddaddy oak of northeastern North America; no other grows larger in the Midwest and Northeast. It is another magnificent tree with a much higher crown than bur oak atop a long dramatic trunk, often clear of branches for 50 feet or more when growing in a forested setting. It is found across the entire Midwest, except for some of the northwestern fringe but can be cultivated well throughout. It is the tree that produces highly desired lumber, a favorite wood flooring of many midwesterners.

HOW TO GROW Northern red oak is found in well-drained mesic soils and is confined to sheltered north- and east-facing slopes on the western part of its range. It is much more easily transplanted than bur oak but not as tolerant of dryness or disturbed soils. It is easily raised from acorns. Young trees grow moderately fast and, when given open conditions, create stout rounded crowns atop what seems an oversized trunk. It is our most shade tolerant oak and a denizen of more mesic or dry-mesic upland woods. It will not tolerate compacted or otherwise destroyed soil structure left by modern construction activities.

LANDSCAPE USE This premier shade tree becomes massive over time, 60–100 feet tall or more). It's very wind firm and long-lived, readily reaching 200 years old. It was widely available as a landscape tree but is currently not very popular. Its large acorns make it one of the premier mast crops to wildlife and the tree hosts a plethora of beneficial insects.

When young, *Quercus rubra* (northern red oak) shows warm fall colors.

Mature trees of *Quercus rubra* (northern red oak), with striking striated trunks, are surrounded by countless young sugar maples poised to succeed the oaks.

ORNAMENTAL ATTRIBUTES Open-grown trees have magnificent rounded crowns while forest-grown trees are often ultimately "king of the forest" with a high mushroom-like crown. The trunks are simply exquisite with silvered bark striped by charcoal-colored fissures. Foliage is readily identifiable with pointed lobes that do not indent much more than half way between the outer edge and the midrib of each leaf. Fall color is variable from fine burnt reds to shades of bronzed oranges and gold. Purchase a tree while it's in fall color if you desire one with more colorful qualities. Flowers emerge in spring with the leaves and are monoecious, comprised of pendant, golden male catkins and tiny female flowers that take two years to produce the distinctive large acorns that have a relatively thin, flat-topped cap.

RELATED PLANTS Shumard oak (*Quercus shumardii*) in the Lower Midwest and sparingly northward into Michigan is closely related and a superb native shade tree with similarly valued lumber. Shumard oak is much more tolerant of heat, dryness, and a variety of soils as its native haunts include drier ridges and almost diametrically opposite mesic floodplains. Its foliage is shiny with deeper lobes, creating a more splayed/divergent look to the lobes. The trunks do not develop as strikingly beautiful striped trunks as northern red oak does, but the acorns are quite similar, sporting a larger, domed/convex cap. I have witnessed many Shumard oaks grown from southern sources whose leaves freeze green on the tree in autumn so it is important to choose locally grown or trees sourced from the northern edge of their midwestern range. It's a great choice for the Lower Midwest and through USDA zone 5b when grown from northern sources.

Robinia pseudoacacia
Black locust

Fabaceae (legume family)
Landscape group: successional trees

Black locust is one of the world's most beautiful flowering trees and beloved for its flowers wherever it can be grown. It has been widely planted for land reclamation, wood use, and for its ornamental nature and subsequently naturalized across the entire Midwest so the exact native range is unknown (shown by dashed lines on many maps). Black locust is considered invasive beyond where native where it is a threat to prairies, savannas, and other open habitats as some strains spread extensively by underground rhizomes and trees release an allelopath that kills many plants beneath. In its native range, it occurs in mixed forests, under control by neighboring trees it has to compete with as it is intolerant of shade and dies when overgrown by other

Quercus shumardii (Shumard oak) has more deeply splayed lobes than northern red oak.

In peak spring bloom, the canopy of *Robinia pseudoacacia* (black locust) is like a cloud of creamy white.

species. There it has important roles as nurse for upland trees and as a legume, providing nitrogen to other plants (though not through root nodules).

HOW TO GROW Black locust is native to the Appalachians and discontinuous westward in the Ozark Highlands and Ouachita Mountains. This range makes it native across southern Ohio and eastern Kentucky, possibly in southern Indiana, the Shawnee Hills of Illinois and southern Missouri into Arkansas and eastern Oklahoma. It's tolerant of a wide range of soils from pure sand to calcareous clay and acidic mine spoils. Easy to germinate from seeds, more often it quickly reproduces by underground rhizomes. Do not plant this tree near any natural land it could invade, as it is very challenging to remove once established. It is not shade tolerant so cannot compete in a closed forest.

LANDSCAPE USE Not recommended anymore because of its invasive, suckering nature and susceptibility to wood and stem borers. In historic and its presettlement native landscapes, it should be maintained. It could be carefully farmed for its productive wood (rot resistant with high BTU fuel value). The insects linked with this tree, including the destructive but beautiful locust borer beetle and the lovely locust underwing month, have followed it wherever it is planted. The tree grows in my neighboring woodland so I get to enjoy its flowers, fragrance, and accompanying insects. Unfortunately, it also grows on private property adjacent to my meadow/prairie restoration and it's a chore to continually remove suckering plants.

ORNAMENTAL ATTRIBUTES The showy flowers are produced in mid to late spring and are pendant clusters of intensely fragrant white pealike flowers. Its bloom time coincides with the emergence of the destructive EAB but it also is what one of my associates looks forward to on her birthday. Stems have true spines. The tree is upright with rugged furrowed bark of light brown and produces small flat pea pods, which mature brown and hang on the tree through winter. The pinnately compound leaves are finely textured and require no raking. A leaf-mining moth often mires the leaves, which in fall freeze green on the tree or drop after turning yellow-green.

RELATED PLANTS There are a few cultivars of this tree readily available including Chicago Blues ('Benjamin'), which is borer resistant with showy blue-green leaves, and 'Frisia', with spring foliage that emerges butter yellow maturing to chartreuse. Two popular nonflowering cultivars include globelike 'Umbraculifera' and Twisty Baby 'Lace Lady'. There are more obscure cultivars, many selected from this native tree in Europe. It has been hybridized with other species with pink to purplish flowers.

Salix spp.
Willows

Salicaceae (willow family)
Landscape group: floodplain/bottomland trees

Willows are fast growing, relatively short-lived trees that germinate from fresh seed and then only on moist to wet, bare ground in sun. This adaptation makes them ideal colonizers after floods or droughts around streams, rivers, and other wetlands. They need to grow fast in their ever-changing habitat and nurse the next wave of succession in a wetland forest that includes related cottonwoods, elms, hackberry, ash, and others. Willows are host to a wide array of insects so are trees of high wildlife value. Their niche does not fit most of the landscapes we create. Only nonnative weeping willows are readily available as shade-tree-sized willows in nurseries—beloved for their graceful pendant branches grown almost as a cliché along a pond or other water body.

There are two native midwestern species: black willow (*Salix nigra*) and peachleaf willow (*S. amygdaloides*). Many large wild willows are nonnative white willow (*S. alba*) and most often, its hybrid with crack willow (*S. euxina*, syn. *S. fragilis*) creating whitecrack willow (*Salix* ×*fragilis*, syn. *S.* ×*rubens*), which was promoted to settlers as quick timber for the open prairie. As the *Flora of North America* states: whitecrack willow is "the most commonly cultivated and naturalized tree-willow." I observe massive whitecrack

Salix amygdaloides (peachleaf willow) is cloaked in its characteristic peachtreelike leaves.

willows dominating farmland swales from northern Illinois through Iowa and Minnesota, and it's documented across all states of the Midwest except Kansas and Oklahoma.

HOW TO GROW Willows are easiest to propagate by simply rooting a stem in water yet also are easy to grow from fresh seed on moist soil. They require full sun or partial shade. Seedlings are drought tolerant when established. Willows have water-seeking roots that can clog septic and sewer lines. Their branches easily snap, making them susceptible to wind and ice storm damage though they quickly recover.

LANDSCAPE USE Willows are fine left where native along a wetland and I have fond recollections of black willows shading a deck built around them at the Oxley Nature Center in Tulsa. They should be planted in stream and wetland restoration projects as an important component of tree diversity and all the life they nurture. Butterflies often congregate on their trunks and many songbirds use their cottony seed fluff for nest building. Willows contain salicylic acid, the pain-relieving ingredient in aspirin.

Salix nigra (black willow) sets the stage for a shaded deck at Oxley Nature Center.

ORNAMENTAL ATTRIBUTES The fine-textured, narrow leaves of native species are a nice contrast to other trees. Willows green up early in spring and hold their leaves late, usually turning yellowish in fall. Flowers of shade tree species are hidden among the leaves; the flowers of male trees have golden-yellow stamens while female plants have even less showy greenish flowers that produce the capsules from which the cottony seeds disperse.

Salix amygdaloides (peachleaf willow) is found across the entire Upper Midwest and southward into the Lower Midwest along the Missouri, Mississippi, and Wabash Rivers. It can be a small tree but does readily mature to 50 feet and much taller. It is aptly named as its leaves look like a peach tree's. It is a more graceful branched tree than black willow and has pendant branches. It is also hardy throughout the Midwest.

Salix nigra (black willow) is the largest and most widespread of our native willows found across the Eastern and Lower Midwest and absent only from the northwestern Midwest beyond eastern Nebraska and central Minnesota. It readily grows to 50 feet and is hardy everywhere in the Midwest.

Sassafras albidum
Sassafras

> Lauraceae (laurel family)
> Landscape group: successional trees

No tree has a more fun name or such fun leaves found in three shapes—simple, mitten (one lobe), and ghost (two lobes)—all of which may be present on a single branch. The tree has a lemony-spicy aroma to its crushed buds, stems, and roots and is, along with bay leaves, in the laurel family.

HOW TO GROW Sassafras is found across most of the Eastern Midwest and westward across the Lower Midwest through the Ozarks—spreading westward into the Osage Plains. It's not fully cold hardy north of USDA zone 5b. Sassafras is difficult to transplant and establish but tenacious where present, usually suckering into a grove. Trees are either male or female and seed can be readily germinated and transplanted from container grown stock. Sassafras requires full sun and just about any well-drained soil.

LANDSCAPE USE Sassafras makes an amazing shade tree or grove in a more naturalistic landscape. Don't be fooled that it's a small tree by young groves of this plant—it'll eventually reach well over 50 feet tall at maturity. It's a good bird and butterfly garden plant as the late summer fruits are relished and dispersed by songbirds and its one of only two host plants for the spicebush swallowtail.

Sassafras leaves, in all three shapes, show well when backlit by the sun.

A clump of *Tilia americana* (basswood) creates the perfect setting for an outdoor deck.

ORNAMENTAL ATTRIBUTES Sassafras has a beautiful winter silhouette with sympodial tiers of branches, the young stems and twigs decidedly green—it actually was used as a substitute Christmas tree before evergreens were planted on the prairies. Bark on mature trees is deeply furrowed. In spring, the greenish yellow flowers stud the upturned branch tips before the leaves emerge and are the perfect complement to blooming redbuds. The foliage is so elegant and turns extraordinary colors in fall from yellow to orange, red, and purplish—all colors often present on a single tree. The fruit are also very unique—a bluish black berry set on a succulent red pedestal, somewhat like a golf ball on a tee, designed to be seen by birds against the green foliage.

Tilia americana

Basswood, American linden

Malvaceae (mallow family)
Landscape group: mesic forest trees

This may be our best shade tree for pollinators, blooming in early summer with intensely fragrant, nectar-rich flowers. Basswood is now placed in the mallow family along with the tropical hibiscus. Old trees over 3 feet in diameter hollow out, creating ideal dens for wildlife. Trees regenerate by suckers around the base, so the tree is often found growing in clumps where there once was a single trunk.

HOW TO GROW Basswood is one of the few trees found wild across the entire Midwest in moist, well-drained soils. It's challenging to grow from seed and is rarely grown by nurseries compared to nonnative lindens. The tree is easy to transplant, grows fast, and is quite shade tolerant.
LANDSCAPE USE Basswood is a fine shade tree for a woodland garden or food forest, it's a premier honey tree. It casts dense shade and has abundant surface roots so when mature, turf does not grow well beneath it. Basswood grows 60–80 feet tall.
ORNAMENTAL ATTRIBUTES The tree's flowers are creamy golden yellow in pendant clusters attached to a leafy bract that acts as a propeller to disperse the seed. The fragrance is heady and noticeable from a distance, intensifying at dusk. The flowers attract insects from bees to butterflies and moths at night. The large leaves contrast nicely with finer-textured trees. Fall color is yellow at best. The tree is usually very upright, having a pyramidal crown when young that fills out with age.

RELATED PLANTS Several cultivars are available with more pyramidal crowns including American Sentry ('McKSentry') from Wisconsin, 'Boulevard' and Frontyard ('Bailyard') from Minnesota, 'Dakota' from South Dakota, 'Lincoln' and Legend ('DTR 123') from Illinois.

In the Lower Midwest the tree is variable and was once split into several species. (*The Woody Plants of Ohio* splits trees into four species). The white basswood (variety *heterophylla*) is most distinct with its large leaves that are silvered on the underside. It's native sporadically across the Lower Midwest but cold hardy at least through USDA zone 5 with a magnificent tree at the Klehm Arboretum in Rockford, Illinois. The national champion tree is easy to observe at the Missouri Botanical Garden. The silvered leaves make it a distinctive tree, especially with night lighting. 'Continental Appeal' is an excellent cultivar of white basswood selected in Illinois for its pyramidal form and clean foliage.

Ulmus americana
American elm

Ulmaceae (elm family)
Landscape group: floodplain/bottomland trees

I almost didn't buy the house I reside in because an American elm arches up over the deck and creates a wonderful space with its leafy canopy overhead. The threat of it dying of Dutch elm disease (DED) and its removal costs crossed my mind. Seven years later, I've lost eight American elms around the wooded yard as they've grown from adolescence to adulthood (they become susceptible to the disease at that age) but the deck elm is still going strong, as are several other wild elms. Like all living things around me, I've decided to enjoy the deck elm's company every day I'm blessed with its presence. No finer canopy could have

A branch of *Tilia heterophylla* (white basswood), dripping with fruit, clearly shows the whitened underside of the leaf.

A classic specimen of *Ulmus americana* (American elm) thrives at the Cranbrook estate outside Detroit, Michigan.

been designed to create such a comfortable shaded space for the deck.

My deck elm made me include American elm as a viable shade tree in this book. I can't recommend planting one, but American elms inherited in a landscape are worth embracing. It also makes me question where horticulture is headed in regards to elms—new nonnative hybrids are exclusively promoted while throughout the Midwest a few classic American elms defy disease. These remnant trees are rarely being tested for disease tolerance or promoted as viable local "legacy" shade trees. Luckily, a few are on the market, 'St. Croix' from Minnesota, one of the few originating in the Midwest. American elm is the state tree of North Dakota. With globalization, DED and our native elms provide a lesson on how vulnerable all plants are from some unforeseen creatures or pathogens from another part of the world. Living things, innocuous where they evolved, when moved into an alien environment may wreak havoc. It's a reminder to always plant a diversity of plants, just in case. Elm yellows (elm phloem necrosis), another deadly elm disease caused by phytoplasma (*Candidatus ulmi*), is unfortunately becoming more prevalent. Learn more about DED at na.fs.fed.us/fhp/ded/.

HOW TO GROW American elm is a floodplain native throughout the Midwest. The historic Edsel Ford estate just outside of Detroit is a place to still see magnificent, mature American elms originally laid out by Jens Jensen. They really care for the trees there and though some have been lost, I'll admit to hugging their remaining American elms towering at least 100 feet up. Flashback to the street I grew up on prior to the DED plague that hit my neighborhood in the 1970s.

LANDSCAPE USE American elm was the quintessential street tree lining the streets of midwestern communities. Yes, it was what's now termed "overplanted," but whoever would have imagined our native elms would be almost wiped out

Early spring seeds of *Ulmus rubra* (slippery elm) are larger and showier than American elm.

by an alien disease? Native elms provide first-of-the-season seeds for wildlife and also host a wide variety of insects, especially moths. The double-toothed prominent (moth) caterpillar is the most astonishing—it's back has the same serrated edge as the elm's leaf. For a creature that can't even see well, this moth has a remarkable back that provides good camouflage against hungry birds.

ORNAMENTAL ATTRIBUTES The American elm's upward branching that cascaded around the edges (described as vase-shaped) formed tunnel-like cathedrals of trees. The effect was beautiful as well as functional, creating a moderated winter and cooler summer.

RELATED PLANTS Slippery elm (*Ulmus rubra*) is found across all but the farthest northwest sliver of the Midwest and is hardy everywhere. It's named for the mucilaginous substance under the inner bark with medicinal value; I use it in lozenges when my voice is strained. The name red elm comes from the reddish wood. The tree is more flat-topped than American elm with un-slippery, raspy surfaced leaves and seeds that are larger and showier than American elm in spring before the tree leafs out. It has the same chromosome count as Siberian elm and they readily hybridize wherever they grow together creating the new "slippery-Siberian" elm (*Ulmus ×notha*).

Rock elm (*Ulmus thomasii*) is a sporadic tree, often growing in gravelly, limestone-based river floodplains and adjacent bedrock outcrops. It's found across the Upper Midwest more sporadically southward and is also hardy everywhere. It's a much more upright than the other elms with younger stems that are wonderfully corky. A magnificent tree two blocks from my boyhood home was a spectacular pyramidal dome 80 feet tall—it died of DED but three other rock elms remain in my old neighborhood whereas all the American and slippery elms along the streets are now gone.

Evergreen Trees

The extremes of weather, past widespread wildfires, and soils all played a part in that the Midwest was not a land of evergreen trees (and shrubs for that matter) when it was settled. Only eastern red cedar is found across the entirety of the Midwest and prior to settlement it was not remotely as widespread as it is today. Fires limited red cedar to areas where fire could not incinerate its highly flammable foliage. Ancient gnarly trees 300 to 400 years and older can be found on cliff faces of the Driftless Area and Ozark Highlands in particular.

Other evergreen species ring the periphery of the Midwest with outlying relics here and there where landforms created microclimates for these species to persist or advance with climate change. It's hard to believe that pollen records show that evergreens were once dominant in glacial times, which in the big scheme of things was not that long ago.

Most of our evergreen trees are conifers; red cedar and arborvitae are in the cypress family (Cupressaceae), while pines, spruces, balsam fir, and eastern hemlock are in the pine family (Pinaceae). Some of these species are of a more northern affinity; these do not tolerate summer heat well. Other species are of a southern affinity; these have a few outlying native populations in the Lower Midwest and are readily adapted all across the Lower Midwest. In fact, shortleaf and loblolly (*Pinus taeda*, not covered in this book) pines are the state tree of Arkansas and are represented in the Greater Kansas City region with Virginia pine readily naturalizing.

The rare southern broadleaf evergreens are even more challenging in the midwestern climate and limited by the current climate to only the Lower and Eastern Midwest. The Upper Midwest simply gets too bitter cold for them to survive more than a few, mild years. American holly (*Ilex opaca*) has naturalized northward into the region since settlement as it once was a popular ornamental, though now largely replaced by alien hybrid hollies that grow fast enough to meet the current economy of nursery production. Sad, as American holly in the holly family (Aquifoliaceae) is still the hardiest and most long-lived holly and broadleaf evergreen for that matter.

Evergreens, however, remain a critical component in a well-designed landscape with extremely important roles in sustainable landscapes that conserve energy. They have been used as windbreaks since the prairies were settled. Their winter foliage is loved by many a gardener to admire through the long winters. Countless creatures find cover in them on cold winter nights. Where would you want to spend a windy night when the temperature reaches -10°F?

In the descriptions that follow, USDA hardiness zone ratings refer to the cold (winter) tolerance of a plant while the American Horticultural Society (AHS) heat zones refer to a plant's heat (summer) tolerance or requirement. Northern affinity evergreens are fully cold hardy but limited in their use southward by their tolerance of summer heat; thus, their descriptions include a heat zone rating. Likewise, southern affinity evergreens may require a certain amount of heat to survive and need a heat zone and a hardiness zone to describe their adaptability.

There are too few evergreen tree species to create a meaningful top dozen list recommended for traditional landscapes.

Abies balsamea (balsam fir) has a striking spirelike crown but usually opens up with age when growing in a less-than-ideal site.

Abies balsamea

Balsam fir

Pinaceae (pine family)
Landscape group: northern affinity evergreens

This is the traditional American Christmas tree and common Northwoods tree extending southward into the Upper Midwest. I grew up near remnant native stands of this tree and its rather uniform spirelike growth also graced a few local landscapes.

HOW TO GROW Balsam fir is native from Minnesota to northeast Iowa and eastward to central Michigan. It's best in the cooler regions of the Upper Midwest in AHS heat zone 5 or cooler. Plant in moist, well-drained soils that are protected from hot afternoon sun and summer winds. The tree is propagated by seed and available from some nurseries.

LANDSCAPE USE A striking evergreen for a cool, afternoon-shaded site, balsam fir works in sometimes-challenging north or east courtyards of homes and buildings. It can be used as a focal evergreen in appropriate cool summer areas with moist soils.

ORNAMENTAL ATTRIBUTES The needles are aromatic, soft, dark green with a silvery underside. The cones are lovely blue-green and up-facing, disintegrating as they mature, leaving a toothpicklike center. The tight narrow crown of the tree is distinctive, but older, lower branches thin out with age.

Ilex opaca

American holly

Aquifoliaceae (holly family)
Landscape group: southern broadleaf evergreens

American holly is the hardiest broadleaf evergreen tree bar none. It's a classic native tree across the Southeast and along the Atlantic coast northward to Cape Cod. Its presettlement native range brushed the south edge of the Midwest, but for more than a century it was widely cultivated across the Lower and Eastern Midwest where it is hardy. Magnificent trees grace old neighborhoods from Kansas City to Saint Louis, Louisville to Cincinnati, and northward across much of Indiana and Ohio into southern Michigan. This classic American evergreen species has naturalized sparingly across this range as far north as southwestern Michigan and has become a significant tree across two-thirds of the Midwest.

HOW TO GROW Only northern strains of this holly should be cultivated in the Midwest. More than 100 selections have

A female tree of *Ilex opaca* (American holly) shines through winter with evergreen leaves and colorful red fruit.

The brilliant red berries of *Ilex opaca* (American holly) are iconic to this broadleaf evergreen.

been made, though they have fallen out of favor for several decades now, as they are slow to grow and homeowners usually want instant gratification. American holly is dioecious and a male tree is required to pollinate female plants. Although the species is difficult to find in nurseries, it far excels the current widely available hybrid hollies that grow with hybrid vigor and readily propagate quickly from cuttings. Hybrids are far less hardy, reduced to paper bag brown by severe winters. American holly can defoliate in extremely cold winters but recovers. Plant it in neutral to acid soil in full sun to light shade in moist, well-drained soil. A site sheltered from fierce winter winds is best and winter shade is necessary for the plant to survive in marginal areas of southern USDA zone 5b. I have seen trees under such conditions as far north as Rockford, Illinois.

LANDSCAPE USE American holly is a premier broadleaf evergreen that is worth the wait to grow and develop. It's a welcome site in winter, especially female trees studded in red (or yellow or golden orange) berries. It's a fine evergreen to flank stately, large homes and because it's shade tolerant, a good windbreak or screen for wooded areas. Use female plants for focal areas but remember a male plant must be present nearby for females to produce fruit.

ORNAMENTAL ATTRIBUTES American holly is most beloved for its wintertime red berries on female trees that epitomize the holidays. Trees are more olive green and not green-green, so gardeners shun them, demanding evergreen plants that are truly green in winter. Most American holly cultivar selections were made for improved leaf color, hardiness, and exceptional fruiting. This holly grows upright and pyramidal, with extremely strong, light-colored wood. The flowers are little noticed by people, but rich in nectar and sought out by a plethora of pollinators.

NOTES The Dawes Arboretum east of Columbus, Ohio, has a superb American holly collection (229 cultivars) that showcases cultivars adapted to the Midwest.

Juniperus virginiana
Eastern red cedar

Cupressaceae (cypress family)
Landscape group: midwestern evergreens

Eastern red cedar has greatly increased across the midwestern landscape since settlement. It is a successional species, colonizing abandoned fields and other disturbed ground as well as open lands in premier natural areas including glades and prairies where it shades out native plants. Luckily, this cedar is the easiest invasive to remove through fire and cutting, and it does not resprout from a stump. The very

Juniperus virginiana (Eastern red cedar) enhances the springtime beauty of Whitmire Wildflower Garden at Shaw Nature Reserve in Franklin County, Missouri.

An old *Juniperus virginiana* (red cedar) grows from Buzzard's Roost Rock in Ohio.

aromatic wood is resistant to decay and repels insects, thus is used in lining cedar chests. The foliage is comprised of sharp prickles on juvenile plants or shaded growth, but is otherwise a scalelike awl.

HOW TO GROW This is the only evergreen native across nearly the entire Midwest and is hardy everywhere. It is easy to grow from the seeds in its berrylike cones. Cuttings are challenging. The plant thrives in any well-drained soil and does best in full sun; it is weak and open in light shade and does not survive in full shade. It is incredibly heat and drought tolerant and grows on rock outcrops with little soil.

LANDSCAPE USE Eastern red cedar is a premier evergreen windbreak and ornamental in challenging soils where other evergreens won't survive. It is a wildlife-friendly plant whose cones are relished by many birds. It's also the sole host of our only green butterfly: the olive juniper hairstreak. The fruits are edible to a degree (juniper is the flavoring of gin); one in a thousand trees have sweet berries with a gin aftertaste.

ORNAMENTAL ATTRIBUTES Eastern red cedar is dioecious, and female trees produce copious amounts of beautiful blue, berrylike cones. Male trees produce abundant orange-brown pollen cones that discolor the tree and produce pollen in spring. The foliage turns reddish brown or orangish olive in winter—unappealing to many gardeners but beautifully in harmony with hues of winter. Trees growing on precarious rocky outcrops are often incredibly beautiful giant bonsais with unique sculpted trunks and branching. Tree trunks are quite lovely with whitish and tannish strips that exfoliate vertically, and often the trunk is adorned with unique white lichens.

NOTES Eastern red cedar is the alternate host to three rusts that affect fruit trees, flowering crabapples, and hawthorns. Removal of eastern red cedars is often listed as a cure, but as they are now so widespread that will have little effect in most areas. It is best to plant rust-resistant fruit tree and crabapple cultivars.

RELATED PLANTS Several female cultivars have been selected for greener foliage in winter including 'Canaertii' from Kansas with a distinctive and picturesque open crown, 'Hillspire' from Illinois with a neat sweeping though pyramidal form, and 'Taylor' from Nebraska that forms a tight cylindrical column. 'Glauca' has bluish green foliage and is also a female.

Picea glauca

White spruce

Pinaceae (pine family)
Landscape group: northern affinity evergreens

White spruce is another Northwoods evergreen barely native on the north edge of the Midwest with an outlying population in the Black Hills of western South Dakota that is actually much more adapted to most of the hotter and drier Midwest. The South Dakota strain of the plant is called the Black Hills spruce (*Picea glauca* 'Densata', sometimes listed as var. *densata*) and is the state tree of South Dakota.

HOW TO GROW Virtually all the white spruces in midwestern nurseries are from the South Dakota form. It is easy to grow from seed and is easy to transplant. It requires full sun or only part shade in moist, well-drained soils and can suffer in the hot, humid Lower Midwest.

LANDSCAPE USE A prime evergreen for a windbreak or focal ornamental. The cones produce copious amounts of seeds that attract winter songbirds. Older trees often lose their lower branches, especially if shaded or in stressful locations.

ORNAMENTAL ATTRIBUTES The full, pyramidal form of the plant is highly desirable; the needles are more bluish green than other evergreen trees, creating a nice contrast. The cones are small and often abundant, adding winter interest.

A mature *Picea glauca* 'Densata' (Black Hills spruce) resides in front of at the historic Vaile Mansion in Independence, Missouri.

Picea mariana
Black spruce

Pinaceae (pine family)
Landscape group: northern affinity evergreens

Black spruce brings to mind wetlands in the Northwoods where its tall narrow crown, along with tamarack and balsam fir, creates a landscape reminiscent of Doctor Seuss. Since settlement it has greatly increased, documented by original land surveyors' witness trees in Minnesota.

HOW TO GROW Native southward into the Upper Midwest in relict stands in bogs, it grows well in Upper Midwest landscapes of AHS heat zone 5 and colder. Black spruce is difficult to procure and is rarely sold at nurseries, but can be grown from seed and does readily transplant. My boyhood neighbor had two perfect, 40-foot-tall focal trees, grown from seedlings off a relative's property in northwestern Wisconsin. Because black spruce is not shade tolerant, the two neighborhood trees were overgrown by squirrel-planted walnut trees by the early 1990s.

LANDSCAPE USE Black spruce makes a very unique ornamental evergreen tree. It's a nice addition to a conifer-themed garden or beside a water or bog garden.

Spirelike *Picea mariana* (black spruce) grows next to *Larix laricina* (tamarack) in full fall color.

ORNAMENTAL ATTRIBUTES The form of the plant is quite striking, especially when highlighted by snow. The cones are also tiny and cute, like pendant ornaments.

Pinus banksiana
Jack pine

Pinaceae (pine family)
Landscape group: northern affinity evergreens

Jack pine ranges across the Northwoods and boreal forests of North America, but there are populations that thrive southward into the heart of the Midwest, usually in areas of sandy or otherwise poor soils. It's usually such a northern plant that it is odd to find it growing close to flowering dogwoods in the Indiana Dunes or with river birch in sandy prairie barrens along the Lower Wisconsin River. It has naturalized in poor soils as far south as Powell Gardens near Kansas City, Missouri, but is susceptible to pine wilt disease in more stressful settings. Virginia pine, its close relative with a similar niche, is a better choice for the Lower Midwest.

HOW TO GROW Jack pine is easily grown from seed. Some strains will not open their cones and release their seeds unless they are burned. Jack pine requires full sun and thrives in nutrient-poor soil. It readily reforests areas that have burned or been disturbed to expose the soil to sunlight.

Pinus banksiana (Jack pine) survives in harsh sites like this sand dune habitat at Muskegon State Park, Michigan.

LANDSCAPE USE Jack pine is often called a scrub pine because it is shorter and bushier and found in successional areas. Because it is not the standard Christmas tree shape desired by most homeowners, it is seldom used in home landscapes. It's a great wildlife habitat tree, producing abundant cones and providing winter cover.

ORNAMENTAL ATTRIBUTES Its bushy nature can make it a fine screen or smaller pine with two wonderfully short needles per bundle, giving it a very fine texture. Winter needles are more golden green and not true green so desired by home-owners, but this makes the tree contrast well with other evergreens.

Pinus echinata

Shortleaf pine

Pinaceae (pine family)
Landscape group: southern affinity pines

This is the most widespread of the southern pines, having significant stands in the Appalachian Plateau of Ohio westward to the eastern and southern Ozarks. A few trees cling to the bluffs overlooking the Mississippi River in the very scenic LaRue Pine Hills of southwestern Illinois. The species is cold hardy only in USDA zone 6 southward and requires the same summer heat zone (AHS zone 6).

HOW TO GROW Shortleaf pine is easy to grow from seed. It is fickle about poorly drained or alkaline soils, but otherwise thrives in poor, acidic soils across the Lower Midwest. Plant only local strains for best results as southern-sourced plants will winter burn or die after harsher winters.

LANDSCAPE USE Shortleaf pine is a fine screen or windbreak tree in the Lower Midwest. It is being restored to some parts of the Ozark Highlands where lack of fire has reduced its numbers and the marvelous open pine savannas it created have almost entirely been lost. Look for beautiful restorations of shortleaf pine savanna at Hawn State Park south of St. Louis in Missouri.

ORNAMENTAL ATTRIBUTES Mature trees have a unique blocky bark pattern of vertical, reticulated rectangular lighter plates surrounded by darker furrows. A walk through mature stands in open savannas is memorable along with the sound of wind through the needles and the rich "southern" pine scent the fallen needles create. I would not call the 4- to 5-inch needles short, but they are in bundles of two and occasionally three.

Pinus echinata (shortleaf pine) thrives on poor, well-drained soils.

Pinus ponderosa (ponderosa pine) is readily identified by its tufts of long, dark needles.

Pinus ponderosa var. *scopulorum*
Rocky Mountain ponderosa pine

> Pinaceae (pine family)
> Landscape group: western conifer reaching midwestern states

The Rocky Mountain ponderosa pine's native range ventures out of the mountains and well into the Great Plains, reaching the tallgrass prairie along the spectacular Niobrara River in northern Nebraska. It's a premier pine for much of the western Midwest in drier, well-drained soils. In the more humid Midwest, it is very fickle in heavy clay soils, so surviving plantings are restricted to loess soils in the Kansas City region, for example.

HOW TO GROW Plant in well-drained soils from loess to sand with good air circulation. This pine thrives in wind and requires full sun. It is grown from seed and available as seedlings from some state nurseries and those specializing in evergreens.

LANDSCAPE USE This is a good evergreen tree for windbreaks and ornamental plantings in sites that may be challenging for other evergreens.

ORNAMENTAL ATTRIBUTES The long needles are dark green, appearing in striking tufts on older trees. The bark of mature trees becomes quite unique, changing from charcoal black to exfoliating to reveal warmer colors.

Pinus resinosa
Red pine

> Pinaceae (pine family)
> Landscape group: northern affinity evergreens

This evergreen tree comprises a major component of the Northwoods but has a few wild populations southward into the Upper Midwest from Minnesota to Michigan, as far south as Starved Rock State Park in Illinois. It is the state tree of Minnesota.

HOW TO GROW Red pine is easy to grow from seed and is offered by many state nurseries including Missouri, where it is recommended only for the northern part of the state. Red pine requires well-drained soils and languishes in wet clays. It also is not tolerant of extreme heat and drought so is not recommended in the Lower Midwest. It is best in AHS heat zones 5 and northward. As far south as AHS heat zone 7, it can survive in appropriate loess or sandy sites with excellent drainage.

LANDSCAPE USE Red pine is used as a windbreak tree and ornamental specimen evergreen, although its screening

Pinus resinosa (red pine) bark is flaky, reddish, and highly ornamental.

value is lost as the tree opens up with age. The tree can be identified by its long needles in bundles of two that readily snap apart rather than bend when folded. It's a prime wildlife plant with good cone crops and evergreen cover in the crown of the tree.

ORNAMENTAL ATTRIBUTES The tree is named after its reddish flaky bark that is quite beautiful as the tree matures. I think of a tree in my boyhood neighborhood planted lone as a sentinel with its gorgeous trunk and upright oval crown of shaggy needles. A winter walk through a stand of mature red pine is alive with birds prying the flaky bark for insects or the cones for seeds.

Pinus rigida
Pitch pine

> Pinaceae (pine family)
> Landscape group: southern affinity pines

Pitch pine is one of the few pines that can burn in a fire or be cut down and sprout adventitiously (from the tufts of needles along young trunks and older branches). The needle tufts along its branches give the look of hairy arms that help identify it readily when driving by in a car. The cones are somewhat squat with prickles, some not opening without fire. Pitch pine has coarse needles in bundles of three. It is found in the Appalachian Highlands of Ohio and Kentucky with a few outliers westward. With lack of fire management, it is relegated to surviving only in harsh sites

The Greater Kansas City champion pitch pine (*Pinus rigida*) shows how the crown of mature trees can be quite spectacular.

Pinus strobus (Eastern white pine) was widely planted in cemeteries across the Midwest.

where it can compete with other trees. It naturally hybridizes with shortleaf pine.

HOW TO GROW This pine is easily grown from seed but very rarely available anymore. It thrives in well-drained, poor, acidic soils. Marvelous mature specimens occur across the entire Lower Midwest from native stands westward to the Kansas City region. It is cold hardy northward into USDA zone 5.

LANDSCAPE USE Pitch pine is an appropriate evergreen for informal or naturalistic screening, wildlife plantings, and historic and native landscape restorations. It is best reserved for soils and sites that provide challenges for other pines.

ORNAMENTAL ATTRIBUTES The irregular crown and spreading outline of older trees with needle tufts along some of the older branches create a striking sculptural sight at maturity. The cones mature rich brown and lie short and squat along the branches. Needles are twisted and more golden green than those of most pines.

Pinus strobus

Eastern white pine

Pinaceae (pine family)
Landscape group: northern affinity evergreens

This pine is a favorite tree of mine. I wish I could have witnessed what virgin stands of eastern white pine looked like. It's hard to imagine the tree once grew over 200 feet tall;

in New England such trees were claimed for the British Navy, helping inspire the American Revolution. Great rafts of white pine timber were steered down the major rivers of the Midwest to be used in building many of our cities and to create great wealth for lumber barons.

Eastern white pine is the state tree of Michigan and the premier tree of the Northwoods, but its native range extended across the entire Upper Midwest with relict stands southward in central Iowa along the Iowa River greenbelt, southeastern Iowa at Wildcat Den State Park, Starved Rock State Park in northern central Illinois, Turkey Run State Park in west-central Indiana, and the Appalachian Highlands of southern Ohio. The tree is in trouble in these outlying stands as deer browse and lack of fire have put an end to its regeneration. My friend Don Miller's essay "The Pine" in his book *Life Afloat* captures the essence of the last single tree standing at Severson Dells Forest Preserve south of Rockford, Illinois, relict of a once-larger stand that lumbermen mined away long ago. This pine is becoming scarcer in the wild almost everywhere in the Midwest.

HOW TO GROW This pine grows fast from seed and unique forms of it can be grafted onto seedling understock. It grows

Pinus virginiana (Virginia pine) has short needles and is always heavily studded with cones.

best in moist, well-drained soils, regenerating in nature in mineral soils in the aftermath of wildfires. Plants struggle in wet clay soils and during extreme dry spells, but are overall the most widespread and adaptable of all the native pines. Eastern white pine simply won't grow in my home's droughty clay soils, so I have several bonsai in containers to satisfy my need to have this plant around. It does make a fine container plant.

LANDSCAPE USE Eastern white pine is the premier windbreak and screening tree across the entire Midwest. Some gardeners and designers don't like that it opens up with age as its horizontal limbs can succumb to snow and ice loads.

ORNAMENTAL ATTRIBUTES The long, soft needles are in bundles of five (w-h-i-t-e as in five letters is often used to remind children how to identify it); it is the only midwestern native pine with that needle arrangement. Each needle has a whitened stripe down its underside creating a beautiful blue-green overall appearance. The tree's needles capture the wind generating a whispering rush like no other. Needles from the previous year turn rust brown and are shed annually in late summer, sometimes causing alarm to homeowners. Shed needles create fabulous and aromatic mulch under the tree. Mature trunks are charcoal black and ascend to the sky with horizontal limbs that create a striking silhouette as the tree ages.

Pinus virginiana
Virginia pine

> Pinaceae (pine family)
> Landscape group: southern affinity pines

Virginia pine is a more southern counterpart to the Jack pine found regularly in the Appalachian highlands of Ohio and Kentucky with outliers into Indiana's unglaciated Shawnee Hills. It is a pioneer species, and blocks of the tree often delineate abandoned fields. It thrives in a variety of poor soils, tolerates summer heat, and grows well across unglaciated parts of the Lower Midwest. It's cold hardy to USDA zone 6 but looked peeked after recent severe cold in the Chicago area and so does best in warmer summer areas of AHS heat zones 6 and 7. It has much lower mortality to pine wilt disease than Jack pine in the Kansas City area, where it readily naturalizes. Without disturbance such as fire or clearing, the tree cannot regenerate.

HOW TO GROW Virginia pine is easy to grow from seed and adapts well in the old clay soils of the unglaciated parts of the Midwest. It thrives in these challenging soils that can be very wet in spring and powder dry in summer.

LANDSCAPE USE This bushy scrub pine has extremely high wildlife value, and its abundant cone set provides more seeds than any other based on observations around Kansas City. It is a great evergreen choice for a screen or windbreak in the Lower Midwest and is superior for gardens designed to attract birds.

ORNAMENTAL ATTRIBUTES The short, dense needles and fine texture separate it from other pines in the Lower Midwest. The dark brown cones are similar to those of Jack pine but have sharp prickles at the end of each bract.

RELATED PLANTS *Pinus virginiana* 'Watt's Golden' has needles that turn yellow in winter and green up again for spring and summer.

Thuja occidentalis
American arborvitae, northern white-cedar

> Cupressaceae (cypress family)
> Landscape group: northern affinity evergreens

The neighborhood where I grew up had a stately mature hedge of American arborvitae that was the perfect place to hide during games of kick-the-can and hide-and-seek. To this day, the unique aroma of arborvitae's foliage brings me to those fond childhood memories. This hedgerow was removed long ago and now deer trim up all American

Thuja occidentalis (American arborvitae) is a striking evergreen on Evening Island, Chicago Botanic Garden.

arborvitae in the old neighborhood, eating every bit of green they can reach. This deer issue is endangering the tree in its relict midwestern stands. Look for American arborvitae in places like the Wilderness Preserve in Adams County, Ohio or along springs in Trout Park, Elgin, Illinois. The foliage is rich in vitamin C and explains why the name *arborvitae* (tree of life) is so fitting. American arborvitae is amazingly long lived, readily reaching 400 years with some trees documented to reach 1100 years along the Niagara escarpment.

HOW TO GROW American arborvitae is common in the Northwoods but has a localized midwestern range from Minnesota to Ohio where relict stands are associated with cool, moist conditions and sheltered escarpments. The species is easy to grow from seed or rooted cuttings and readily naturalizes near many plantings until deer became sedentary and abundant. American arborvitae prefers moist soils that are neutral to alkaline, often growing on or near wet sites but only where the roots are elevated in well-drained soils with a constant supply of moisture. The tree is more challenging to grow south of AHS heat zone 5 in the western Lower Midwest as it is intolerant of excessive heat and drought. Only a few trees remain in the Kansas City area (without irrigation) after recent historic droughts. Make sure to plant this arborvitae in moist sites in the western Lower Midwest where it is sheltered from summer wind and afternoon sun.

LANDSCAPE USE American arborvitae is a popular wind-break tree in the Upper Midwest. It was formerly a popular sheared hedge and foundation plant, and in many instances has grown into tree size. It makes a lush evergreen screen and is tolerant of light shade. Pyramidal, yellow-foliaged, and other unique cultivars create exceptional focal points in a landscape. Many mature arborvitaes, including several cultivars, grace old cemeteries across the Upper Midwest. Towering columnar 'Pyramidalis', twisted foliaged 'Spiralis', and massive rounded 'Globosa' have resisted heavy snow over decades and are simply spectacular; though those cultivars are no longer available, they should be maintained in historic landscapes.

ORNAMENTAL ATTRIBUTES The foliage is made up of tiny awl-type leaves pressed to the twigs, which splay out into elaborate and undulating fan-shaped growth. The foliage is stunning upon close inspection. The bark of old trees peels off into thin vertical strips of beige to whitish gray that adds to the trees appeal. Cones are small, up-facing, and produced in clusters. They start out green and mature to brown and are a favorite winter seed source for several finches especially pine siskins.

RELATED PLANTS Cultivars are readily available at nurseries across the entire Midwest but native-sourced trees are hard to find and available only at specialty native nurseries. Many cultivars of the tree exist with columnar, pyramidal, globular, or dwarf forms and with foliage colors ranging from yellow new growth to those that stay truer green in winter.

Tsuga canadensis
Eastern hemlock, Canadian hemlock

Pinaceae (pine family)
Landscape group: northern affinity evergreens

Our native Eastern hemlock is a soft and fine conifer and sadly under threat from an imported pest: the Asian hemlock woolly adelgid spreading from the East Coast. Eastern hemlock thrives in mesic forests mainly to the north and east of the Midwest but is found throughout the Appalachian Plateau and Lake Erie Drift Plains and in more localized and rare relict scattered sites in beech-maple and rarely in maple-basswood forests as far west as northeastern Minnesota, the Driftless Area of Wisconsin, and the Shawnee Hills of Southern Indiana. The tree grows well across most of the Midwest as there are fine historic trees in easternmost Kansas and central Iowa. Visit the tree in the wild at almost any of Michigan's state parks along Lake Michigan where it grows delightfully in forests sheltered between sand dunes. Relict sites include the Hemlock Cliffs of Hoosier National Forest in southern Indiana and at Wildcat Mountain State Park in southwestern Wisconsin. Birders flock to historic Mount Mora Cemetery in St. Joseph, Missouri, as a rare place to observe white-winged crossbills feeding on Eastern hemlock cones in winter.

HOW TO GROW Plant Eastern hemlocks in moist, well-drained soil as they are intolerant of wet feet or too dry of a site. They need to be sited out of hot summer winds and to be shaded from the hottest afternoon sun in much of Lower and western Midwest. Plant only local-sourced, cold and drought tolerant trees in USDAS zone 4 and much of the western Midwest. Eastern hemlock is not a good choice for prairie sites.

LANDSCAPE USE Eastern hemlock is one of the few shade-tolerant evergreens so it makes a great evergreen screen in moist, shady sites. It is also tolerant of shearing and makes a marvelous hedge. It was formerly planted as a sheared foundation plant (many of which have grown into trees). Eastern hemlock is a favorite browse of deer which are endangering the plant in its spotty midwestern range and makes its use limited to an evergreen tree above deer reach in areas where deer are overpopulated. It was once a popular evergreen planted in cemeteries and Victorian landscapes.

ORNAMENTAL ATTRIBUTES Our most graceful evergreen, Eastern hemlock has growth reminiscent of water cascading over a cone: from its nodding leader to the way the branches splay downward and outward like falling water. The cones are cute, rarely more than 1½ inches long, and grow green but mature brown. They adorn the branches as pendant ornaments through winter and attract many songbirds to feast on their seeds.

Tsuga canadensis (Eastern hemlock) is shade tolerant and relict in sheltered sites in the Midwest.

Small Trees and Large Shrubs

The rose family (Rosaceae) is the most important group of small trees and large shrubs to the midwestern web of life and includes some of most iconic and inspiring plants: crabapples (*Malus* spp.), hawthorns (*Crataegus* spp.), plums and cherries (*Prunus* spp.), ninebark (*Physocarpus opulifolius*), and serviceberries (*Amelanchier* spp.). Together these species are among the region's most ornamental flowering trees, and most have edible fruit. They host a wide diversity of insects, only exceeded by the oaks.

I wrote an essay on the value of wild plum and cherry trees. Give me these over any Japanese maple so beloved by most gardeners. The first wild plum I planted in my landscape, I placed outside my bedroom window. When it bloomed, I left that shade and window open to fully experience the magical and fragrant white, luminous flowers. I bought it as an American plum but it turned out to be a wild goose plum. The native species are not easy to tell apart and often mixed up in nurseries. From fragrant springtime flowers to hosting a marvelous array of insects, followed up by delicious aromatic fruit, give me a wild plum.

Why are wild plums not planted in every landscape? Like all plants rich in fauna, their leaves are not perfect (chewed and clipped, webbed with silk or otherwise imperfect) and since the tree bears fruit, it's deemed messy.

The birch family (Betulaceae) is the second most important group to the web of life for you to add to your landscape or garden. Four important midwestern small tree and large shrubs are hophornbeam (*Ostrya*), hornbeam (*Carpinus*), hazelnut (*Corylus*), and alder (*Alnus*).

Elderberries (*Sambucus*) and viburnums, once considered part of the honeysuckle family (Caprifoliaceae), are now placed in the moschatel family (Adoxaceae). They produce flat clusters of white flowers followed by colorful fruits. Two species of elderberries become large shrubs but can be pruned into small trees. Three species of viburnums readily become small trees while one species usually remains a very large shrub.

Prickly-ash (*Zanthoxylum americanum*) and wafer-ash (*Ptelea trifoliata*) are not true ashes in the olive family (Oleaceae), but are related to citrus in the rue family (Rutaceae). Both plants are the sole host plants of our largest butterfly, the giant swallowtail.

Sumacs (*Rhus*) and their relative the American smoketree (*Cotinus obovatus*) are in the cashew family (Anacardiaceae), which also includes poison ivy. Frank Lloyd Wright created stained glass designs inspired by our native sumacs. These shrubs are the epitome of fall in the Midwest and so beautiful they are hard to describe in words and yet we rarely include them in our landscapes. Three species run like bamboo, with older stems reaching tree size at over 15

The bright red fruit of *Prunus munsoniana* (wild goose plum) is both ornamental and edible.

feet tall. All are dioecious, the female plants producing cone-like clusters of velvety red berries atop their stems that color up nicely in late summer and last through winter. Fruits are dry but rich in vitamin C, some tasting like sweet tarts (flavor varies from plant to plant) and making a sweet-tart tea. Birds use sumac's fruit as emergency late winter food, especially important in severe conditions including ice storms and late snows. So whether you pronounce their name like Sue and summer or sure and sugar (soo-mack or shoe-mack), sumacs should be visible in every midwestern landscape.

SMALL TREES FOR TRADITIONAL LANDSCAPES

Amelanchier arborea (downy serviceberry)
Amelanchier ×grandiflora (apple serviceberry)
Amelanchier laevis (Allegheny serviceberry)
Carpinus caroliniana (American hornbeam)
Cercis canadensis (eastern redbud)
Cornus alternifolia (pagoda dogwood)
Cornus florida (flowering dogwood)
Cotinus obovatus (American smoketree)
Crataegus crusgalli (cockspur hawthorn)
Crataegus phaenopyrum (Washington hawthorn)
Crataegus viridis (green hawthorn)
Hamamelis virginiana (common witchhazel)
Malus ioensis 'Bechtel' (prairie crabapple)
Ostrya virginiana (eastern hophornbeam)
Prunus americana (American plum)
Prunus canadensis 'Princess Kay' (Canada plum)
Viburnum lentago (nannyberry)
Viburnum prunifolium (blackhaw)
Viburnum rufidulum (rusty blackhaw)

ORNAMENTAL LARGE SHRUBS FOR TRADITIONAL LANDSCAPES

Cornus amomum (silky dogwood)
Cornus sericea (red-osier dogwood)
Hamamelis vernalis (Ozark witchhazel)
Ilex verticillata (winterberry)
Lindera benzoin (spicebush)
Physocarpus opulifolius (ninebark)
Rhus aromatica (fragrant sumac)
Salix discolor (pussy willow)
Salix humilis var. *humilis* (tall prairie willow)
Sambucus pubens (red-berried elder)
Styrax americana (American snowbell)
Viburnum trilobum (American highbush-cranberry)

Aesculus glabra
Ohio buckeye

Sapindaceae (soapberry family)

Every year I go visit this tree in early spring. It is the first to leaf out and the buds burst forth with unique, dramatic palmate foliage comprised of five to seven leaflets. The classic buckeye fruits that ripen in the fall are a beautiful polished dark brown with a light brown "eye." I guess I am superstitious because I always put a few fruit in my pocket as well as in the car and in luggage as they are symbols of good luck. The fruit are poisonous to people but relished by squirrels. Ohio buckeye is the state tree of Ohio.

HOW TO GROW Ohio buckeye grows an immense, carrotlike taproot with little top growth when young so it does not meet efficient nursery production and thus is rarely seen for sale. It is difficult to transplant unless grown in a container or root-pruned in a nursery. It is easy to grow from the nuts but must be cloned by grafting. The tree thrives in full sun or shade in well-drained soils, most often found wild in mesic floodplain woods in rich deep soils that are not inundated by water for long. Ohio buckeye is native across the entire Lower Midwest northward to southeastern Michigan and southeastern Iowa. It has naturalized near where planted, blurring the native-naturalized line, and is hardy throughout the Midwest.

LANDSCAPE USE This tree is a great choice for a woodland garden and planting under the canopy of existing trees. It is not the best for a formal landscape because most plants are bare before fall. The tree is hard to classify, as it can be a small shade tree, a small understory tree, or a large shrub. Champion trees reach over 50 feet tall and the tree hybridizes with yellow buckeye and other more southeastern species to the point they are hard to identify. Some people find buckeye's fallen nuts to be messy though they are quickly hoarded and literally squirreled away. In some regions, specific trees or selections hold their foliage well into autumn but I've learned from experience that does not necessarily hold true elsewhere. 'Autumn Splendor', for example, holds well in Minnesota but not in Kansas City. I recommend growing a local strain that shows that characteristic. Note that winter buds are very sharp pointed.

ORNAMENTAL ATTRIBUTES Ohio buckeye is early to leaf out and celebrated for that after a long winter. This leaf strategy helps it capture spring sunshine before overhead canopy trees leaf out but often the tree is the first to lose its leaves in late summer, which is a drawback if used in a formal landscape. Its bright green, early spring foliage is lit up even more by the candles of nectar-rich green-yellow to

Aesculus glabra (Ohio buckeye) in full bloom makes an ideal wildlife tree at Montrose Point, Chicago.

Alnus incana subsp. *rugosa* (speckled alder) usually grows in dense wetland thickets.

yellow flowers. The flowers are visited by bees and butterflies as well as songbirds like the hummingbird, Baltimore and orchard orioles. The spring foliage and flowers are the perfect complement to blooming redbuds, an unforgettable pairing. Fall color can be beautiful golden yellow to almost burgundy and blends thereof.

RELATED PLANTS 'April Wine' is a cultivar with wine red new leaves. Texas buckeye (*Aesculus glabra* var. *arguta*) is a smaller tree (rarely to 20 feet tall) or large shrub with leaves comprised of 7–11 leaflets. Texas buckeye is found in the western edge of the Midwest in the Flint Hills and Osage Plains from Kansas City west and south.

Alnus spp.
Alders
Betulaceae (birch family)

Virtually every alder planted in the midwestern landscape is a European black alder (*Alnus glutinosa*) which has become an invasive exotic in the Upper Midwest.

HOW TO GROW Two species are native to the Midwest; one northern, the other eastern and southern and they come together in northeastern Ohio where they hybridize. They are easily grown from seed but seem quite demanding of a moist site in cultivation and are not drought tolerant. Smooth alder is not found naturally in limestone soil areas and prefers an acidic soil while speckled alder grows in various wetlands including limestone areas.

LANDSCAPE USE Preserve alders where they are native. They are quite interesting and special trees for pondside, streamside, or other wetland gardens. Like a legume, they fix nitrogen to soil.

ORNAMENTAL ATTRIBUTES Alders don't have a gimmick or colorful nature that entices a gardener or landscaper to plant one. The conelike fruit and catkins adorn the plant all through winter and they come alive early in spring with long pendant male flowers while the female flowers look like mini-cones of stunning magenta. You can identify the two species by their leaf shape and leaf edges.

Alnus incana subsp. *rugosa* (hazel alder, speckled alder, syn. *A. rugosa*) is a northern species extending southward into the Upper Midwest in cool wetland habitats. Look for it from North Dakota southeastward to northeastern Iowa,

Alnus serrulata (smooth alder) has tooth-edged, simple leaves.

Amelanchier arborea (downy serviceberry) is nearly leafless when it sports its white flowers.

northern Illinois, northern Indiana, northern Ohio, and points north. This multistemmed large shrub reaches small tree status on occasion but often forms dense thickets. Distinguish it from smooth alder by its heavily speckled, dark twigs and leaves that have somewhat doubly toothed edges.

Alnus serrulata (smooth alder, hazel alder, common alder) is a southern and eastern alder extending northward into the Lower Midwest from the Ozark Highlands across southern Illinois, Indiana, and southern and eastern Ohio and a disjunct population reported in northwest Indiana. Distinguish it from speckled alder by its smooth trunks more reminiscent of an American hornbeam with leaves that have simply toothed edges.

Amelanchier spp.
Serviceberry trees
Rosaceae (rose family)

When the ground has finally thawed across the American north, serviceberries burst forth with bloom as one of the first flowering trees. That's how they got their name: in earlier times, when serviceberries bloomed, graves for the deceased could finally be dug. Serviceberries grow across most of temperate North America but are most diverse across the northeastern part of the continent. They are self-fertile and hybridize as well, creating a difficult (if not impossible) to identify menagerie of plants from small shrubs to small trees. I have observed trees that bloom like

one species, a leafless cloud of flowers one year, then have coppery leaves at bloom time the next spring.

HOW TO GROW Serviceberries can be easily grown from seed but self-sow on extremely rare occasion. They are more likely to resprout from stumps after logging, fire, or other disturbance than grow from new seedlings. Rhizomatous forms can be divided and select plants can be grown from rooted cuttings off soft wood. They are so relished by deer in my area that there is no chance of any plant surviving beyond their reach. Rabbits also savor small plants, so tree guards, tree wrap, netting, or fencing is required as critter protection through winter.

LANDSCAPE USE Treelike serviceberries are suitable for formal landscapes as they don't self-sow, sucker, or otherwise misbehave. They make ideal small trees (growing 15–30 feet) near a front entryway or around more intimate outdoor spaces. Most are multitrunked so can act as a perfect baffle screen between spaces. They are native to woodland understory and woodland edge habitat so look great in woodland gardens. The more clumping and suckering types are not invasive and make good shrubs for back of the border, screening hedges, and edges of woodlands. Some species can suffer from foliar maladies some years but others have clean foliage through most of the growing season. Their delicious fruit make them perfect in edible landscapes and food forests but birds find them just as delectable so they're integral in bird gardens.

ORNAMENTAL ATTRIBUTES The white flowers in early spring are a welcome sight after winter but are very brief, lasting

Fall color on *Amelanchier ×grandiflora* (apple serviceberry) is spectacular, with leaves in full sun being the reddest.

Amelanchier interior (inland serviceberry) usually has well-formed leaves at flowering time.

Amelanchier fruit ripens in late spring or early summer and is delicious.

The white flowers of *Amelanchier laevis* (Allegheny serviceberry) contrast with its coppery new leaves.

only a few days in warm, windy weather. Serviceberries are among the first plants to fruit in early summer, but are rarely messy as birds quickly strip them. Fall color can be one of the best: rich blends of warm yellow to apricot, orange to burnt red, and purplish red. Some plants have all these colors with leaves in full sun the reddest. The trunks are highly ornamental being smooth gray, developing dark striations of charcoal with full maturity.

Amelanchier alnifolia (Saskatoon), a small clumping tree, is native only to the far northwestern part of the Midwest in the Dakotas, northern Nebraska, westernmost Minnesota, and into northwestern Iowa. I recall it there growing in thickets on the edges of prairies or with groves of bur oak, green ash, basswood, and quaking aspen. It grows 6–8 feet and occasionally 12 feet tall and has the largest fruit that was an important food for Native Americans. There are several cultivars available for good fruit production and the

plant grows well across much of the Upper Midwest. The cultivar Standing Ovation ('Obelisk') has an upright form that is attractive in all seasons and makes a good hedge.

Amelanchier arborea (downy serviceberry) is aptly named as it is the most arborescent (tree-sized) of the serviceberries in most areas. The most spectacular trees I've witnessed grow in central Iowa along the Des Moines River bluffs. There are mature trees with trunks reminiscent of an old apple tree. In most areas, downy serviceberry is simply a cloud of white in bloom, the woolly emerging leaves tiny at bloom time, not masking the flowers whatsoever. This small tree is a common site in the Ozarks in early spring and the only tree serviceberry in Missouri. The winter buds are thin and sharply pointed, more so than any other serviceberry.

Amelanchier ×grandiflora (apple serviceberry) is a hybrid between the downy and Allegheny serviceberries. There are many horticultural cultivars including 'Autumn Brilliance' selected in Illinois. It does well throughout all the Midwest and lives up to its name with consistent rich orange-red to purplish fall color.

Amelanchier interior (inland serviceberry) is a Midwest endemic species carried by a couple of nurseries. Smaller than *A. ×grandiflora* or *A. laevis*, it is the perfect size for smaller gardens as it usually just reaches tree size at around 15 feet with typical smooth gray bark. It's the only tree serviceberry that I can find in my hometown's county despite being on range maps for downy and Allegheny serviceberries which are growing locally just to the east along the Mississippi River bluffs.

Amelanchier laevis (Allegheny serviceberry) has smooth, coppery new leaves that contrast exquisitely with the sparkling white flowers and give this understory tree its own beauty. Its berries are the most succulent and delicious of our midwestern tree species.

Amelanchier sanguinea (round-leaved serviceberry) is a small clumping tree. I actually divided a section of this off a wild clump growing on a rocky woodland opening at the golf course. I always wondered how it got its botanical name but accidentally nicked the cambium and it was "sanguine" bloody red: ah hah (though it's probably because the fruits are red before ripening). It readily grows around 8 feet tall, a denser thicket in full sun, rather open in competition or shade. The leaves are rounded with coarse teeth (less than 27) mostly around the top half of the leaf but toothed to the base. The typical white flowers in midspring above the new leaves are followed by delicious purplish fruit in early summer. The fall color is yellow through burnt orange.

Amorpha fruticosa
Indigobush
Fabaceae (legume family)

I have known this shrub since childhood and finally was able to identify it with the *Newcomb's Wildflower Guide*. The range listed in the book did not include where I was so I thought it was a rare anomaly. Its native range actually includes nearly the entire Midwest and it is hardy throughout.

HOW TO GROW Easily grown from seed, indigobush is not a standard nursery plant but is occasionally available as seedlings from state nurseries or native plant nurseries. It requires full sun or nearly so and is native in well-drained soils from sandy or rocky floodplains usually growing on sunny banks (that can flood but not for long) to dry ridges on the edges of prairie or meadows. Rejuvenate indigobush by removing old stems, but it blooms on old wood so flowering is sacrificed on any regrowth.

LANDSCAPE USE Indigobush is a marvelous large shrub, usually over 8 feet tall, for a natural setting but has too many foliar and twig issues for most formal landscapes. It makes a choice shrub along a stream or pond or at the edge of a prairie or woodland. At Powell Gardens, it provides just enough shade to shelter two benches in our meadow. I also recall a lovely plant shading a gazebo at Fernwood in Michigan.

ORNAMENTAL ATTRIBUTES This shrub sports 4-inch spikes of beautiful deep indigo flowers and contrasting orange stamens at the tips of all sunny twigs in late spring. The

Contrasting yellow-orange stamens highlight the dark indigo-colored flowers of *Amorpha fruticosa* (indigobush).

Asimina triloba (pawpaw) produces flowers in midspring before the tree leafs out.

flowers are fragrant to some noses, not so much to me, with a hint of bitter peanut butter. The flowers are followed by seedpods (actually legumes) that change from golden green to brown, maturing gray, and adorn the plant into winter. The pinnately compound, grayish green leaves are finely textured but disheveled by several insects including silver-spotted skipper butterfly caterpillars that can web the leaves together for their protective nests. Galls, deemed unsightly by some, can swell twigs: created by amazingly intriguing life forms (gall wasps), they are purely cosmetic issues that do not kill or otherwise harm the plant.

Asimina triloba
Pawpaw
Annonaceae (custard apple family)

This little tree looks tropical and that's because, in a way, it is. It's the sole hardy member of the tropical family Annonaceae, which includes custard apple and cherimoya. Pawpaw's edible fruit ripen in early fall and have a unique flavor, aroma, and texture called "Nebraska" or "Michigan" banana. Be sure to eat only ripe fruit. Shake the tree and if any drop, ripen them indoors—otherwise the raccoons and opossums will beat you to them.

HOW TO GROW Pawpaw grows wild across all of the Lower Midwest and all but the north edge of the Eastern Midwest. It grows northward into the Upper Midwest along the Missouri River into southeast Nebraska and the southwest corner of Iowa and has a very spotty range up the Mississippi River valley into east central Iowa and northern Illinois. It is cold hardy in USDA zone 5. The tree is native in alluvial woods, mesic floodplain forest, and mesic ravines into dry upland woods. It grows in either full sun or shade but requires rich soil and is amazingly heat and drought tolerant once established. It's easy to grow from seed, but select varieties grown for superior fruit are grafted. The tree suckers into the proverbial pawpaw patch but is very difficult to divide. Two seedlings or different cultivars are required for pollination and fruit production.

LANDSCAPE USE Pawpaw can be a focal ornamental where it forms a dense pyramidal tree in full sun. It's also lovely as a thicket-forming grove in a shade garden. It's an essential plant for a food forest or an edible landscape. It's also integral to a butterfly garden as the only host plant of the zebra swallowtail butterfly, the sole hardy member of the tropical kite swallowtails.

ORNAMENTAL ATTRIBUTES The large leaves are the most handsome aspect of the plant, smaller and droopy in full sun, but larger and held horizontal in a shady setting to

The trunks of *Carpinus caroliniana* (American hornbeam) create the perfect understory tree for a woodland garden.

capture available light, illuminating their character. Fall color can be spectacular yellow and occasionally a unique white-yellow. The flowers bloom in spring before the leaves emerge and look like purple trilliums: maroon purple with three petals. The buds are brown and fuzzy in winter and are a sight to watch in spring as they enlarge and become velvety purple to green before opening.

RELATED PLANTS There are many cultivars of pawpaw selected for superior fruit size, flavor, and quality. The ripe fruit bruise easily and have no shelf life so will never be a standard fruit available at local grocery stores. It's worth it to grow your own and savor the flavor and the season of this native.

Carpinus caroliniana
American hornbeam, musclewood, blue-beech
Betulaceae (birch family)

This is the "other" ironwood, related to the hophornbeam with equally durable wood.

HOW TO GROW Propagation is the same as for hophornbeam but American hornbeam is found in more moist conditions and is more shade tolerant, growing wild in mesic floodplain to maple-basswood and beech-maple woods. It is less tolerant of drought and is absent from the western third of the Midwest with isolated populations growing in sheltered valleys westward to Central Iowa along the Des Moines River. It will grow in full sun in moist soils where native. It's cold hardy throughout the Midwest but needs proper siting away from hot, drying winds and scorching sun in the western Midwest.

LANDSCAPE USE Here's a marvelous small ornamental tree for a shadier site, wonderful on the north or east side of a home or building. American hornbeam makes a beautiful understory tree under existing trees. It's a great native substitute for Japanese maple.

ORNAMENTAL ATTRIBUTES Cultivate multitrunked trees to show off their fluted, muscular-looking bark that is smooth and bluish gray. The ornamental trunks are eye-catching in the winter landscape. The pendant, pagoda-looking seed heads are light green and contrast well with the summertime foliage, and then mature to brown and hang on the tree through winter. Fall color is one of the best with shades from yellow to orange and scarlet red, often all colors on a single tree glowing like a flame as the outer branches that receive more light have redder fall color.

Celtis tenuifolia
Dwarf hackberry

Ulmaceae (elm family)

Shade tree-sized hackberry, sugarberry, and elms along with the small tree-sized dwarf hackberry are all in the elm family. I wish nurseries would embrace this delightful little tree but it is nigh impossible to procure one. There is no finer plant for a butterfly garden because it is host to no less than five species and fits into any landscape as it usually grows 8–15 feet tall.

HOW TO GROW This tree is easy to germinate from seed and readily transplants. It's sparingly native to dry, rocky woods including glades across the Lower Midwest and northward into southern Michigan and on sand dunes locally in the Indiana Dunes. It thrives in full sun to light shade and tolerates extreme heat and droughty conditions. Cold hardiness is listed as USDA zone 4, but I've never witnessed a plant in that zone.

LANDSCAPE USE Dwarf hackberry is a fine little tree for a natural garden with butterfly or bird themes. When grown in good soils, it grows quite fast into a vase shape but trees in harsh native conditions often have marvelous gnarly growth and crowns. It is usually single-trunked so hard to call a shrub but can flower and fruit when only a few feet tall.

ORNAMENTAL ATTRIBUTES The foliage is clean and neat, not displaying nipple galls or witches' broom common on hackberry trees. The reddish purple fruit delightfully adorn the tree into winter and are edible—though it's just a sweet coating on a hard seed. Many songbirds relish the winter

fruit. Mourning cloak, question mark, American snout, tawny and hackberry emperor butterflies' caterpillars feed on the foliage. Fall color is yellow. Bark is smooth with corky warts and quite interesting.

Celtis tenuifolia (dwarf hackberry) has distinctive leaves with teeth on the upper half of the leaf.

Spherical, showy white flowers adorn *Cephalanthus occidentalis* (buttonbush) in midsummer.

Cephalanthus occidentalis
Buttonbush
Rubiaceae (madder family)

Most of the woody plants in the madder family are tropical. Buttonbush is an amazingly hardy plant and one of a very few native from Minneapolis to Miami. When this plant blooms in midsummer, proudly displaying its clusters of round white flowers, it often gets a "what is that?" question. People are just as surprised to hear it's a native plant growing along floodplain swamps and other wooded wetlands.

HOW TO GROW Buttonbush is native in all midwestern states except the Dakotas but is found as far north as along the St. Croix River in eastern Minnesota so is cold hardy in USDA zone 4 when grown from northern strains. It grows wild along floodplain swamps and oxbows, doing best where it gets light at the edge or opening in the forest. It tolerates all but poor, excessively dry soils and is drought tolerant. The plant is fuller with more flowering in full sun but it will survive in light shade. It is easy to grow from seed, cuttings, or by layering stems.

LANDSCAPE USE Buttonbush readily grows 8–15 feet tall and occasionally taller, mature plants can be very wide spreading—some measured to 35 feet across. I find it looking best when pruned into a multitrunked small tree, but some strains are smaller and shrubby. It's an excellent plant for a butterfly garden, rain garden, along a pond or stream garden.

Spring bloom on *Cercis canadensis* (Eastern redbud) occurs before the tree leafs out and is the most cherished aspect of the plant.

ORNAMENTAL ATTRIBUTES The foliage is often dark green and glossy but the summertime flowers steal the show and look otherworldly—the cultivar 'Sputnik' describes them well. The flowers, which are very nectar rich and visited by many pollinators including butterflies, form similar-shaped clusters of fruits that age rosy to finally ripening brown and remaining on the plant into winter as they disintegrate. Many songbirds including waterfowl eat the fruits.

RELATED PLANTS 'Sugar Shack' is a smaller shrub variety selected in Michigan while 'Magical Moonlight' grows 5–8 feet tall and is from Ohio.

Cercis canadensis
Eastern redbud
Fabaceae (legume family)

The spring-flowering trees of North America are almost universally white blooming, and then there's redbud. I always make time to drive through countryside and neighborhoods with outstanding floral displays of redbud. The color is vibrant raspberry sherbet and equally refreshing—just what the doctor ordered after a long winter. Redbuds are legumes in the senna subfamily, which includes some of the world's most flamboyant flowering trees such as the cassias, paloverdes, and royal poinciana. Eastern redbud is the state tree of Oklahoma.

HOW TO GROW Redbuds are at their best in the Lower Midwest and native northward throughout all of Ohio, southern Michigan, central Iowa, and southeastern Nebraska. The tree is magnificent in Oklahoma where it is appropriately the state tree. Redbuds are cultivated everywhere in the Midwest with recent Minnesota strains cold hardy through USDA zone 4. Redbud loves our rich, calcareous soils as long as they are not compacted and are well drained. It grows best in full sun but will grow in high shade, though more open with less flowering when shaded. Redbud is easy to grow from seed while cultivars are usually budded on rootstock as it's really challenging to root from cuttings. It grows fast and lives about as long as we do—a tree that meets instant gratification. Removing declining trunks and allowing new basal shoots to replace dying trunks can rejuvenate old trees. Redbud is extremely susceptible to verticillium wilt so use it carefully in areas where that disease is prevalent and know where any mulch you use came from.

LANDSCAPE USE Redbud is widely cultivated as an ornamental flowering tree. It's at its best along the edges of woodlands or as the right-sized tree to grace an entryway, porch, deck, or patio. The main trunks of old plants often lean, sometimes on the ground. The tree's branching pattern and form are among the best of all living sculptures, and landscapes that preserve such trees with groundcovers planted beneath should be lauded. Spring wildflowers like Virginia bluebells, woodland phlox, polemonium, and violets look stunning beneath redbuds. Redbuds also have edible flowers and young pea pods (actually legumes), with nectar-rich flowers so they are good plants for food forests and insectaries gardens.

ORNAMENTAL ATTRIBUTES The spring bloom, before the tree leafs out, is the most cherished aspect of the plant—tufts of flowers occur on the trunk and along every twig. The peapod fruit is sometimes thought of as unsightly, but I find the light green pods beautiful as they become crimped while they mature, turning brown when ripe, and add interest as they hang on the tree in winter. The form of the plant is unparalleled with each a work of art and the bark of older trees often exfoliates to reveal reddish younger bark. Outside of polluted urban environments, the tree's trunk is cloaked in blue-gray and chartreuse lichens.

RELATED PLANTS There is a flood of new redbuds on the market with most selected from the American South. It's essential to grow locally proven strains and to grow proven hardy cultivars. I must admit the purple and chartreuse foliaged cultivars are striking and a much healthier choice for the environment than Japanese or Norway maple cultivars. The cultivar 'Alba' has white flowers, as does hardier 'Royal White' selected from Illinois. The double-flowering

'Flame' redbud, also from Illinois, is cold hardy only to USDA zone 6—its flowers don't show well but are literally extra delicious. Some people find the blue-pink flower color harsh and bright cherry to peeked pink-flowering cultivars are also available. There also are weeping and variegated cultivars.

Chionanthus virginicus
American fringetree

Oleaceae (olive family)

American fringetree is in the same family with swamp-privet and ash trees. As a close relative of ash trees threatened by EAB, American fringetree is a potentially important plant to aid many native insects tied solely to ash (they may face extinction). I have seen the beautiful caterpillars of fawn sphinx on my American fringetrees, as well as on my ash trees so I recommend planting it where ash trees are now missing. Unfortunately, EAB has now been found in some American fringetrees. Barely a tree and really more like a huge shrub, American fringetree is much wider than tall at maturity. It is dioecious; male trees appear to grow taller with slightly showier bloom while females are decidedly wider than tall. As members of the olive family, female plants produce a fruit very similar to an olive and can actually be prepared like an olive to be edible.

HOW TO GROW American fringetree grows mainly across the Southeast but ranges northward into Ohio and Missouri and grows well across almost all of the Midwest and has naturalized northward. Northern plants have more rounded leaves, often with fuzzy new growth and that form is best (fully hardy) for the Upper Midwest. Fringetree grows slowly to moderately, rarely more than 12 inches a year. It's native to woodland understory but will grow in full sun in rich, well-drained soils. When stressed in poor or compacted soils, it is very susceptible to lilac-ash borers that will kill infested trunks. Removing damaged trunks can easily rejuvenate fringetree and new basal shoots will emerge and quickly replace them. This may be a way to maintain a fringetree against EAB since that borer prefers larger stems. American fringetree is easy to grow from seed but requires a double dormancy so wild seedlings emerge after two winters.

LANDSCAPE USE American fringetree is best utilized in a woodland garden or woodland edge or as a focal plant under high shade of large trees. It can be pruned to make a nice hedge too, but the beauty of its flowering is destroyed. It is a great choice for a white or moon garden or near outdoor seating spaces.

The white flowers of *Chionanthus virginicus* (American fringetree) give the plant its common name and light up the forest when the plant blooms in late spring.

ORNAMENTAL ATTRIBUTES I write this in late spring, the week of walking beneath our fringetrees in full bloom at Powell Gardens and at home. The delicate, pendant tufts of white flowers tucked beneath the bursts of new growth and the soft and sweet perfume filling the air are nothing short of sublime. Fringetree blooms between the hurrah of spring and the greens of summer. The fruit fill out green over the summer, turning purplish from the bottom up and finally bluish black in late summer, sometimes hanging into winter if robins don't eat them all. Fall color is at best yellow, mainly yellow-green. The gray trunks develop bark striated with charcoal when they mature.

Cornus spp.
Dogwoods
Cornaceae (dogwood family)

Dogwoods make some of the most popular small trees and large shrubs for landscaping. They all have whitish flowers and colorful fruits in fall, but none of them hold their fruits into winter. Flowering and pagoda dogwoods are the most treelike; roughleaf and gray dogwoods have stems that readily become the size of small trees, but they are thicket forming, while silky and red-osier dogwoods are more shrublike.

Cornus alternifolia
Pagoda dogwood, alternate-leaved dogwood
Cornaceae (dogwood family)

I grew up with this dogwood as a common understory tree and transplanted a couple of plants into my parents' yard where their splendor in the garden became apparent and their top attribute—berries loved by the birds in late summer—brought in species not seen in the yard until then. Pagoda dogwood grows fast when young, adding a whole new tier of branches each season.

HOW TO GROW Pagoda dogwood is native across northeastern North America westward into central Minnesota, central Iowa, and the eastern Ozarks. It is hardy throughout the Midwest. It demands well-drained soils and struggles in heavy clay or compacted soil. It can be grown from seed, cuttings, or by layering. It is native in moist woodlands so prefers protection from hot afternoon sun and even in the Upper Midwest will languish in a hot dry site. It will grow in full sun in a moist, sheltered site but prefers part shade to high shade. It can self-sow where happy. Note that two separate trees are needed for good fruit set. Old trunks can suddenly die and turn orange, but can be quickly rejuvenated.

Distinctive horizontal tiers of branches give *Cornus alternifolia* (pagoda dogwood) its common name.

Simply remove dead trunks and encourage new stems to grow from the base (root crown) of the plant.

LANDSCAPE USE The distinctive horizontal tiers of branches (sympodial branching) give it the name pagoda and make it a fine ornamental tree for a wooded landscape. It also is useful at the corner of a house where it softens harsh edges, and it is exquisite in and on the edge of a woods. It's a premier bird garden plant.

ORNAMENTAL ATTRIBUTES My favorite aspect is the horizontal branching that curves upward at the twig ends, so beautiful in all seasons. The flowers are clusters of creamy white set above the foliage in late spring. The fruit ripen bluish black in late summer and are high in lipids and relished by a wide array of songbirds. Pagoda dogwood brought the first bluebirds to our yard but attracted everything from catbirds and robins to red-headed woodpecker and red-eyed vireo. The foliage is clean and neat and turns yellow to burgundy red in the fall, the richer colors produced in full sun or on the outer parts of the plant that receive more light.

The bluish fruit of *Cornus amomum* (silky dogwood) in late summer into fall distinguishes it from other native dogwoods.

Cornus drummondii (roughleaf dogwood) fruit attracts songbirds in fall like no other plant.

RELATED PLANTS Several new cultivars have recently become available and I am all for them if it helps get this great plant into more midwestern landscapes: Golden Shadows ('W. Stackman') is yellow variegated, 'Gold Bouillon' has golden leaves, Big Chocolate Chip ('Bichozam') shows a chocolate burgundy cast to mature leaves, French Vanilla ('Frevazam') has frothy good bloom with orange fall color, and 'Pistachio' has burgundy new growth.

Cornus amomum
Silky dogwood

Syn. *Cornus obliqua*
Cornaceae (dogwood family)

This dogwood is truly a shrub, never a tree, with a multistemmed base of greenish to reddish stems.
HOW TO GROW Silky dogwood is found across the Eastern and Lower Midwest and in the eastern Upper Midwest, westward to central Minnesota, and southeastern South Dakota. It grows on the edges of wetlands but also can be found in swales in upland prairies and along springs, streams, and other riparian areas. It can be grown from seed, cuttings, or layering. It thrives in full sun to part or light shade and is tolerant of all but the driest soils. It's fully hardy anywhere in the Midwest.

LANDSCAPE USE It makes a fine shrub for massing, especially in problematic wet areas, and it works well in rain gardens as it can take being wet or dry. It readily grows to 8 feet tall. Removing older stems keeps the plant looking tidier.
ORNAMENTAL ATTRIBUTES The stems are somewhat ornamental in winter, usually greenish to reddish brown, more colorful on younger stems. Flowers appear in late spring and are clusters of creamy white followed by uniquely colored, bluish berries that ripen in late summer. Sometimes the berries are fused blue on white like porcelain. The fruits drop in the fall. Fall color is usually very good from red to burgundy, with yellows and oranges on the shaded leaves.
RELATED PLANTS Powell Gardens has selected a red-twigged form from a unique population in its native prairie remnants. It is available under the trademark name Red Rover ('Powell Gardens') and if rejuvenated shows good red winter twigs and dark burgundy fall color.

Cornus drummondii
Roughleaf dogwood

Cornaceae (dogwood family)

No plant in my yard attracts more wildlife than this thicket-forming large shrub. I have set up a chair near it when it's in full fruit in September and watched various resident and migrant songbirds come in to snatch its lipid-rich berries—including phoebes and pewees, various vireos and warblers normally insectivorous. This species is mixed up in the nursery trade with the Upper Midwest's gray dogwood (*Cornus racemosa*), which is similar but overall smaller with smooth, flatter leaves unlike the raspy cat's-tongue surface to aptly named roughleaf dogwood. Roughleaf dogwood also holds its leaves with the edges curled up showing its lighter underside.

HOW TO GROW It readily germinates from seed but is easiest to propagate by dividing off plants from a thicket. It's found across the Lower Midwest (absent from eastern Ohio) and northward sparingly into northeastern Iowa, northern Illinois, and southeastern Michigan. It's a brushy, mainly edge-of-the-woods and successional species, dying out as a woodland matures and shades it out. It's very heat and drought tolerant and cold hardy through USDA zone 4.

LANDSCAPE USE It is one of the finest screens and hedgerow plants but is ornamental enough to be a focal ornamental plant if pruned and its suckers removed. It can be trained into a little tree too.

ORNAMENTAL ATTRIBUTES The white berries, set atop red stems, float like clouds over the shrub in late summer into fall—a phenomenal contrast to red and burgundy fall color of the foliage. The berries do not last into winter; they wither and drop if not eaten by birds. White-throated sparrows and other ground-feeding birds seek out the seeds of fallen fruit. The creamy white flower clusters bloom in late spring after the plant has leafed out and are visited by masses of pollinators.

RELATED PLANTS Gray dogwood (*Cornus racemosa*, syn. *C. foemina* subsp. *racemosa*) is another thicket-forming shrub. Where I grew up, it harbored the marvelous thumping dance of the ruffed grouse. It's found across the Upper Midwest from North Dakota south to northern Missouri (also sparingly in the Ozarks) and eastward to all but southernmost Ohio. It has the same niche, cultural requirements, and landscape use as roughleaf dogwood but is more tolerant of wet soils. Its fall color is more burgundy to purple with a polished leather look to the leaves and is finer and of smaller stature than roughleaf dogwood. It is hardy throughout the Midwest.

155

Cornus florida
Flowering dogwood

Cornaceae (dogwood family)

This is the longest-lasting, showiest of our native flowering trees and the state tree of Missouri. A leafless tree foliated in its classic four-bracted flower clusters in midspring is like a stunning white cloud of floating butterflies. Woodlands and communities with exceptional displays are tourist attractions. It was widely planted wherever hardy and is by no means an easy plant to grow. When dogwood anthracnose came along and destroyed many trees, flowering

Flowers of *Cornus florida* (flowering dogwood) create a heavenly cloud of white.

The white autumn fruit combined with wine-purple fall color make *Cornus racemosa* (gray dogwood) a stand out in the fall.

dogwood became a popular target for horticulturists who offered substitutes and almost made this integral component of the American flora a villain.

HOW TO GROW Flowering dogwood is native across the Eastern Midwest and westward across the Lower Midwest through the Ozarks and spreading into the Osage Plains. It's cultivated northward through USDA zone 5b and has actually naturalized near plantings as far north as Rockford, Illinois, where it is sparingly cultivated in the shelter of the Rock River and the city's urban heat island—a good example of how locally adapted this plant can be. It's always a good idea to grow regionally sourced or locally proven cultivars. Easily grown from seed and easily cloned by layering (pinning down lower branches and allowing them to root), it's commercially propagated by bud grafting on seedlings. Flowering dogwood requires well-drained soil, light or high shade to full sun. In areas with dogwood anthracnose, open locations with good air circulation are best. Afternoon shade is often beneficial in the western part of its range when cultivated, though it grows wild in full sun on hot, rocky glades near Branson, Missouri. It may be challenging to transplant while a bird-planted seedling can be tenacious.

LANDSCAPE USE Flowering dogwood is a marvelous small tree for a woodland garden, woodland edge, or as a focal ornamental plant in a traditional landscape. It's a must for a bird-themed garden as its fruit are rich in fats to fuel bird migration.

ORNAMENTAL ATTRIBUTES The flowers are tiny yellowish clusters at the center of the four showy white bracts. The bracts emerge a marvelous creamy to lime green, sometimes blushed with pink, and mature milky white. When viewed from below and illuminated by sunlight, they are unmatched. The bracts hold as the leaves emerge and drop after about 10 days. The fruits ripen in early fall, becoming scarlet red in contrast to the foliage, an advertisement to birds that disperse their seed. The fall color is one of the longest lasting and consistently showy: purplish tinges start after the autumnal equinox and become redder until they are pure red before dropping in late fall. The tree's winter form reveals layers of branches made up of upturned twigs studded with flower buds. The charcoal bark of old trees gets pebbly to reticulate and the wood is heavy and strong.

RELATED PLANTS There are many cultivars of this plant, most selected from south of the Midwest and these are mostly not reliably hardy in the northern part of the range. Seedlings can bloom on alternate years, while cultivars are selected for annual bloom and special attributes that include larger flowers, pink or red bracts, and variegated foliage. The University of Tennessee has introduced several anthracnose-resistant cultivars.

Cornus sericea
Red-osier dogwood

Syn. *Cornus stolonifera*
Cornaceae (dogwood family)

Cornus sericea is our native red-twigged dogwood whose cultivars are readily available at most nurseries across the Midwest. If you are ever in doubt about the difference between this and the silky dogwood, cut a twig—silky dogwood has brown pith while red-osier's pith is white.

HOW TO GROW Red-osier dogwood is found in cool, moist sites and wetlands across the Upper Midwest from Nebraska, northern Iowa, and eastward across the northern edge of the Corn Belt to Ohio and points north. This dogwood is actually stressed under hot and dry conditions and often suffers many foliar maladies and stem cankers to the point that it's not a long-term landscape plant outside its native range. It's best in AHS heat zone 5 and colder; no healthy plants survive at Powell Gardens (in AHS heat zone 7). Removing older stems keeps the plant tidier as well as maintains the best red stems—the red color is lost on old wood.

LANDSCAPE USE It's a fine large shrub (readily growing 8 feet tall) for mass plantings in swales or other wetter sites and works in a rain garden.

The red-stems of *Cornus sericea* (red-osier dogwood) add color on a dreary winter's day.

ORNAMENTAL ATTRIBUTES It really stands out in a snowy winter landscape with its red stems, some plants more colorful and almost coral red. The flowers are creamy white clusters in late spring. Berries are white on red stems in late summer and they also drop before winter. Fall color can be quite beautiful in shades from yellow and orange to dark red on foliage that receives full sun.

RELATED PLANTS The cultivar 'Cardinal' has extra colorful red stems and is more heat tolerant than most, doing well southward to the St. Louis area. Some cultivars have yellow, rather than red stems, and there are also variegated and golden-leaved clones. There are also selections of smaller stature that fit in the shrub category including Firedance ('Bailadeline') that grows just 3–4 feet tall by 4–5 feet wide and was found in Minnesota.

Corylus americana (American hazelnut) once formed dense thickets across the Midwest and is best used in a natural garden, where it can spread.

Corylus americana

American hazelnut

Betulaceae (birch family)

When one reads the Midwest's historical flora records, this shrub was once one of the most abundant species. It formed dense thickets in prairies, savannas, and oak-hickory woodlands. What happened to it? It's a mere vestige of what it once was and became infrequent in most areas, originally absent only from southwest Minnesota and adjacent northwest Iowa and northeast South Dakota. Yes, it's the native counterpart to the hazelnuts (*Corylus avellana* and hybrids) sold in grocery stores with very similar, though smaller nuts.

HOW TO GROW American hazelnut is easily grown from nuts but can be divided off or layered from an existing plant. It tolerates any well-drained soil in full sun to light shade and is hardy across the entire Midwest.

LANDSCAPE USE This lightly to strongly suckering shrub gets larger than often attributed—6 feet tall is a small plant and it will readily grow 12 feet tall. The champion is in Washtenaw County, Michigan, at an astounding 35 feet tall and wide. The plant is best in natural and wildlife-themed gardens, natural area restorations, edible landscapes, and food forests. It is a favorite browse for rabbits and deer as well as many species of caterpillars that feed on related birch.

ORNAMENTAL ATTRIBUTES It's often the first plant to bloom in spring with tiny female flowers that look like fuchsia-red spiders on the buds, while male flowers are quite handsome golden catkins that drape abundantly from the twigs to release their pollen in the wind. Bracts that encase the nuts turn light green in summer, contrasting with the dark green leaves. Fall color is one of the best, turning oranges to reds.

RELATED PLANTS Closely related beaked hazel (*Corylus cornuta*, syn. *C. rostrata*) is found across the northern edge of the Midwest and southward through the Driftless Area where it is restricted to cool, mesic forests. It also produces edible nuts that look more like a filbert (*Corylus maxima*) in that they are enclosed in a bract that forms a beaklike projection. Beaked hazel is widespread across North America, also growing in the Pacific Northwest but only native forms should be cultivated in the Upper Midwest.

Cotinus obovatus

American smoketree

Anacardiaceae (cashew family)

There's nothing like seeing a tree in the wild to learn its true colors. On a college field trip through Missouri, I first saw this tree in the wild at Caney Mountain Natural Area in the Ozark Highlands near Branson. Here, smoketrees were growing out of rock on the edges of a glade and they grew larger than I imagined (some 30–40 feet) with such beautiful mature trunks. American smoketree is like the common nonnative smoketree (*Cotinus coggygria*) but on steroids.

HOW TO GROW Our smoketree's largest area of native range is in the Midwest—in the southern Ozark Highlands into adjacent northeastern Oklahoma. It's also found in southeastern Tennessee and adjacent Alabama and in the Texas

Hill Country. It's relatively easy to grow from seed, grows quickly, readily transplants, and is fully cold hardy through USDA zone 4. Its major requirement is well-drained soil and full sun and its adaptations shrug off extreme heat and drought.

LANDSCAPE USE American smoketree makes one of the finest small trees (usually 20–30 feet) for a hot, dry locale and grows well on exposed limestone and gravelly sites. If you have a scorching hot, west-facing outdoor space, try planting this tree there.

ORNAMENTAL ATTRIBUTES The tree has steely, rounded leaves that will take your breath away in autumn as they turn brilliant yellow to orange and fiery red. It's related to sumac. Older trees develop a flaky light gray-brown bark and a sculptural character. The flowers are silky clusters that develop into feathered seeds that give the look of small clouds of smoke over the foliage in summer.

The fall color is the most spectacular ornamental asset of *Cotinus obovatus* (American smoketree).

Crataegus spp.
Hawthorns
Rosaceae (rose family)

Hawthorns are the state flower of Missouri. At one time they were a small tree beloved by the Midwest's pioneering native plant gardeners and designers, but today most landscapers, professional designers, and homeowners loathe them. Most hawthorns have wide-spreading, horizontal branching so were embraced as trees that mimicked the horizontal lines of the midwestern prairies. Haw foliage suffers from many maladies resulting in leaves often speckled with yellow, orange, and brown, and dropping prematurely by late summer in many species. Modern landscapes do not allow for plants with any defect, even though it may be purely cosmetic and not harm the plant in the long run. Thorn has also become a legal liability.

Hawthorns, related to apples, were once common across much of the country, in open woodlands, woodland edges, savannas, pastures, hedgerows, and abandoned fields. Their diversity of foliage was astounding, with a mind-boggling array of regional variations once all named as species. With our woodlands becoming dense forests that shade out hawthorns, our pastures becoming rare, and our fields farmed fence to fence, hawthorns are in severe decline. The increase in red cedars, alternate host to cedar-quince (*Gymnosporangium clavipes*) rust, the worst of three cedar rust maladies that afflicts hawthorns, is probably also to blame. Missouri flora author George Yatskievych has looked at herbarium specimens and documented that older specimens show few signs of the rust, while it's prevalent in more recent collections.

Cedar-quince rust can disfigure hawthorn fruit—in this case, *Crataegus crusgallii* (cockspur hawthorn).

Cedar-apple rust causing leaf spotting on *Crataegus mollis* (downy hawthorn).

I have watched magnificent mature hawthorns, planted by the likes of O. C. Simonds, die and not be replaced. Do we want this chapter of our botanical and horticultural history to end? A recent visit to the Edsel Ford estate in southeastern Michigan was uplifting; Jens Jensen's original and spectacular hawthorns still stand there. A mature dotted hawthorn (*Crataegus punctata*) northeast of the home was simply one of the most awe-inspiring small trees I have ever set foot under. Yes, its leaves were imperfect but the tree's horizontal form and branching pattern atop a fluted, twisted trunk, created nature's finest living sculpture. Thank you, Mr. Jensen, for your design here and thank you to the estate's conservators for their preservation of this historic landscape's plants.

Midwestern hawthorns all bloom in spring, almost exclusively with white flowers that are malodorous and pollinated by flies and beetles. While in bloom, they're a great place to observe flycatchers and other hungry migrating birds. In autumn, the plants produce reddish fruit, which in some species persist on the tree through winter. The fruit are edible, make the finest preserves, and contain compounds beneficial to the human cardiovascular system. The fruit also feed countless birds and small mammals, so they are a must in a wildlife-themed garden. Mature trees can have wondrous horizontal forms and branching patterns, simply spectacular in the winter landscape.

Hawthorn's Latin name *Crataegus* originates with the Greek word that signifies strength and invincibility. Let's hope we don't sterilize our modern landscapes or neglect wildlands to the point we lose these iconic trees. On my small acreage, I'm saving the three wild species present by removing competing vegetation or transplanting wild seedlings into more light. I have also planted four added species. I tried trimming off cedar-quince rust but found the plants do better following my mom's advice to "let it be." Removing my red cedars would be moot too as spores can travel for miles. Give nature a chance and I'm sure hawthorns will find their way. Horticulturists should always be on the lookout for disease-resistant trees to propagate.

HOW TO GROW Dozens of hawthorn species are found throughout the Midwest with three species widely cultivated (Washington, cockspur, and green). They are easy to grow from seed but most landscape plants are grafted on understock that may be another species of hawthorn, apple, and even quince (*Cydonia oblonga*). Hawthorns need full sun or just part or high shade. Various species are tolerant of any soil—wet to dry.

LANDSCAPE USE Hawthorn is no longer a tree for the traditional landscape but should be preserved in native, natural, and historic landscapes where it makes a beautiful sight in bloom and when leafless through our long winters. Its thorny nature can make it a premier barrier plant yet thornless varieties of many species are available. They can be planted on the edges of food forests for their fruit and are great additions to bird gardens as prime nesting habitat and for their insect-attracting flowers and fall-to-winter fruit. Because of their dropping fruit, downy and dotted hawthorns may not be suitable for pristine gardens.

ORNAMENTAL ATTRIBUTES The winter form of the trees is their greatest asset, some with beautiful exfoliating bark. Their spring bloom can be quite showy as with their fall fruit. Some species hold their colorful red fruit into winter.

Crataegus crusgalli (cockspur hawthorn) is native across the Eastern and Lower Midwest and northward into southern Iowa. It is cultivated and hardy everywhere in the Midwest. It has classic horizontal branching and holds its red fruit beautifully into winter. It is one of the spiniest species, making an impenetrable barrier hedge. Variety *inermis* is

A close-up of *Crataegus crusgalli* (cockspur haw) shows the dull red winter fruit and cockspur thorns.

The large red fruit of *Crataegus mollis* (downy hawthorn) makes the finest preserves.

thornless and widely planted. The cultivar Crusader ('Cruzam') also is thornless, smaller statured, and more disease resistant.

Crataegus mollis (downy hawthorn, redhaw) is a magnificent hawthorn found across most of the Midwest except for much of eastern Ohio. It blooms early in midspring and has large fruit that make the finest redhaw preserves. The fruit drop after ripening, but mature trees in winter are spectacular sculptures even without colorful fruit. The leaves are often spotted yellow from cedar-apple rust, and I'll be the only one to say I think it's beautiful—others would find the colors attractive only on a coleus.

Crataegus phaenopyrum (Washington hawthorn) is native across the Lower Midwest from the Ozarks to southern Ohio and south. It is the most widely cultivated hawthorn in the Midwest but is cold hardy only through USDA zone 5 and into the warmest parts of USDA zone 4, doing well as far north as La Crosse, Wisconsin. I never saw cedar-quince rust on this species in Ames, Iowa, or Rockford, Illinois, and was shocked to see trees mired with the rust when I moved to the Kansas City region in 1996. Since then, cedar-quince rust has invaded northward and I've seen the rust in Minnesota and Wisconsin. It is still one of the most disease resistant haws with a twiggier upright crown, very late spring and malodorous flowers, excellent red to burgundy fall color, and very showy, small red fruit which last into winter until stripped off by hungry birds.

Crataegus punctata (dotted hawthorn) is from central Minnesota south across Iowa and Illinois eastward throughout the Eastern Midwest and into southeastern Canada and New England. It produces the finest, horizontal branching pattern of any hawthorn. The younger stems can be very light gray, the name "white" hawthorn used in some

This awe-inspiring *Crataegus punctata* (dotted hawthorn) once grew in Sinnissippi Park in Rockford, Illinois. It blew down in a storm in 1992.

regions. As a teenager, I successfully transplanted one from a pasture. The fruit drop after ripening. The cultivar 'Ohio Pioneer' is thornless.

Crataegus viridis (green hawthorn) is native across the Lower Midwest from the Osage Plains eastward to southern Indiana. It is hardy and planted northward through USDA zone 5. This hawthorn is also quite disease resistant with the most beautiful gray bark that exfoliates to reveal cinnamon-colored fresh bark, which ages back to gray. This species can have the showiest red fruit that hang into winter. The cultivar 'Winter King', selected from southwestern Indiana, is widely planted.

Crataegus phaenopyrum (Washington hawthorn) holds its red fruit into winter—the fruit remain colorful through below zero cold.

Crataegus viridis (green hawthorn) has vibrant red fruit that hold into winter and a more wide-spreading growth habit than Washington hawthorn.

The bark of *Crataegus viridis* (green hawthorn) exfoliates to reveal cinnamon-colored younger bark.

Euonymus atropurpureus
Eastern wahoo

Celastraceae (staff vine family)

Say *Euonymus* and it conjures up the invasive evergreen wintercreeper (*E. fortunei*) and burning bush (*E. alatus*). Other Eurasian small treelike species have escaped too (*E. europaeus*, *E. bungeanus*) and muddy the identification of our native version. So how do you identify the native? It's the only one with madder red or, as it is named, "dark (*atro*) purple (*purpureus*)" flowers.

HOW TO GROW Eastern wahoo is found across almost the entire Midwest except North Dakota and hardy everywhere. It's easy to grow from seed, cuttings, or divisions. It will grow in full sun to shade, found most often in mesic floodplains (not continually saturated) to moist upland woods. Oldest stems reach 4 inches in diameter and 15 feet tall, usually dying at that point but the plant living on through younger stems. I wonder if trees with large trunks reaching 30 feet tall are true to type or hybrids with the nonnatives.

LANDSCAPE USE Wahoo is best in a natural or shade garden because the plant occasionally suffers severely from euonymus scale and mildew, suckers into a thicket, and grows rather unkempt. It's too open for use as a screen but nice mixed with other understory shrubs and trees. It is a good plant for a bird garden; eastern bluebirds eat the fruit in fall into winter.

ORNAMENTAL ATTRIBUTES The fruit are shockingly beautiful on close inspection. What a clash of colors—light violet-pink (capsule) paired with vermillion red (aril surrounding the seed). The capsule bleaches salmon-colored by winter and doesn't hold its color like related bittersweet when cut. The fall color ranges from purplish to orange-red and is most noticeable in fruit after leaf drop. Stems are greenish and stand out with the typical grays of winter. Plants are highly variable, some strains more suitable for a traditional landscape than others, but no cultivars have been selected.

The fruit capsule of *Euonymus atropurpureus* (eastern wahoo) shows it's related to well-known bittersweet.

Forestiera acuminata
Swamp-privet

Oleaceae (olive family)

This little-known small tree is one tenacious plant with nothing showy; it's been ignored by the landscape industry.

HOW TO GROW Swamp-privet is native in river floodways from southeastern Kansas across southern and eastern Missouri into central Illinois along the Illinois River, into southwestern Indiana along the Wabash River, and western Kentucky along the Ohio River. It is heat, cold and drought resistant but is susceptible to lilac-ash borer when really stressed. It remains to be seen if emerald ash borer will infest it. It's easy to grow from fresh seed and also grows well from cuttings. It grows fast and is tolerant of a wide range of soils as long as they are not too dry. It grows in full sun or high shade.

LANDSCAPE USE Swamp-privet is a small tree for a naturalistic shade, wetland, or rain garden and can also be grown as a large screen plant. My guess is it would make a better hedge than privet.

ORNAMENTAL ATTRIBUTES The flowers are cute in early spring: paired yellow tufts of stamens adorn branches on male plants while green spidery flowers line in tandem the stems of female plants. It's one of the first plants to set fruit for birds and wildlife, ripening purplish elongated (sometimes cashew-shaped) drupes in late spring. The foliage is small, creating a fine texture, and stays clean and neat all growing season before turning yellowish in the fall. Winter

Early spring male flowers (shown here) of *Forestiera acuminata* (swamp-privet) are showier than female flowers.

plants stand out along riverways with their lighter-colored, tan bark.

NOTES Swamp-privet is the sole host of a sweet little moth (*Philtraea monillata*) that has white wings adorned with orange bands and circles that are outlined with black polka dots—a good reminder of the unseen value of every plant.

Hamamelis spp.
Witchhazels
Hamamelidaceae (witch hazel family)

There are two species of midwestern native witchhazel whose flowers mark both the beginning and the end of the floral season. Common witchhazel is widespread in the Midwest while Ozark witchhazel is endemic to the Ozark Highlands.

Butter yellow leaves of *Hamamelis vernalis* (Ozark witchhazel) may dry brown and linger through winter on some plants.

Hamamelis vernalis
Ozark witchhazel
Hamamelidaceae (witch hazel family)

Winter-flying noctuid moths have a friend in this, the earliest flowering native plant in the Midwest—often blooming in winter in the Lower Midwest. In the Upper Midwest, it will bloom during a late winter thaw but usually no earlier than March.

HOW TO GROW This very large shrub is endemic to rocky streamsides of the Ozark Highlands and adjacent Ouachita Mountains in Oklahoma, Arkansas, and Missouri. It's widely cultivated wherever it is hardy, through all of USDA zone 5, and will survive in USDA zone 4 when sited in a warm location. It has naturalized sparingly near plantings as far north as Rockford, Illinois, and at Powell Gardens, Missouri, outside its native range. It's also easy to grow from seed, dispersed by the same explosive capsule as the common witchhazel.

LANDSCAPE USE Ozark witchhazel is a popular traditional landscape plant used in shrub borders and shade gardens. It rarely achieves small tree status at maturity, mainly a large shrub 8–12 feet tall and usually growing taller than wide to start with but eventually spreading wider than tall. Site plants where they can be viewed from indoors and are backlit by the sun to highlight the flowers.

ORNAMENTAL ATTRIBUTES The small flowers are the first of the season when gardeners are starved for bloom and are very fragrant to boot. Flowers vary from yellow to purplish and all combinations in between. There are many cultivars reflecting various flower size and color. I recommend

picking out seed-grown plants at the nursery while in bloom to choose a color and fragrance that suits you but the foliage also can be quite variable from dark green to bluish green and some plants hold their leaves marcescent. Fall color is always beautiful but also variable from pure yellow to fiery yellow-orange to red, often multiple colors on a single leaf.

Hamamelis virginiana
Common witchhazel
Hamamelidaceae (witch hazel family)

This is the understory tree that announces that the flowering season is at its end. In late September (north) into November and December (south) it cloaks its twigs with spidery yellow flowers as the forest goes dormant.

HOW TO GROW Witchhazel grows across most of the Eastern Midwest, northwestward through the Driftless Area and southwestward into the Ozark Highlands and is hardy throughout the Midwest. It grows wild in mesic forests, and west of Lake Michigan, it's usually found in sheltered ravines, usually on the north or east side of steep bluffs. It's easy to propagate from its shiny black seeds that grow fast and transplant well. Seed burst forth from the capsule when it snaps open in fall. I recall a capsule on my office desk in Illinois opening and the seed ricocheting down the hallway and into another office. Wow, the engineering in the seed dispersal mechanism of this plant!

LANDSCAPE USE It's a small tree 15–25 feet tall suitable for a traditional landscape but is seldom cultivated, probably because no one is at the nursery when it's in bloom. It does

Hamamelis virginiana (common witchhazel) flowers glorify a leafless, late-fall woodland garden.

best in shade or at least shade from the hot afternoon sun in Lower and western Midwest. The tree lightly sends up new shoots from the base to create a multistemmed tree; older trunks often lean—I've paced off mature trees in the understory of old growth woods at 55 feet across.

ORNAMENTAL ATTRIBUTES Witchhazel is a graceful understory tree. I see it as a ballet beneath the canopy. The end-of-season bloom is subtle but simply beautiful against a blue sky or a tree trunk. The tree has rich yellow fall color and plants that lose their leaves before blooming are highly desired so they don't mask the flowers. Sometimes weather dictates flowering and leaf fall. Some plants are marcescent, holding their leaves into winter.

Ilex spp.
Hollies
Aquifoliaceae (holly family)

The holly family includes several species found in the Midwest but only two, winterberry and possumhaw, are the size of a small tree to large shrub and native across larger swaths of the region. Both are dioecious (plants are either male or female) as well as deciduous and are some of our showiest fruiting plants in fall into winter.

Ilex decidua
Possumhaw, deciduous holly
Aquifoliaceae (holly family)

Possumhaw is a Lower Midwest specialty that appears to be blooming fiery red after leaf drop in autumn. Female trees can be loaded with bright red berries and last for at least 2 months and occasionally through winter—providing longer color than any flowering tree. Subzero cold can discolor fruit and birds can strip the berries but that's rarely earlier than mid December.

HOW TO GROW Possumhaw is widespread from the American Southeast but grows up the Mississippi Valley well into the Lower Midwest where it is found in the Osage Plains, Ozark Highlands, Interior River Valleys and Hills, and the Shawnee Hills of southern Illinois. Easy to grow from seed or cuttings it may be one of the most soils tolerant of all plants, growing from floodplains to dry rocky glades. It's dioecious so plants are either male or female; both are needed nearby for pollination. It will grow in light shade but flowers and fruits heaviest in full sun. This plant is hardy into USDA zone 5b but best in the Lower Midwest with its hot summers. The showiest plants from Powell Gardens are under trial in Michigan.

LANDSCAPE USE It can be shrublike but will always grow larger than 15 feet with a wider crown at maturity. As with its smaller cousin the winterberry, possumhaw should be grown where its bright fall and winter color can be observed from indoors. It cheers up any landscape in that season but is a highly variable plant. Rarely are two plants alike; some have suckering bases, others have upright open branching, while still others may form dense rounded crowns. Possumhaw is essential for a bird garden as it attracts fruit-eating birds, its larder often guarded by a mockingbird. The whitish flowers are tiny but nectar rich and make a fine honey so it's also a good plant for an insectaries or bee garden.

In the winter landscape *Ilex decidua* (possumhaw) provides longer lasting seasonal color than a spring-flowering tree.

Red hot fruits of *Ilex verticillata* (winterberry) warm a dreary winter day. This large shrub can be trained up into a multistemmed, tree-like plant.

ORNAMENTAL ATTRIBUTES There is no showier plant for late fall and winter color in the Lower Midwest than a female possumhaw fully decked in red fruits. The foliage is small and dark green, creating a fine texture with little color change in the fall.

RELATED PLANTS Several cultivars have been selected from plants from southwestern Indiana, including 'Red Cascade', a female with undulating somewhat pendulous branching and reaching 20 feet tall and wide; 'Sentry', a female that is more columnar reaching 20 feet tall but only 10 feet wide; and 'Red Escort', a fine male selection.

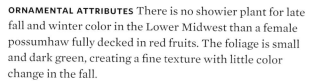

Ilex verticillata
Winterberry

Aquifoliaceae (holly family)

Winterberry is a rather common landscape shrub grown for its outlandish red berries that brighten up female plants after they have dropped their leaves in fall. Over time, all but the most compact cultivars grow larger and larger and many can be pruned into multistemmed treelike plants. I've paced off wild shrubs with a spread of 27 feet but winterberry usually grows 6–8 feet tall and wide.

HOW TO GROW This eastern North American shrub has a divergent midwestern range. Found almost everywhere in the Eastern Midwest, it grows northwestward into Minnesota and southwestward sparingly into the eastern Ozark Highlands. It is native in more acidic sandy or peaty wetland soils, but will grow in rich upland soils and is quite drought tolerant. It's easy to grow from seed, producing plants that are unknown male or female while cuttings will clone showier female or tidier male plants. Plants flower and fruit more heavily in full sun, partial or light shade. They suffer in alkaline or excessively dry soils.

LANDSCAPE USE Winterberry works well in a shrub border, and a female plant can be a focal ornamental shrub. It's great in bird and bee gardens as the berries attract many birds and the tiny flowers are nectar rich. It's also beautiful along streams, in wetlands, ponds, and rain gardens.

ORNAMENTAL ATTRIBUTES This shrub is a loud hurrah at the end of the growing season when its fruits ripen bright red and remain after leaf fall on the bare shrub until birds gobble them all up or until bitter cold discolors them. The foliage is clean and neat, turning a bronzy yellow, but lackluster compared to the fruit. I would not be without this plant as it shines in November, often the grayest and cloudiest month in much of the Midwest. Winterberries are a landscape cure for seasonal affective disorder (SAD).

RELATED PLANTS Virtually all plants in nurseries are cultivars selected for their spectacular fruit displays. Be sure to plant an appropriate pollinator male plant that blooms at the same time. Later-blooming 'Southern Gentleman' (8 feet plus) and earlier blooming 'Jim Dandy' (6 feet plus) are the most widely available male cultivars. Wildfire ('Bailfire') is a 6- to 7-foot selection from Minnesota; Berry Heavy ('Spravy') and Berry Nice ('Spriber'), both 6–8 feet, are from Michigan; vermillion or orangish berried 'Afterglow' (10 feet), scarlet red 'Winter Red' (9 feet) and its sport 'Winter Gold' with golden, pinkish-orange blushed berries are from Indiana. 'Sprite' is a female cultivar with large, abundant fruit on a compact shrub.

Larix laricina
Tamarack, American larch
Pinaceae (pine family)

HOW TO GROW This deciduous conifer grows as far north as any tree and as far south as the Upper Midwest. It is found only in boggy wetlands southward to central Minnesota, Volo Bog in Chicagoland, northern Indiana, and northern Ohio. It can be grown from seed and transplants well into average to wet organic soils. It is probably best in areas of AHS heat zone 5 and colder.

LANDSCAPE USE This tree grows as tall as 75 feet and lives over 300 years old but is a much smaller and shorter-lived tree in the Midwest. It usually grows to about 35 feet tall here and makes a fine addition to an Upper Midwest wetland garden, pondside, or lakeshore.

ORNAMENTAL ATTRIBUTES Tamarack is lovely upright tree of finest "furry" texture of blue-green summer needles. The needles emerge vibrant chartreuse in spring and turn golden yellow late in the fall. The tiny cones are simply lovely and hang on the bare tree into winter.

NOTES Look for lovely tamarack trees at the Eloise Butler Wildflower Garden in Minneapolis or at the Matthaei Botanical Garden in Ann Arbor, Michigan. The Dawes Arboretum east of Columbus, Ohio, is conserving relict Ohio tamaracks *ex-situ* (out of their native site) by propagating and cultivating them.

Lindera benzoin
Spicebush
Lauraceae (laurel family)

The spicy aroma of all parts of this plant, especially in the dry twigs and stems give this plant its common name. My friend Marjory Rand always reminds me it would be a fine men's cologne. The spicebush, along with the bay leaf we use for cooking, is a member of the laurel family.

HOW TO GROW Spicebush is found across the entire Eastern Midwest and westward throughout the Ozark Highlands. It grows mainly in moist floodplain forest and other mesic woodlands that are not densely shaded. It's not cold hardy north of USDA zone 5b so not suited to most of the Upper Midwest. I've read historic accounts of it being killed back to its roots after extreme winters where native, but I have never observed that. It is also very regionally adapted so grow only locally sourced plants to be successful. It's easy to germinate from seed and is best grown in shade, or at least afternoon shade.

With its finely textured needles, *Larix laricina* (tamarack) looks almost furry.

The glossy red berries on female plants of *Lindera benzoin* (spicebush) are its finest asset and important fuel for migrating songbirds.

LANDSCAPE USE This shrub is often listed as growing 6–12 feet tall, but with age and in ideal conditions it becomes almost a small, multistemmed tree over 15 feet tall. I have paced off spicebush with a spread of 30 feet in old growth forest in the Indiana Dunes Nature Preserve. It's a magical addition to a woodland garden or along a stream or pond garden. It's the host of the spicebush swallowtail so a must in a butterfly garden, and birds seek the lipid-rich fruit that help fuel their migration.

ORNAMENTAL ATTRIBUTES The yellowish tiny flowers in early spring are a delight on the tree's bare stems, the foliage is clean and rich green, folded into leaf tacos that are the daytime shelter of the snakelike-appearing spicebush swallowtail caterpillar, which ventures out to feed only at night. The berries on female plants are glossy lipstick red in late summer and early fall, followed by fall color that can't be beat and is a consistent rich yellow.

NOTES Spicebush is a deer-resistant plant so is increasingly common in woodlands where deer are prevalent. It hosts two iconic insects—spicebush swallowtail and promethea moth—but unfortunately it is does not host a rich variety of insects as many other plants do.

Magnolia tripetala
Umbrella magnolia

Magnoliaceae (magnolia family)

I first saw this tree as a teenager at the newly opened Minnesota Zoo and wow, what a tropical looking plant growing outside in the temperate Twin Cities. I've since seen its phenomenal foliage floating in its native haunts—the understory of rich mesic forests.

HOW TO GROW Umbrella magnolia grows easily from seed but takes five or six years to bloom. It requires a sheltered site in moist, well-drained soil that mimics its mesic forest haunts. It will grow in full sun with moisture in all but the westernmost Midwest. At settlement, it was mainly found in the southeastern United States, but relic populations grow northward into the Midwest in Ohio and more frequently in the southern Ozarks. Unlike the other American magnolias, its evolutionary history begins in Asia so it would have quite a story to tell if it could speak. It has naturalized near plantings at the Klehm Arboretum in Rockford, Illinois, and is cold hardy through USDA zone 4.

LANDSCAPE USE This unique ornamental tree is a conversation piece for an outdoor seating area where its large floating whorls of leaves capture the breeze and the imagination. It shines in night lighting as the undersides of the leaves reflect light. It is most open, graceful, and airy when

Young *Magnolia tripetala* (umbrella magnolia) in bloom displays its characteristic whorls of large leaves.

167

The fragrant flowers of *Malus ioensis* (prairie crabapple) are exquisitely pink.

growing in the shade of larger trees. I also like the tree where you can look out on it from above or into its canopy. **ORNAMENTAL ATTRIBUTES** The large umbrellas or whorls of huge 18-inch-plus leaves held horizontally are its greatest asset and are lightly silvered on the underside. The flowers are ivory white and set atop the maturing leaves in late spring. First-day flowers hold their tepals (they are neither petals or sepals) up like a closed tulip, but on the second day they splay outward, 8 inches across, and on the third day, age to tarnished brown. The flowers are malodorous but I have to share that one ladies' gardener group I was having sample the smell described it as "dark chicken in lemon sauce." The fruit also are beautiful in late summer into fall, looking like popsicles that turn pinkish and burst with waxy red seeds. The fall color can be a marvelous blend of gold to brown and is quite beautiful from below as the leaves are then translucent which enhances the colors. The naked tree is quite coarse and open, more full and pyramidal when grown in full sun.

Malus ioensis
Prairie crabapple
Rosaceae (rose family)

I should have known that this was the tree that inspired Jens Jensen's interest in native plants. It was heavily planted by the prairie-style landscape architects in parks and gardens, but has been widely shunned since then in favor of disease-resistant species. For years, I routinely drove by a unique gnarly grove of this tree in a historic park in downtown Rockford. Then one day it was gone, replaced with "anywhere U.S.A." disease-resistant Asian ornamental "flowering" crabapples. Aldo Leopold's quote—"another little episode in the funeral of our flora"—was the first thing that came to mind. My first recollection of the tree is that of being blown away by its clear pink blossom color and intense fragrance.

HOW TO GROW As a teenager, I tried to dig out a 6-foot-tall tree somehow spared by road crews. I dug down more than 2 feet and there still was no lateral root, so I let it be (later it was sprayed and bladed off). Needless to say, any wild plant that has not been root-pruned to make its taproot branch is nearly impossible to transplant, though it's easy to transplant container-grown plants. This species germinates readily from seed and is sometimes grafted on apple rootstocks. Prairie crabapple requires well-drained soils and full sun to flourish, but will live with partial shade. Some plants spread by rhizomes to form thickets. It's heat, drought, wind, and ice impervious but its foliage is host to many insects, rusts, and other *Malus* maladies. The tree is a savanna and woodland edge species, often found in pastures across the core of the western Midwest with sporadic outlying populations into South Dakota and Oklahoma. It is hardy everywhere in the Midwest.

LANDSCAPE USE This premier small tree readily grows 15 feet tall, rarely to 30 feet, and is best for restoration of native or natural landscapes, and historic parks. It is shunned in

A beam of sunlight illuminates the large, multishaped leaves of *Morus rubra* (red mulberry).

and up the Missouri River bluff lands to Kansas City. I was worried that other trees were shading out a wonderful grove of wild sweet crabapples at a local conservation area, when a beaver systematically cut all the surrounding trees but left the crabapples. I have read that the cambium of this tree contains cyanide so that is possibly why the grove was left alone. A larger, double-flowering form 'Charlottae' exists in older plant collections. Wild sweet crabapple is cold hardy to USDA zone 4 and growing at the Minnesota Landscape Arboretum.

Morus rubra
Red mulberry
Moraceae (mulberry family)

Mention mulberry and it usually conjures up images of messy purple fruit and weak, short-lived, weedy trees. That describes the widespread and invasive white mulberry (*Morus alba*), but our true native mulberry is not the same. Red mulberry is in trouble, being naturally hybridized out of existence (it readily cross-pollinates, creating intermediate trees) by imported white mulberry. Red mulberry is a much more magnificent and tastier fruited tree. It's a denizen of the understory of mesic floodplain forests and rich woodlands where its large leaves create almost a tropical feel. The tree can grow quite large into the shade tree category but prefers to grow beneath the canopy of large trees. It's dioecious and the fruit on female trees are large and delicious, a new superfood rich in antioxidants. Mulberries don't keep or ship so you need a tree to imbibe their goodness that appears as one of the first ripe fruits in early summer. Male trees produce copious amounts of windborne pollen.

HOW TO GROW Red mulberry is listed as cold hardy through USDA zone 4 but that is not true; it grows as far north as extreme southeastern Minnesota near the Mississippi River but is tender beyond where native. I know of lovely wild trees from Ames, Iowa to Rockford, Illinois. Grow it from local, wild sources if that is possible. I inherited this tree in my woodland. One female tree in particular that I put my compost pile beneath has incredibly large and flavorful fruit. I'd like to introduce it as a new cultivar 'Alan's Compost Pile' but unlike the invasive white mulberry, our native won't root from cuttings and needs to be grafted. It's very hard to find this tree in cultivation, though it transplants readily when dormant as we have dug wild saplings at Powell Gardens for our Living Off the Land component of the Heartland Harvest Garden. It will grow in any rich, well-drained soil that can be saturated during short periods

formal landscapes and by almost all the nursery industry because it does suffer from foliar disorders so the foliage prematurely drops and is never perfect after spring.

ORNAMENTAL ATTRIBUTES Its fragrance is what first comes to mind, as sweet as violets, perfuming a wide area. The flowers are as true a pink as any, with orange stamens, and they are set against the felty, gray-green leaves. If the tree blooms during hot weather, the flowers may be quite pale. The exfoliating bark and picturesque gnarly form speak eloquently to our region. The fruit mature green and are about 1½ inches across. Tart and sour but edible, they allegedly make the finest pickles but I have yet to sample one.

RELATED PLANTS Several natural forms have been selected and propagated including sumptuous double-flowering forms 'Bechtel' and 'Klehm's Improved' that are sterile with petaloid stamens. The selection 'Fimbriata' has delicately fringed flower petals.

Having half the chromosomes of prairie crabapple, wild sweet crabapple (*Malus coronaria*) has more refined branching and smooth or nearly smooth leaves and twigs. It's named after its fragrant blossoms, just as pink but possibly more intensely fragrant than the prairie crabapple. This tree otherwise is very similar to the prairie crabapple in cultivation but described as having shallow roots. I've never transplanted this one and have both species wild on my three acres. Its native range is across the Eastern Midwest and sporadically westward to near Rockford, Illinois, but will not grow in wet or compacted soil. Sun or shade is equally fine. This tree is far more fickle than the Asian weed tree; it dislikes any trunk damage and is slow to recover. My two favorite trees are in slow decline from construction damage when my house was built.

LANDSCAPE USE Red mulberry is a premier edible landscape, food forest tree, or natural wildlife tree if you could ever find one in a nursery.

ORNAMENTAL ATTRIBUTES Red mulberry's coarse, large leaves are a striking contrast to the smaller, finer leaves of other trees. This feature makes it a standout, especially in the forest understory where its horizontal layers of leaves are illuminated by ambient light. The fruit are showy for a brief time in early summer when they turn from green to red and purple when ripe. The trunks develop orangish or tannish rugged bark when mature.

Ostrya virginiana
Eastern hophornbeam, ironwood

Betulaceae (birch family)

If there ever was a tree that was a sleeper, our native hophornbeam is it. It's the plain Jane of trees with elm-like leaves and flowers that aren't showy to most gardeners' eyes. I often request that people observe the delicate branching of a mature tree in winter—the fine twigs and branches make a reticulated pattern as detailed as the veining in a leaf—all the better to capture the indirect light in its under-the-canopy, shady home. This tree has such strong wood, it's oblivious to snow and ice load, only broken by canopy trees falling on it. Its colloquial name "ironwood" fits it to a tee. Pollen records show it was one of the first deciduous trees to recolonize the Midwest after glacial times.

HOW TO GROW It's difficult for mainstream nurseries to produce and sell this tree, so it almost never makes it into our landscape even though it's native across nearly the entire Midwest and it's as sturdy and long-lived as any plant. Fresh green seed at the end of summer will germinate, but let it turn brown and it requires a double dormancy. Also watch the skin-irritating hairs on the hoplike fruit. I painstakingly transplanted a woodland sapling as a young man and, with proper siting and care, it grew faster than I ever imagined. A moist to seasonally dry, well-drained upland site is all that's needed. Though native to the shady woodland understory mainly in oak-hickory woods, it thrives in full sun too. What kills our hophornbeam are compacted or poorly aerated, wet soils. Eastern hophornbeam is hardy throughout the entire Midwest.

LANDSCAPE USE This species is sometimes listed for use as a street tree, but I have never seen one live for long under such conditions. It makes a superior smaller shade tree or tree for a confined space or under the canopy of other native trees like oaks and hickories. Its winter leaves make it a good windbreak or winter screen in areas where an evergreen cannot be utilized.

ORNAMENTAL ATTRIBUTES The sum of its subtle beauty through the seasons equals one stellar tree. It produces pendulous golden male catkins in spring before the leaves fully emerge, and the female flowers form pale green fruit that look like hops and mature to brown and hang on the tree into winter. Cardinals and purple finch eat the seeds of the fruit. The foliage is clean and disease free through the growing season and the fall color is golden brown, often persisting as rich brown marcescent foliage through the winter. I've observed cedar waxwings settle in for the night snuggled in its winter leaves. Young trees have a pyramidal form but with age they spread, creating unique sculptural forms under a canopy of trees as they search for available light. The delicate branching pattern is exquisite in winter along with the rugged nature of mature trees and their finely longitudinal-striped bark.

Sturdy and long-lived, *Ostrya virginiana* (Eastern hophornbeam) bears clean, disease-free foliage throughout the growing season.

Physocarpus opulifolius

Ninebark

Rosaceae (rose family)

Colorful foliage selections of this widespread native plant have become popular shrubs across commercial and residential landscapes. It's usually grown as a small to medium-sized shrub, but the real deal does become a very large shrub 8–12 feet tall and eventually wider than tall. I've heard the common name comes from the number of herbal uses the plant had or from the bark that curls off the stem in the shape of the number nine.

HOW TO GROW Ninebark has a widespread midwestern range surrounding the Great Lakes, westward to eastern Minnesota, and scattered as far west as South Dakota and Nebraska south through the Ozark Highlands and Bluegrass Region of Kentucky. This shrub is quite easy to grow from seed and cuttings and is hardy throughout the Midwest. It's found in moist, well-drained open woods usually on embankments or bluffs where it receives more sunlight. It's also regularly found northward on rises in and around wetlands.

LANDSCAPE USE Ninebark is one of the finest large screening and hedge plants. It's beautiful on the edge of a woodland but usually too open and sparse in a shade garden.

ORNAMENTAL ATTRIBUTES The winter shrub is quite attractive with bark on the older stems that exfoliates into pale tan papery strips. Brown clusters of four-parted seed capsules (botanically known as dehiscent follicles) crown the shrub's fountain of arching stems. It blooms in late spring after the leaves emerge with clusters of nectar-rich white flowers that attract plenty of beneficial insects. The flowers produce capsules that turn from pinkish to rich rosy red shades before maturing brown in the fall. Fall color is usually lackluster yellowish to purplish.

RELATED PLANTS There are many purple, bronze, coppery, and golden-leaved cultivars along with selections that are more compact with tinier leaves.

Physocarpus opulifolius (ninebark) produces white flowers in mid to late spring.

Fruit of *Prunus americana* (American plum) look as delicious as they taste.

Prunus americana

American plum

Rosaceae (rose family)

Prunus americana is the most widespread plum found across almost the entire Midwest, absent only in some parts of Michigan. I make time to go visit them each spring when in bloom, 10 April in Missouri, 1 May in northeast Iowa. They are alive with pollinators and springtime butterflies, but it's their fragrance that brings me back home, remembering how they grew along the fencerow at my maternal grandparents' farm—my grandmother's laugh and loving hands part of the recollection. They epitomize the importance of fragrance in our lives and how it triggers memories more than any of our other senses.

HOW TO GROW All wild plums are easy to grow from their plum pits, but root suckers also can be separated from the plant. Special varieties must be grafted on seedling rootstock. Wild plums are easy to transplant and require full sun or only part shade, being open and weak in full shade. They are tenacious, surviving in poor to rich soils as long as they are not continually wet.

LANDSCAPE USE Most trees are rhizomatous-forming thickets, and they host so many creatures that their foliage can look tattered by midsummer. Interior twigs can develop into thorns, some plants nearly impenetrable. This makes them best suited to natural landscapes including sunny woodland borders and fencerows or the edges of food forests. Selections of single-trunked forms could improve their use in formal landscapes.

ORNAMENTAL ATTRIBUTES The fragrant, white spring flowers with yellow anthers are iconic too, blooming with redbuds in the Lower Midwest. The flowers are rich in nectar for a whole host of pollinators. They bloom during neotropical migration in the Upper Midwest so they often have orioles drinking from their blooms there. The fruits in early autumn are quite beautiful blends from yellowish through rose with purplish tinge. Fall color can be good from apricot yellow to red. Older plants develop nice bark with striking curled edges, and their winter form can be uniquely gnarly. Look for large cecropia cocoons on the plant in winter.

RELATED PLANTS Canada plum (*Prunus nigra*) is an Upper Midwest specialty growing across northern Minnesota, Wisconsin, and Michigan southward into northeast Iowa and northern Ohio. The anthers of its flowers emerge white but age to pink, giving a more pink look to the flowers which was how I could locate and distinguish them as a young man in Iowa. The gorgeous flowering cultivar 'Princess Kay' has been selected from wild Minnesota stock for superb double flowers equally as exquisite as any oriental flowering cherry but providing no nectar for pollinators.

Wild goose plum (*Prunus munsoniana*) is a Lower Midwest specialty, possibly a bit more refined with more uniform, narrower leaves than American plum but easy to distinguish in fall as its smaller fruits ripen red and look just like red paint balls.

Hortulan plum (*Prunus hortulana*) and Mexican plum (*P. mexicana*) are two additional species found across the Lower Midwest.

Prunus pensylvanica
Pin cherry

Rosaceae (rose family)

I nominated the former Illinois champion of this tree at the Klehm Arboretum in Rockford, Illinois, but soil compaction from construction equipment near the tree killed it prematurely. Pin cherry is a pioneer species that suckers into a thicket with the oldest tree in the center ultimately reaching 30 feet.

Like all wild plums, *Prunus munsoniana* (wild goose plum) is a cloud of white when it blooms in early to midspring.

The pinkish floral parts of *Prunus nigra* (Canada plum) make it stand out when in full bloom.

The ripe fruit of *Prunus pensylvanica* (pin cherry) look like mini sweet cherries.

Prunus virginiana (chokecherry) is early to leaf out and blooms by midspring.

HOW TO GROW Pin cherry grows wild across the Upper Midwest from scattered sites in North Dakota southward to Iowa, Illinois, northern Indiana, and northeastern Ohio. It can be grown from its fruit's pit (which can lie dormant in the seed bank for 50 years) or by dividing off a suckering stem. It requires full sun so gives way to maturing woods that shade it out. It probably is best cultivated solely in the cooler Upper Midwest.

LANDSCAPE USE This forgotten tree is perfect for the edge of a woodland garden or in a bird-themed garden. It's best to let it become a multistemmed thicket, but suckers can easily be removed. The small red cherries in midsummer are edible and tart; birds feast on them as soon as they ripen.

ORNAMENTAL ATTRIBUTES Mature trees are strongly upright, their trunks encased in classic cherry bark with a somewhat horizontal polished look with lighter lenticels. The white flowers in dome-shaped clusters are showy in midspring before all the leaves fully emerge, and the fall color is energetic as it turns shades of red.

Prunus viginiana
Common chokecherry

Rosaceae (rose family)

As a boy, the common name of this shrub was a dare to taste its cherries, which I knew were edible. Yep, the fruit has an aftertaste that coats the tongue but there is no sense of choking from it. The juice is rather rich and flavorful. Chokecherry is usually a suckering shrub to small tree and I sought out thickets of it as the favorite haunts of beautiful cecropia cocoons (the largest moth in North America).

HOW TO GROW Chokecherry is native across the Upper Midwest, sparingly southward, but it's hardy everywhere in the Midwest. It grows in a variety of habitats from full sun to light shade but dies out in dense shade. It can be grown from its pits or from divisions of a thicket. Once established, a plant is tenacious.

LANDSCAPE USE Common chokeberry is a fine tree for a natural landscape or at the edge of woods where it can be allowed to become a clump or thicket. Older trees have an irregular crown. Chokeberry has a very high wildlife value from numerous insects to birds relishing the fruit. On occasion, the tree will become quite large and not sucker; the biggest one in my old neighborhood was over 30 feet tall. In some areas it is susceptible to black knot disease that disfigures the stems with black growths.

ORNAMENTAL ATTRIBUTES The lacey white flowers in bottlebrush-shaped clusters in midspring are quite lovely and occur with emerging new leaves that can be green, bronze, or purplish shades. Fall color is bronze to apricot yellow and sometimes red.

RELATED PLANTS Cultivar 'Schubert' and its sport 'Canada Red' have leaves that turn rich burgundy purple after they mature and are quite showy—usually they are the only chokecherries available at a nursery.

Ptelea trifoliata
Wafer-ash, hoptree
Rutaceae (rue family)

Here's a tree more loved as a garden plant in Europe than in its heartland home.

HOW TO GROW Wafer-ash is easily grown from seed. It's native to woodland edges, hedgerows, and open woods across the core of the Midwest, oddly in a diagonal swath from southwest to northeast (southeast Kansas to southern Michigan). It readily naturalizes wherever planted, blurring the line of native, and is cold hardy to USDA zone 4 so grows well almost everywhere in the Midwest. The little tree is very heat and drought tolerant.

LANDSCAPE USE A must for a butterfly garden, it makes an ideal natural garden small tree on the edge of woods or in light shade. It's not for formal landscapes as it can have foliar maladies that create premature leaf drop though not all plants are affected. Giant swallowtail caterpillars are so numerous on my plants that they defoliate them and have even killed some plants because every leaf is repeatedly eaten. I rarely found these caterpillars on wild wafer-ash in the Rockford, Illinois, area.

ORNAMENTAL ATTRIBUTES Plants are dioecious with wonderfully fragrant tufts of creamy to golden light green flowers visited by many pollinators. Female plants produce the waferlike fruit that adorn the plant; light green in summer they ripen brown and remain on the tree into winter. Plants are single trunked but often low branched. The form of the tree is somewhat like an umbrella with a rounded outer crown of foliage and open branching underneath.

RELATED PLANTS The cultivar 'Aurea' with golden green spring foliage is sometimes available at specialty nurseries. The Indiana Dunes has its own form of this plant (forma *lanata*) with woolly twigs, an apparent adaptation to the extreme conditions in the sandy environment.

Quercus prinoides
Dwarf chinkapin oak
Fagaceae (beech family)

If your landscape is too small for most oaks trees but you still want an oak for its wildlife and nature value, then dwarf chinkapin oak is for you.

HOW TO GROW This small oak is native across the Lower Midwest and northward to southeastern Nebraska and southern Michigan. It is usually found on sandy or acidic ridges, dry ridges westward. It can be easily grown from acorns and readily transplants, preferring full sun to part or light shade. It is hardy to USDA zone 4.

The yellow-green flowers of *Ptelea trifoliata* (wafer-ash) are pleasantly fragrant.

Quercus prinoides (dwarf chinkapin oak) leaves have fewer teeth on their margins (usually less than a dozen).

174

Rhamnus lanceolata (lanceleaf buckthorn) fruit ripen by late summer and soon disappear.

Frangula caroliniana (Carolina buckthorn) has elegant, polished foliage that turns orange in the fall.

LANDSCAPE USE This plant is variable, but trees I have seen in cultivation form a small treelike plant reaching 8 feet tall in a dozen years. Some plants are shrubs and spread by underground rhizomes while others can grow to 25 feet. Small tree-type plants make lovely ornamentals on a hot, dry site while shrubbier types are better in a sunny shrub border—ask about the seed source of this plant.

ORNAMENTAL ATTRIBUTES Older plants have a rugged look and they usually have great burnt orange to red fall color.

175

Rhamnus lanceolata
Lanceleaf buckthorn

Rhamnaceae (buckthorn family)

Speak the word *buckthorn* and the invasive exotic species are what immediately comes to mind. Grrrr! A few species are native to the Midwest, and lanceleaf buckthorn has a rather widespread range though is now rare in many areas. Powell Gardens has healthy populations of this small tree on its wildlands so I have gotten to know it well. I could never find it in the Rockford, Illinois, area though it was collected there historically.

HOW TO GROW Lanceleaf buckthorn readily germinates from fresh seed, grows fast as a young plant, and transplants easily. It's found in upland successional areas on the edge of prairies and woodlands but never is present abundantly like the nonnative buckthorns, which germinate in a carpet, shade out and smother everything as they grow.

LANDSCAPE USE Lanceleaf buckthorn makes a fine large shrub or little tree 8–15 feet tall. It takes shearing well so could be used as a trimmed hedge (certainly a better choice than exotic privets and currants) but also works well on the edge of a woodland garden or in a shrub border.

ORNAMENTAL ATTRIBUTES Everything about this plant is subtle and that's why it's not been cultivated and is usually overlooked. It leafs out earlier than most shrubs while sporting tiny four-petaled, though pleasantly fragrant, yellow-green flowers (flowers are either male or female on the same or separate plants). The fruit ripen black in late summer but do not hold on the plant like the exotic buckthorns. Fruit soon are missing from the shrub and ground beneath, but I haven't observed if birds or nighttime mice eat them, or where they end up. The leaves look very much like those of a black cherry and hold a bit later than surrounding plants in fall, then turn yellow before dropping.

RELATED PLANTS Carolina buckthorn (*Frangula caroliniana*, syn. *Rhamnus caroliniana*) is our other small tree-sized native buckthorn found across the Lower Midwest from the Ozarks to southern Ohio. It makes a refined small tree or large shrub usually 8–15 feet tall and is cold hardy to USDA zone 6. The elegant foliage and fruit in late summer through fall are its best attributes. The fruit turn from green to red to black, often with all colors present, and may remain on the plant after leaf drop.

Rhus aromatica

Fragrant sumac

Anacardiaceae (cashew family)

Fragrant sumac doesn't have underground rhizomes that run uncontrollably, but many midwestern strains do become sprawling large shrubs. Plants native in Missouri and adjacent regions reach 8–12 feet (rarely taller) and spread wider than tall. I've paced some off at 35 feet across. Low fragrant sumac (*Rhus aromatica* var. *arenaria*) from northern Illinois into Michigan and northern Ohio is a small shrub rarely more than 3 feet tall (see description in chapter on small shrubs).

HOW TO GROW Fragrant sumac is found nearly statewide in Missouri but more fragmented in all other midwestern states, absent wild only from North Dakota and Minnesota. This shrub can be grown from seeds or cuttings. It grows in full sun to light shade in well-drained, limestone-based soils from pure sand and gravel to clay. Appropriate strains are cold hardy to USDA zone 4.

LANDSCAPE USE This popular shrub is occasionally used as a focal ornamental but mainly for back of the border or as a screen plant. The edible fruit have become trendy for flavorings and teas and are in demand by chefs working with local ingredients.

ORNAMENTAL ATTRIBUTES The flowers are yellow, but nothing like forsythia that blooms at the same time in early spring before they leaf out. Fragrant sumac's tiny flowers are dripping with nectar and relished by the first insects of spring, their bloom coinciding with the emergence of the spring azure butterfly. Forsythia offers only a feast for the eye. The flowers of male fragrant sumac plants are showier but female plants produce stunning red, fuzzy fruit that ripen in late spring into midsummer when they are one of the most ornamental features of any plant, contrasting perfectly to the green foliage. Fruit discolor to dark reddish brown in late summer and occasionally persist into winter. Fall color is exceptional on plants originating locally or regionally, but can be disappointing if you cultivate plants from outside of your region. Winter twigs are tipped with the flower buds for next spring, almost looking like mini birch catkins, and are a favored browse of deer and rabbits.

The yellow, early spring flowers of *Rhus aromatica* (fragrant sumac) are subtle but rich in nectar for early pollinators.

The flowers of *Rhus copallinum* (flameleaf sumac) are in squat clusters compared to other sumacs, and the winged leaf stem (rachis) between leaflets is distinctive.

RELATED PLANTS Poison sumac (*Toxicodendron vernix*) is a related shrub found only rarely in swamps and bogs in the Upper Midwest and is absent from most of the Midwest. It looks somewhat similar to true sumac but has white berries.

Rhus copallinum
Flameleaf sumac, winged sumac
Anacardiaceae (cashew family)

Flameleaf sumac is mainly a species of the American Southeast but found across the Lower Midwest and scattered northward to central Wisconsin and Lower Michigan.
HOW TO GROW Flameleaf sumac is tolerant of the hottest, driest, and poorest soils so is a good solution for tough sites.
LANDSCAPE USE This sumac doesn't run quite as much as the other two species making it a better choice for traditional landscapes.
ORNAMENTAL ATTRIBUTES Fall color is a true red, vibrantly stunning when backlit. This sumac flowers in late summer, long after the other species.

Rhus glabra
Smooth sumac
Anacardiaceae (cashew family)

This is the only woody plant found in all of the 48 lower states and adjacent provinces, but the heart of its range is in the Midwest where it is found throughout. It's the prairie invader, hated by managers and restorationists because it is challenging to keep from overtaking native prairie remnants and plantings.

HOW TO GROW Smooth sumac is found in well-drained soils in roadsides, power line cuts, hedgerows, and any successional area on the edge of grasslands to dry upland woods.
LANDSCAPE USE This sumac makes a fine mass planting on steep embankments or against a woodland or tree planting. Keep in mind, it runs in all directions except towards shade.
ORNAMENTAL ATTRIBUTES Fall color has intense and saturated reds often with a pinkish overtone. The flowers appear in early to midsummer, are light green, and are noticeably fragrant. Many insects and woodland butterflies visit the flowers. The leaves are dark blue-green above with lighter bluish, luminous undersides reflective of night light that make this plant shine in such situations.
RELATED PLANTS 'Laciniata' is a cut-leaved, female cultivar.

Rhus typhina
Staghorn sumac
Syn. *Rhus hirta*
Anacardiaceae (cashew family)

This is the largest and most treelike of the running sumacs. It has coarse stems that are felty like a stag's horn.
HOW TO GROW This sumac is native to northeastern North America as far west as central Minnesota, northeast Iowa, and southward to northern Indiana, all but southwestern Ohio, and scattered in Kentucky. All sumacs can be grown from seed or divisions off male or female clones. Staghorn

Rhus glabra (smooth sumac) often shows vivid red fall color.

Rhus glabra (smooth sumac) displays inspiring luminous and patterned leaves, crowned by cones of reddish fruit.

Foliage of *Rhus typhina* (staghorn sumac) creates an interesting herringbone pattern.

sumac is hardy and grown throughout the Midwest. It is found on the edges or openings in mesic woods in moist, well-drained soil. It requires full sun or part shade. Cutting them back to the ground can rejuvenate plantings of any sumac that gets too tall or has other issues.

LANDSCAPE USE Staghorn sumac is the largest, reaching 35 feet or more, and quickly colonizes open ground as a short-lived nurse tree for advancing woodland. It dies out when shaded. It makes an excellent edge-of-the-woods or hedgerow planting in large landscapes.

ORNAMENTAL ATTRIBUTES Sumac is best known for its flaming fall color, and staghorn sumac's is yellow, orange to red, often all on the same plant. The female plants of all sumacs are showier than the males as their fruit ripens red in late summer and lingers through winter, exceptionally so when adorned with snow. Staghorn sumac has the showiest fruit, as it's fuzzy with metallic hairs somewhat like its velvety young stems that are also quite noticeable and usually redbrown in winter.

RELATED PLANTS Staghorn and smooth sumac readily hybridize where they grow together (*Rhus ×pulvinata*), and these are mixed in the nursery trade, so are occasionally found in plantings far from where native as, for example, in Kansas City.

Two cultivars of staghorn sumac are widely available: 'Laciniata' (cutleaf staghorn sumac) has leaflets that look cut, creating a fine texture though the herringbone pattern is lost. Tiger Eyes ('Bailtiger') is the golden-leaved form of cutleaf staghorn sumac.

Salix spp.
Willows
Salicaceae (willow family)

Willows may be the most underutilized of our native woody plants. The smaller species are much more structurally sound and can easily be rejuvenated if they grow too big or problematic. There's an astonishing array of insects tied to willow so they are premier wildlife plants. They are also dioecious so plants are either male or female. The immature male catkins are the silky pussy willows that truly flower when they're adorned with yellow, pollen-rich stamens. The female flowers are less showy and produce the capsules that burst with cottony parachuted tiny seeds. Many birds collect the cottony seed to line their nests. The nectar and pollen produced from willow flowers is always busy with pollinators, an especially important food source on the early spring blooming species. Willows also hybridize, making it nigh impossible to identify some plants.

Seven midwestern willows for garden use are as follows, from the most treelike to the most shrublike: *Salix caroliniana*, *S. discolor*, *S. bebbiana*, *S. eriocephala*, *S. humilis* var. *humilis*, *S. interior*, and *S. myricoides*. Several other willow

The catkins of *Salix bebbiana* (Bebb's willow) are slightly smaller than pussy willow.

Salix caroliniana (Ward's willow) makes a handsome small tree with a wide crown.

species become large shrubs, all of value in their appropriate native natural areas but little utilized as landscape plants. The shining willow (*S. lucida*) with its aptly named lustrous glossy and leathery leaves and showy yellow flowers may be one of the best of these but I have little experience with it. The silky willow (*S. sericea*) with glistening silvery hairs on the leaf undersides has been an unkempt landscape plant at Powell Gardens. Heart-leaf or sand dune willow (*Salix cordata*, syns. *S. syrticola*, *S. adenophylla*) is a specialty of the Great Lakes' dune lands.

HOW TO GROW All willows are easy to root from cuttings placed in water or grown from fresh seed that is simply pressed on soil and kept moist. Seedlings will be of unknown sex until they bloom. You can clone favorite male or female plants by rooting cuttings from them.

LANDSCAPE USE Willows are fine additions to rain gardens and around water gardens. Many are important plants to stop erosion in problematic sites or to stabilize streambanks. They should be in every butterfly, insectaries, or bird-themed garden. All host the viceroy butterfly (monarch mimic), whose caterpillar overwinters on the plant inside the remnant of a leaf it has rolled into a sleeping bag, the leaf's petiole tethered by silk to the plant so it won't drop with the other leaves in fall.

ORNAMENTAL ATTRIBUTES Early spring flowers of most species are quite beautiful, especially the pussy willow catkins of the male plants. The cottony seeds of female plants also are interesting. The subtle colors and unique texture of the foliage make willows handsome plants, especially on plants that have been rejuvenated by cutting them back. Be sure and cut them back right after flowering so they will produce flower buds on new growth and bloom the next

The early spring catkins on male plants of *Salix discolor* (pussy willow) are adorable.

The catkins of *Salix eriocephala* (Missouri willow) look like woolly (*erio*) heads (*cephala*).

The narrow, wispy leaves of *Salix interior* (sandbar willow) are very distinctive.

The bluish underside of *Salix myricoides* (blue-leaf willow) foliage is striking all through the growing season.

season. If not cut back, willows usually get twiggy older trunks and are shorter lived. Their fall color is late and usually yellowish.

Salix bebbiana (Bebb's willow, beaked willow) also looks like a pussy willow with smaller catkins and more netted or veiny, gray-hairy leaves that are less strikingly discolored from upper to underside. It's an Upper Midwest willow found in various wetlands from the Dakotas eastward across the northern parts of the Corn Belt states. It becomes a wider-spreading little tree, sometimes with unique forms on older plants. I'm fond of a little grove of these mini-trees on an algific talus slope along the Oneota trail in Decorah, Iowa. Because our flora is so little understood, these trees are hacked on as weedy brush when they never would be problematic to the trail. Who can distinguish a unique small willow from a large-growing species?

Salix caroliniana (Ward's willow, Carolina willow, coastal willow, syn. *S. wardii*) is found across the entire Lower

Midwest in rocky streamsides and southward through the American Southeast. It is a good choice for the Lower Midwest as an umbrella-crowned small tree 15–30 feet tall and wide. This willow is probably hardy northward but I never see it beyond where native. It flowers with leaf emergence and has distinctive round leaflike stipules (appendages where the leaf petiole meets the stem).

Salix discolor (pussy willow) is a better landscape plant than the nonnative European goat willow (*Salix caprea*), the source of most pussy willows at nurseries, because it doesn't grow nearly so large (usually 10 to rarely 20 feet tall) but its catkins are smaller. It's found in peaty wetlands across most of the Upper Midwest (except Nebraska), and from northeastern Missouri it ranges southeastward into Kentucky. It's better in AHS heat zone 6 and northward as we have not been able to grow it at Powell Gardens. Pussy willow has lovely smooth, whitened (discolored) leaf undersides that make it quite showy; it also has the largest early spring catkins of our native species. It forms an upright crown, usually multistemmed with a thick base, and responds well to being cut back, growing long stems of catkins.

Salix eriocephala (Missouri willow, diamond-leaf willow, heart-leaf willow, syn. *S. rigida*) was the first pussy willow I knew as it grew along the banks of the Upper Iowa River just across the street and behind the dike from where I grew up. As a boy, I would pick the little pussy willow stems in early spring. The catkins are smaller than those of pussy willow but cute; their stem is hairy and the later leaves green beneath. Missouri willow is found along gravelly rivers and streams across the entire Midwest except for

The male catkins of flowers on *Salix humilis* var. *humilis* (prairie willow) illuminate an early spring day.

Sambucus canadensis (American elderberry) is a spectacular sight when in bloom during the summer solstice.

Oklahoma. A missed opportunity, this large shrub (under 6 feet) or little tree (to 15–18 feet) is not cultivated but is the perfect plant for a streamside or smaller garden.

Salix humilis var. *humilis* (tall prairie willow) is an underutilized willow available from native plant nurseries. When pruned to keep it vigorous, it is a very handsome landscape shrub with foliage that stands out in a garden. The shrub looks gray-green with a unique texture as the leaves are untoothed, gray-hairy, and with visible leaf veining. Tall prairie willow is the most upland of our willows, often growing in dry glades and prairie openings. The catkins are delightful and the smallest of the spring pussy willows here, yet open to wonderful flowers abuzz with hordes of early pollinating insects including honeybees. The shrub easily grows 8–12 feet tall but can be kept smaller by annual rejuvenation after flowering.

Salix interior (sandbar willow, syn. *S. exigua* subsp. *interior*) is nature's streamside flood control engineer, germinating on fresh sandbars and colonizing them with a rhizomatous thicket above ground and a strong network of roots below that squelch erosion. It is too invasive for most properties. Maybe it could be called bamboo willow. It is certainly less invasive than any of the *Phyllostachys* bamboos and more easily controlled though deciduous so not a dense screen in winter. I've wondered if it would thrive in a large container to control the spread of its roots, as it is such a lovely plant. This upright, wispy willow has the narrowest, most finely textured leaves of any so is attractive through the growing season. It often develops a witches' broom cluster of twigs that standout in winter. It does not have springtime catkins and blooms after it leafs out in summer.

Salix myricoides (blue-leaf willow, syns. *S. glaucophylla*, *S. glaucophylloides*) has such beautiful leaves all growing season with a strongly gray-blue underside that is eye-catching in our midwestern breezes. I have yet to try growing or to see it in a botanical garden's collections. Look for it in wetlands around Lake Michigan, Huron, and Erie, scattered southward in Illinois and Ohio. Beautiful plants occur at Illinois Beach State Park. I simply love this plant, which grows into a large shrub 10 feet tall at maturity. Every time I see this willow I think, why is this not cultivated? Is this willow one that defies cultivation or have we simply neglected it?

Sambucus canadensis
American elderberry
Adoxaceae (moschatel family)

No native shrub blooms so spectacularly through summer's solstice. When the greens of June cloak plants, elderberry bursts with white dinner plates of flowers that shine in the intense sunlight. Mature shrub size is usually 8–15 feet with multiple stems.

HOW TO GROW American elderberry is hardy everywhere in the Midwest but native in the Eastern and Lower Midwest and northwestward to central Minnesota and southeastern South Dakota. It's easy to grow from seed or from divisions of plants. It grows fast and is easy to transplant but does require well-drained soil and full sun to only part shade to flower and fruit well.

LANDSCAPE USE Elderberry is not common in traditional landscapes because it tends to sucker, its stems may randomly dieback for various reasons, and we have become afraid of anything with a potentially messy fruit. It is a fine plant for natural landscapes, bird gardens, edible landscapes, and food forests. The fruit (and flowers) are edible, making excellent jellies, syrups, and preserves. Elderberry is rich in antioxidants and used by some as a nutritional supplement. Cutting out old or dead stems easily rejuvenates this shrub. Why some plants become almost treelike without a single suckering stem is beyond me.

ORNAMENTAL ATTRIBUTES The huge, flat clusters of lacey white flowers in midsummer are gorgeous and often rebloom into summer before becoming abundant displays of fruit that ripen to dark purplish black and are attractive in late summer into fall. Note that stems and foliage of elderberry are highly toxic.

RELATED PLANTS Several cultivars have been selected for fruit production including 'Aurea', which has golden spring foliage and more treelike stature.

Sambucus pubens
Red-berried elder

Syn. *Sambucus racemosa* var. *pubens*
Adoxaceae (moschatel family)

This shrub has been lumped with its Eurasian counterpart, but it's important to grow the midwestern version, best adapted here.

HOW TO GROW Red-berried elder may have a Northern Hemisphere range, but it is found in the Upper Midwest

Sambucus pubens (red-berried elder) produces striking red fruits in midsummer.

from scattered locales in North Dakota, across all of eastern Minnesota, into northern Iowa, Illinois, Indiana, and eastern Ohio. It has relict stands in northeastern Missouri too. It's found in cool, moist forests and does well southward into the Lower Midwest if sheltered from hot summer sun and wind and if sourced from the southern populations. It's easy to grow from seed and readily transplants.

LANDSCAPE USE This charming large shrub for a shady woodland garden or bird-themed garden thrives in the shade on the north or east side of buildings in the Lower Midwest.

ORNAMENTAL ATTRIBUTES The large swelling buds burst with foliage early in spring and bloom fully by midspring. The puffy flower clusters emerge from red buds and open to creamy white; they have an unusual scent that reminds me of a freshly opened package of rubber bands. The fruit ripen by midsummer when they turn brilliant red so birds can easily spot them in contrast to the greens of late June. I recall rose-breasted grosbeaks feasting on their berries at the Eloise Butler Wildflower Garden in Minneapolis, Minnesota.

Staphylea trifolia
American bladdernut

Staphyleaceae (staff tree family)

If nursery professionals would ever discover a nonsuckering form of this large shrub or small tree, I believe it would finally hit the mainstream. When I show flowers of it in programs, it is usually thought of as a silverbell (*Halesia* spp.).

HOW TO GROW Bladdernut is found across most of the Eastern and Lower Midwest and northwestward up the Mississippi and Minnesota River valleys well into central Minnesota and up the Missouri River into western Iowa and eastern Nebraska. It's hardy everywhere in the Midwest. Though mainly found wild in mesic forest slopes, it's one of our most drought- and shade-tolerant woodland understory shrubs. It's relatively easy to grow from seed or from dividing off a suckering piece of the plant. Plant it in almost any well-drained soil with at least afternoon shade.

LANDSCAPE USE It makes an excellent shrub for massing along a woodland edge or in a woodland setting, where it makes a fine boundary plant in shade. Some plants sucker less and can be used as a focal ornamental.

ORNAMENTAL ATTRIBUTES Its best feature is its namesake fruit that are inflated bladders. Once ripe, the seeds rattle inside the three-chambered capsule, which is light green in summer and brown in winter. The capsules were used as baby rattles by Native Americans and audibly jangle on the tree in winter wind as the pods hold well all winter. The lovely white pendant flowers emerge after the leaves so are somewhat hidden. The leaves comprise three leaflets like the hoptree but are serrated around the edges and usually clean and neat with little or no color in the fall. Stems are striated tan and green and subtly beautiful as they mature.

Styrax americana
American snowbell
Styracaceae (storax family)

Snowbells and silverbells are part of the storax family, which is known for its white, bell-shaped flowering trees from around the world.

HOW TO GROW This shrub is native mainly in the American Southeast, just reaching the southern edge of the Midwest. It has a disjunct range in the core of the Midwest along the Kankakee River in northwestern Indiana and into adjacent Illinois. It is cold hardy through USDA zone 5 and is easy to cultivate. Snowbell needs moist, well-drained soil that is neutral to slightly acidic and thrives well in rich prairie soils as far west as Kansas City where a mass can be seen growing at the Missouri Department of Conservation's Discovery Center (these were grown from the Kankakee River population). Snowbell does best in the shade of canopy trees.

LANDSCAPE USE A refined shrub for a shrub border, snowbell also can be used as a focal ornamental when trimmed into a somewhat multitrunked-looking mini-tree. It usually grows

no higher than 8 feet but can reach 12 to even 15 feet in its search for light in shade.

ORNAMENTAL ATTRIBUTES The midspring, pendant white, starlike flowers emerge below the leaves and many bees visit them. Songbirds and wild turkeys eat the hard greenish gray, maturing brown fruits (actually drupes) in late fall into winter.

Viburnum spp.
Treelike viburnums, nannyberry, blackhaws
Adoxaceae (moschatel family)

183

Three treelike viburnums readily reach small tree size and stature in the Midwest: one northern—nannyberry (*Viburnum lentago*), one midland—blackhaw (*V. prunifolium*), and one southern—rusty blackhaw (*V. rufidulum*). The three species do not grow wild together, but blackhaw does grow with its northern and southern counterpart where their ranges overlap.

HOW TO GROW All three germinate after long dormancy of the seed. They also all can be grown from cuttings, layering, or dividing off a basal sucker from the main plant. They tolerate a wide range of soils. Nannyberry is the most wet tolerant while rusty blackhaw prefers well-drained soils. All are heat and drought tolerant once established.

LANDSCAPE USE All are usually suckering large shrubs that develop a main trunk and grow 20–35 feet tall at maturity. They are fine edge-of-the-woods plants or for large screens.

The white, bell-shaped flowers of *Staphylea trifolia* (American bladdernut) are beautiful on close inspection.

The white flowers of *Styrax americana* (American snowbell) are simply delightful with petals that curve back like a shooting star.

All will grow in light shade of open woods but flower heaviest in full sun. They are occasionally used as focal ornamental plants; certain plants are single trunked and can be pruned up as little trees.

ORNAMENTAL ATTRIBUTES Showy clusters of white flowers with yellow stamens are produced with or after leaf emergence in midspring, later spring in rusty blackhaw. Fruit set is best with two or more noncloned plants; berries turn lovely sage green when full sized then gradually to pinkish and blue, all ripening a blue-black, drying on the tree, and holding into winter. The fruit are edible and taste a bit like a musky or bitter raisin. Fall color is good with some trees having exceptional blends from yellow to red or uniformly red. The tree trunks form an interesting blocky texture with age.

Viburnum lentago (nannyberry) is distinguished by long-beaked, taupe winter buds and is native from New England westward across the Great Lakes states and sparingly into the northern Great Plains of the Dakotas as far west as Saskatchewan. It is often found in wetlands but also in open, dry upland woods. It is hardy throughout the Midwest, growing well southward. Powell Gardens has fine plantings that survive without irrigation. I have measured a single-trunked plant with a spread of 30 feet at the Edsel Ford estate. Thickets of the plant can spread 50 feet.

Viburnum prunifolium (blackhaw) has short-beaked, light brown winter buds and is native from the Mid-Atlantic states westward across the eastern and central Corn Belt, across Missouri and into eastern Kansas. It is a highly variable species, some with a more treelike form of horizontal branches reminiscent of a hawthorn. Some blackhaws display stunning red fall color and the species is hardy throughout the Midwest.

The lacecap flower of *Viburnum trilobum* (American highbush-cranberry) stands out against the fresh foliage of late spring.

The beaklike terminal bud of *Viburnum lentago* (nannyberry) caps a stem in peak fall color.

The blue-black fruit of *Viburnum prunifolium* (blackhaw) are quite showy in fall and hang on the tree into winter.

The polished leaves of *Viburnum rufidulum* (rusty blackhaw) turn flaming red in the fall.

Zanthoxylum americanum (prickly-ash) has handsome rich green, pinnately compound foliage.

Viburnum rufidulum (rusty blackhaw, southern blackhaw) is distinguished by its rusty, dark brown winter buds and exceedingly lustrous, almost lacquered leaves. The species is found across the American Southeast from southeast Virginia to the Hill Country of Texas; it grows northward into the Lower Midwest and is found in the Bluegrass Region of Kentucky and adjacent Indiana and Ohio and is quite common throughout the Ozark Highlands and sparingly in adjacent regions. I was surprised to find a native clump in my backyard woods, as it's a localized and rare shrub in the Osage Plains. It thrives in any well-drained soil and is extremely heat and drought tolerant. It is cold hardy through USDA zone 5, growing well northward to the Iowa State University campus and Morton Arboretum outside Chicago.

Viburnum trilobum

American highbush-cranberry

Syn. *Viburnum opulus* var. *americanum*
Adoxaceae (moschatel family)

American highbush-cranberry is a more shrublike viburnum but may have tree-sized stems. It's named for its vibrant red berries that are a rare hot color in a bitter cold, snowy landscape. Winter birds such as cedar waxwings seek the fruit in wintertime after the berries have frozen and thawed and probably fermented. Fruit is edible, but very tart.

HOW TO GROW This viburnum is native in the Upper Midwest, from eastern North Dakota southward in cool, moist woodland microhabitats into northeastern Iowa, northern Illinois, Indiana, and Ohio. It's easily propagated from rooted cuttings or by layering stems, and it readily transplants into moist, well-drained soils. It will grow throughout the Midwest but must be in shaded, cooler sites on the north or east sides of buildings in warmer climates such as those found in AHS heat zones 6 and 7.

LANDSCAPE USE A magnificent 8- to 12-foot large shrub of upward branches that arch outward, this viburnum makes a good back-of-the-border or edge-of-the-woods plant. It also can stand alone as a stunning ornamental. It can grow taller, with fewer stems reaching for light in its native haunts or in woodland gardens.

ORNAMENTAL ATTRIBUTES I simply love the late-spring, lacecap flowers in stunning white. The foliage is an elegant, three-pointed lobed leaf reminiscent of a maple. The fruit turn vibrant red when ripe and hold their color through subzero weather. Fall color is also noteworthy in a range from yellow to red and burgundy purple on sun-exposed foliage.

RELATED PLANTS Most wild highbush-cranberries in the Midwest are the escaped European species (*V. opulus*), which is bushier than the American native species and more tolerant of a variety of habitats.

Zanthoxylum americanum

Prickly-ash

Rutaceae (rue family)

In 1972, I saw my first giant swallowtail butterfly. As a young butterfly enthusiast, I learned prickly-ash was its only host plant in northeast Iowa, so thus I got to know and grow the lowly prickly-ash.

HOW TO GROW This species is found across the Corn Belt and adjacent regions, essentially throughout the entire Midwest, and is hardy everywhere. It's easy to grow from seed and can be readily divided from a grove. It's native in oak-hickory and second-growth woods. It suckers into dense thickets, but older stems can become treelike (15 feet or taller). If suckers are removed, it can be maintained as a small tree. It thrives in moist to summer-dry soils in full sun to light shade.

LANDSCAPE USE This is a good plant for a hedgerow, barrier, or thicket in a naturalistic wildlife and butterfly garden.

ORNAMENTAL ATTRIBUTES The flowers are tiny in clusters along the stem as the leaves emerge. The pinnately compound foliage is rich, dark green, occasionally dropping early or turning yellow in the fall. The fruit ripen in late summer, turning red and opening like a Pac-man spitting out a black seed. The fruit is aromatic and can be used as a pepperlike flavoring but has a numbing quality.

Evergreen Shrubs

Evergreen shrubs are a rare type of plant in the Midwest. They are more prevalent in regions north, south, east, and west. The Midwest's rich soils are the primary reason why deciduous shrubs are almost totally dominant—plants can afford to regrow their leaves after winter's dormancy. Evergreen shrubs remain a beloved element in a traditional landscape because most gardeners like to see some green in winter. In most cases, evergreens are inappropriate for native restorations but do assist in creating more sustainable landscapes.

188

Most evergreen shrubs in the Midwest are conifers: two species of junipers in the cypress family (Cupressaceae) and one yew in its namesake family (Taxaceae). Dwarf conifers are in two families: the junipers and arborvitae in the cypress family and the hemlock, pines, and spruces in the pine family (Pinaceae).

For our purposes, dwarf conifers are genetic dwarfs or witches' brooms of native evergreen trees. Virtually all of our native conifers produce ball-shaped tufts of dwarfed growth called witches' brooms. These brooms can be propagated by grafting them onto seedling rootstock to create shrub versions of what otherwise would be trees. Some dwarf conifers are true genetic dwarfs while a few are sports of petite growth also grafted onto seedling rootstock. A harsh environment naturally dwarfs some native conifer trees due to where they grow (often on bedrock outcrops), but these plants would become tree-sized if growing in a better site.

Abies balsamea 'Piccolo' (balsam fir) has bright lime green new growth.

All dwarf conifers require the same conditions as their parent species but are usually more easily cultivated because their small size limits their soil and microclimate demands. Northern species may be planted in sites north or east of a structure giving protection from hot summer winds. Native conifers also may be grown in containers for decades, naturally dwarfed by the constraints of the container's size and soil. The ancient art of bonsai is another way to grow conifers as ornamental shrubs.

Most dwarf conifers become really interesting plants with age, creating a rugged, natural look in a small space. They are a better alternative than readily available nonnative evergreen shrubs in a traditional landscape including as foundation plantings. Dwarf conifers create year-round screening where a full-sized tree would be too large, blocking a view or casting too much winter shade. They provide good winter cover for songbirds. They are a good choice where smaller-sized landscapes limit use of a full-sized tree.

Two evergreen shrubs in the heath family (Ericaceae) are found sparingly in the Lower and Eastern Midwest: rosebay rhododendron (*Rhododendron maximum*) and mountain laurel (*Kalmia latifolia*). Both are mainly Appalachian and eastern species but represent some of the most popular broadleaf evergreens in gardens.

Four mainly Northwoods bog specialty evergreen shrubs—all in the heath family (Ericaceae)—are restricted to bogs and poor fens, requiring organic "muck" soils and not easily cultivated unless one is reconstructing a bog garden: bog rosemary (*Andromeda glaucophylla*), leatherleaf (*Chamaedaphne calyculata*), bog laurel (*Kalmia polifolia*), and Labrador tea (*Rhododendron groenlandicum*). All are available at specialty nurseries.

There are several other unconventional evergreen shrub types. From the West are several cacti (Cactaceae) and the yuccas in the asparagus family (Asparagaceae). From the South are mistletoe in the mistletoe family (Loranthaceae) and switchcane in the grass family (Poaceae). Rare unique evergreen shrubs include paxistima related to *Euonymus* in the staff tree family (Celastraceae).

Although cacti are regarded as an element solely of the flora of the American Southwest, one species is native across much of the Midwest and three others are found in the western Midwest. They are comprised of succulent stems known as pads that remain evergreen, though dehydrate in the winter, often looking very wrinkled and near death when dormant. They miraculously rehydrate and look normal by springtime's warmth.

There are too few species of native evergreen shrubs to generate a meaningful top dozen list for a traditional landscape.

Abies balsamea cultivars
Dwarf balsam fir

Pinaceae (pine family)
Landscape group: dwarf conifers

Dwarf balsam fir cultivars are mainly witches' brooms and genetic aberrations. The typical species will grow on mossy rocks and stumps but usually in moist, sheltered conditions.

HOW TO GROW Plant dwarf balsam fir cultivars in moist, well-drained soils where they are protected from hot afternoon sun and summer winds. Balsam fir is shade tolerant, thriving in indirect light but remains challenging to grow in the summer heat of the Lower Midwest. See *Abies balsamea* in "Evergreen Trees" for more details.

LANDSCAPE USE Balsam fir cultivars are elegant evergreens for a cool, afternoon-shaded site, including challenging north- or east-facing courtyards and narrow shady spaces between homes and buildings. They can be used as a focal ornamental evergreen, low screen or backdrop in these often-challenging settings.

ORNAMENTAL ATTRIBUTES 'Nana' is often used as a catch-all cultivar name for dwarf selections of all conifers. 'Nana' balsam fir grows 2–3 inches per year into a rounded, compact form. 'Piccolo' is an extra dwarf globe with smaller twigs and needles. 'Weeping Larry' is an upright dwarf with pendant branches growing 3–4 feet in ten years; its fine-textured, luscious dark green needles keep their color in all seasons.

Andromeda glaucophylla
Bog rosemary

Ericaceae (heath family)
Landscape group: Northwoods heaths

HOW TO GROW Bog rosemary is found in the Northwoods from central Minnesota southeastward to northeastern Illinois, northern Indiana, and northern Ohio. It does not tolerate the summer heat of the Lower Midwest. It can be cultivated in bog gardens created from half sand and half peat with a constant supply of moisture and mulched with pine bark and sphagnum moss. Plants are easily layered or grown from cuttings. They are also available at specialty nurseries.

LANDSCAPE USE Bog rosemary is a diminutive shrub growing under 18 inches tall. I often see this plant for sale at big box stores when in bloom in spring, but buyer beware, it is nigh impossible to grow without a bog garden.

ORNAMENTAL ATTRIBUTES Bog rosemary has foliage that looks remarkably like a large-leaved rosemary, often blue

Andromeda glaucophylla (bog rosemary) looks like its namesake plant in late fall at Cedarburg Bog, Wisconsin.

Arundinaria gigantea (switchcane) provides the perfect habitat for a winter bird feeding station.

or grayish green and whitish and pubescent underneath. The spring flowers are clusters of adorable pinkish white downward-facing urns. The seed capsules face upward and are present into winter.

Arundinaria gigantea
Switchcane, giant cane
> Poaceae (grass family)
> Landscape group: southern origin evergreens

HOW TO GROW Switchcane is the only midwestern native bamboo. It is a Lower Midwest specialty that grows wild from the southern Ozarks eastward to southern Ohio and the Kentucky Bluegrass Region. It is hardy to around -15°F so will grow northward into USDA zone 5b. Severe cold will burn the leaves even in its native range but it quickly recovers. As with all running bamboos, the horticultural adage sleep, creep, and leap describes how this plant grows. It is slow to establish but once it does, it can grow vigorously into a thicket. Switchcane grows readily from seed or divisions. It prefers deep rich and moist soils and is often found in the understory of mesic floodplain forests.

LANDSCAPE USE Switchcane usually grows 6–8 feet high in the Midwest, occasionally 12 feet or taller. It makes a dense thicket that works as a fine screen, especially useful in shady locations where choices are limited, and it is deer resistant. It takes shearing well and can be trimmed to any desired height. Be prepared to control its rhizomatous growth. Switchcane creates premier cover for songbirds and other wildlife so is a great choice for planting near bird feeding stations. The seed is edible and can be used as a grain, so the plant also is appropriate for a food forest. It does not flower abundantly, set seed and then die as with many nonnative bamboos.

ORNAMENTAL ATTRIBUTES Individual stems are straw colored and are elegantly upright with pendant leaves off the finer stems. Foliage is medium to yellowish green, often burning tan around the edges or completely burning tan after a severe winter. It blooms in spring but the flowers are little noticed.

Chamaedaphne calyculata
Leatherleaf
> Ericaceae (heath family)
> Landscape group: Northwoods heaths

HOW TO GROW Leatherleaf is found from central Minnesota southeastward to northeastern Illinois, northern Indiana, and northern Ohio. It can be cultivated in bog gardens created from half sand and half peat with a constant supply of moisture and mulched with pine bark and sphagnum moss. Plants are easily layered or grown from cuttings. They are also available at specialty nurseries. Unlike other evergreen heath shrubs from the Northwoods, leatherleaf is tolerant of the Lower Midwest's summer heat.

LANDSCAPE USE Leatherleaf can only be grown in organic "muck" soils and thus is limited to container gardens or re-created bog gardens.

Chamaedaphne calyculata (leatherleaf) blooms consist of little bell-shaped flowers lined along horizontal stems.

Tiny *Coryphantha vivipara* (common pincushion cacti) shouts its existence while in flower.

ORNAMENTAL ATTRIBUTES Leatherleaf forms a thicket of upward-arching stems as tall as 3 feet, adorned with a line of pendant white bells—one at each leaf along the end of the stem. Each flower forms a seed capsule fruit that faces upward and opens up like petals when fully ripe.

Coryphantha vivipara
Common pincushion cactus, spinystar

Syn. *Escobaria vivipara*
Cactaceae (cactus family)
Landscape group: cacti and succulents

HOW TO GROW Common pincushion cactus is found mainly in the High Plains or western Great Plains but extends into the Midwest region in Nebraska, the Dakotas, and western Minnesota. It grows in lean, rocky sites in dry upland prairies and is easy to cultivate in similar situations in full sun. It may be propagated by seed, but its offshoot viviparous stems also may root and form new plants.

LANDSCAPE USE This is a tiny cactus growing just 3–4 inches tall and seldom more than 2 inches in diameter. It adds interest to a dry rocky site where little else will grow but is also an excellent container subject or can be grown on top of a rock wall where it's more easily viewed. Common pincushion cactus flowers are nectar- and pollen-rich for insects.

ORNAMENTAL ATTRIBUTES The plant looks like a living pincushion and bursts with a crown of bright fuchsia flowers in early summer.

Juniperus communis var. *depressa*
Low common juniper

Cupressaceae (cypress family)
Landscape group: conifers

HOW TO GROW Low common juniper is found wild across the Upper Midwest, usually growing on rock outcrops or sandy shores where a harsh environment and protection from fire provide a niche for it. The plant is denser in full sun, but works fine in partial shade. It grows in any well-drained soil including good garden soil or those of poor fertility from gravel to sand. It's an extremely heat and drought tolerant plant and is hardy throughout the Midwest.

LANDSCAPE USE Low common juniper makes a choice evergreen for a harsh setting including scorching south- and west-facing slopes and stonewalls. It grows only 2 feet tall but readily spreads outward in all directions forming a low bowl-like mound 5–10 feet or more across. The berrylike cones are the preferred juniper flavoring for culinary use.

ORNAMENTAL ATTRIBUTES This juniper has only prickle-type needle foliage and so offers a coarser texture than other native junipers. Each needle has a bluish waxy longitudinal band giving the plant a contrasting silvery look, and the foliage turns earthier tones in winter. Low common juniper is dioecious and female plants produce blue, berrylike cones that are larger than red cedar cones.

RELATED PLANTS Cultivars of low common juniper are more available than the species. Blueberry Delight ('AmiDak') has extra silvery needles and abundant cones; it requires a male plant for pollination. Copper Delight ('ReeDak') is a male cultivar that turns coppery bronze in the fall through winter.

Juniperus horizontalis
Creeping juniper

> Cupressaceae (cypress family)
> Landscape group: conifers

HOW TO GROW Creeping juniper is found across northern North America and sparingly southward into the Upper Midwest, mainly on sand dunes of the western shore of Lake Michigan or along bluffs and sandy terraces of the Upper Mississippi and Lower Wisconsin Rivers; it is found wild sparingly in Minnesota, Iowa, Wisconsin, and Illinois in the Midwest. It grows well in any well-drained soil and is tolerant of very poor soils from gravel to sand. It prefers full sun but will grow in partial shade and is hardy throughout the Midwest. Nurseries propagate it by cuttings.

LANDSCAPE USE This evergreen can be used as a groundcover as it grows in a prostrate form rarely more than 1 foot tall but easily spreading 3 to often 5 feet. It functions best as an evergreen that spills over a dry and sunny embankment or wall and provides good erosion control for scorching hot, rocky slopes.

ORNAMENTAL ATTRIBUTES This juniper is dioecious: female plants produce colorful small bluish, berrylike cones. and males have the distinctive browner pollen cones like a red cedar. Creeping juniper is found in almost every nursery in the Midwest and there are many cultivars selected mainly for attractive bluish needles. Wild strains are hard to come by and most tarnish to earthier tones in winter. The feathery, prostrate form of creeping juniper is enhanced when it spills like water over a wall and it produces only finely textured, scalelike awl foliage. It is at its best when growing in a matrix with companion plants like little bluestem or hairy grama, redroot, Arkansas wild roses, bearberry, and its coarser cousin the low common juniper.

RELATED PLANTS *Juniperus horizontalis* Prairie Elegance ('BowDak') is only 8–10 inches tall and has bright green foliage.

Juniperus communis var. *depressa* (low common juniper) thrives on the Lake Michigan sand dunes of Illinois Beach State Park.

Juniperus horizontalis Prairie Elegance ('BowDak') is one of creeping juniper's many cultivars on display at the Minnesota Landscape Arboretum.

Juniperus virginiana cultivars
Dwarf eastern red cedar

Cupressaceae (cypress family)
Landscape group: dwarf conifers

Many palisades and other rock outcrops throughout the Midwest were the original home of the eastern red cedar. In these harsh, soil-less sites, red cedar is often just a shrub and may be more than a century old. Shrub-sized genetic dwarfs of this tree are rare.

HOW TO GROW Dwarf eastern red cedar cultivars thrive in any well-drained soil in full sun. They are incredibly heat and drought tolerant and will grow on rock outcrops with little soil. See *Juniperus virginiana* in "Evergreen Trees" for more details.

LANDSCAPE USE These dwarf plants are excellent evergreen low screens or focal ornamentals and thrive in challenging soils where other evergreens won't survive including harsh south- and west-facing facades.

ORNAMENTAL ATTRIBUTES The female cultivar 'Falling Water' was found in Detroit and is an upright pendant-branched form looking like falling water and growing just 8 feet tall. The male cultivar 'Pendula' is a larger growing weeping form. Several junipers like 'Grey Owl' are listed as eastern red cedars, but they are actually of hybrid origins.

Kalmia latifolia
Mountain laurel

Ericaceae (heath family)
Landscape group: eastern heaths

HOW TO GROW Mountain laurel is native in the Appalachian Plateau of eastern Ohio and in the Shawnee Hills of Kentucky and southern Indiana. Like rosebay rhododendron, it, too, can become tree sized in sheltered ravines. It is cultivated throughout the Eastern Midwest in well-drained acidic soils and sparingly westward across the Lower Midwest. It is fickle to grow and not recommended in the rich, heavy soils of the Corn Belt region. It is cold hardy to USDA zone 5, but not adapted to most of the western Midwest's climate and soils.

LANDSCAPE USE Mountain laurel is an excellent foundation plant and can be used for naturalizing in wildlife gardens. It goes well with rhododendrons, azaleas, firs, hemlocks, and pines.

ORNAMENTAL ATTRIBUTES This shrub is beloved for its showy flowers that are usually pink budded but open to white flowers, larger but nearly identical to its cousin the bog laurel.

Juniperus virginiana 'Pendula' (weeping eastern red cedar) makes an extraordinary dwarf conifer.

The pink buds of *Kalmia latifolia* (mountain laurel) open to white flowers.

194

RELATED PLANTS There are many mountain laurel cultivars with compact forms and red, pink, or white flowers, some with patterns inside the flowers.

Kalmia polifolia
Bog laurel

> Ericaceae (heath family)
> Landscape group: Northwoods heaths

HOW TO GROW Bog laurel is mainly a Northwoods plant found along the northern edge of the Midwest. It is best grown in the central and eastern Upper Midwest as it does not tolerate the Lower Midwest's summer heat. It can be cultivated in bog gardens created from half sand and half peat with a constant supply of moisture and mulched with pine bark and sphagnum moss. Plants are easily layered or grown from cuttings. They are also available at specialty nurseries.

LANDSCAPE USE Bog laurel's function in the landscape is limited to container gardens or recreated bog gardens.

ORNAMENTAL ATTRIBUTES Sparse ascending stems reach about 2 feet tall. In spring the stems are adorned with clusters of showy rosy-purple (sometimes pink) flowers that are shaped like shallow, five-sided bells. The flowers produce cute little up-facing capsules that ripen brown and hold through winter.

Ledum groenlandicum
Labrador tea

> Ericaceae (heath family)
> Landscape group: Northwoods heaths

HOW TO GROW Labrador tea is primarily a Northwoods plant found along the northern edge of the Midwest, but it also grows southward into eastern Ohio. Because it does not tolerate the Lower Midwest's summer heat, it is best grown in the central and eastern Upper Midwest. It can be cultivated in bog gardens created from half sand and half peat with a constant supply of moisture and mulched with pine bark and sphagnum moss. Plants are easily layered or grown from cuttings. They are also available at specialty nurseries.

LANDSCAPE USE Like the other Northwoods heath shrubs, Labrador tea can only be grown in container gardens or recreated bog gardens.

ORNAMENTAL ATTRIBUTES This open, upright shrub to 3 feet tall bears 1- to 2-inch-long emerald green leaves. Terminal stems are crowned with a tuft of upward-facing, five-petaled white flowers in late spring. The flowers produce seed capsules that mature brown and open up into five sections looking somewhat like mini tulips.

195

Showy bell-shaped flowers adorn *Kalmia polifolia* (bog laurel) in a bog garden.

The evergreen foliage of *Ledum groenlandicum* (Labrador tea) is as fine as any rhododendron.

Opuntia humifusa (eastern prickly pear) shines in flower at Minnesota Landscape Arboretum.

Opuntia humifusa

Eastern prickly-pear

> Cactaceae (cactus family)
> Landscape group: cacti and succulents

HOW TO GROW Eastern prickly-pear is native across much of eastern North America from the Great Plains to Ontario and southern New England and from Texas to Florida. In the Midwest it is native everywhere except the Dakotas and Minnesota, confined to appropriate habitat in the Upper Midwest, while more widespread in the Lower Midwest. It is found in sand prairies and savannas, rocky or gravelly outcrops, and hill prairies, growing in almost any lean, well-drained soil from sand to gravel, but flowers and looks best in full sun. It is most easily propagated by breaking off pads and rooting them, but is also easy to grow from seed.

LANDSCAPE USE Because eastern prickly-pear has sharp spines with tufts of annoying and irritating glochids at their base, it should only be placed in the landscape where you won't come in contact with it and regret planting it (plants are difficult to remove once established). Hot, dry, rocky sites where little else grows suit prickly-pears well. Plants rarely grow more than 12 inches tall, spreading along the ground as much as 3–5 feet. They should be a component of an edible landscape because the pads and fruits are edible if the spines are removed. Many small to large mammals relish the fruits too, but the plant is a staple for the ornate box turtle. The flowers produce an abundance of nectar and pollen for insect pollinators. The plants look best spilling over rock walls or out of large containers, which also provide good viewing, ease of maintenance, and harvest.

ORNAMENTAL ATTRIBUTES Eastern prickly-pear pads are adorned with many 3- to 4-inch gorgeous, true yellow flowers in midsummer. The flowers produce colorful fleshy fruits that ripen red in the fall.

RELATED PLANTS Bigroot prickly-pear (*Opuntia macrorhiza*) has a more western range but is native eastward into Wisconsin, Illinois, and Arkansas. It is very similar to eastern prickly-pear but more often has two or more spines together and flowers that are usually more golden with a more reddish base to the petals. It appears more tolerant of extreme conditions, including cold temperatures and clay soils, than Eastern prickly-pear but otherwise the two can be used interchangeably in a landscape.

Paxistima canbyi

Paxistima, Canby's mountain-lover, cliff green

> Celastraceae (staff tree family)
> Landscape group: southern origin evergreens

Paxistima is a rare relict from a prior time, clinging to life in a few locations off the extinct preglacial Teays River that flowed north out of the Appalachians through what is now Columbus, Ohio, and taking a turn west into Indiana and following the current Wabash River basin. It's a premier small evergreen shrub that is readily available to gardeners helping to conserve this rare species.

Opuntia macrorhiza (bigroot prickly-pear) flowers often have an orange base to their petals.

HOW TO GROW Paxistima is native to a few relict spots in Ohio and Kentucky and in the central Appalachian Mountains, where it grows in full sun to light shade in well-drained soils. It grows wild on limestone in Ohio and is hardy throughout the Midwest. It is usually grown from cuttings.

LANDSCAPE USE This fine small evergreen reaches only 12 inches tall but can spread slowly over time into wide mats. It is often listed as a groundcover but functions better as a unique evergreen for a rock garden, rock wall, or planted on a rock outcrop.

ORNAMENTAL ATTRIBUTES The evergreen foliage is as pretty as boxwood but with longer-toothed leaves. The early spring flowers are four-petaled and rich dark red, reminiscent of a wahoo flower.

Phoradendron flavescens (mistletoe) crowns a red maple at Cave Hill Cemetery in Louisville, Kentucky.

This lush planting of *Paxistima canbyi* is at the Missouri Botanical Garden in St. Louis.

Phoradendron flavescens
Mistletoe

> Loranthaceae (mistletoe family)
> Landscape group: southern origin evergreens

The state flower of Oklahoma, mistletoe is also the native version of the holiday garland beloved for its amore.

HOW TO GROW Mistletoe is a parasitic evergreen shrub growing on the branches of trees across the Lower Midwest from southeast Kansas to southern Ohio and all points south. I have never heard of anyone cultivating the plant, but it should be kept where native.

LANDSCAPE USE Mistletoe rarely is an issue to the host tree. It produces evergreen balls of foliage 1–3 feet in diameter.

ORNAMENTAL ATTRIBUTES Mistletoe provides winter adornment with its foliage becoming visible in winter as it usually grows on deciduous trees. It produces white fruits in late fall through winter that are devoured by birds and spread through their droppings to new sites.

Picea glauca cultivars
Dwarf white spruce

> Pinaceae (pine family)
> Landscape group: dwarf conifers

Most dwarf cultivars of white spruce are known as "Alberta" spruces but there are many other genetic dwarfs and very small witches' brooms.

HOW TO GROW White spruce cultivars require full sun or part shade in moist, well-drained soils. They can suffer in

197

True to its name, *Picea glauca* 'Little Globe' (dwarf white spruce) forms a compact ball in a rock garden.

the hot and humid Lower Midwest where they prefer more shade, protection from summer winds, and more-consistent moisture. See *Picea glauca* in "Evergreen Trees" for more details.

LANDSCAPE USE These spruces are delightful focal points in the landscape, but may create unique low screens and bed backdrops in borders of smaller perennials.

ORNAMENTAL ATTRIBUTES These cultivars come in a variety of forms. 'Eagle Rock' is a witches' broom selection with a flat, nested look, 'Little Globe' is cushion- or bun-shaped, and 'Witches Broom' is a very small and dense mounded cultivar. All three grow just 2–3 feet in 10 years. 'Wild Acres' is a flat, pyramidal intermediate form growing large enough to be used as an evergreen screen. White spruce's small needles are fine textured, and dwarf selections more readily show their needles' two-toned dark upper and lighter undersides.

Picea mariana cultivars
Dwarf black spruce
Pinaceae (pine family)
Landscape group: dwarf conifers

HOW TO GROW Dwarf black spruce cultivars are more tolerant of partial shade, especially in the Lower Midwest. See *Picea mariana* in "Evergreen Trees" for more details.

LANDSCAPE USE Like the species, the cultivars make a striking specimen plant in the landscape. They can be planted with other "northern" evergreens to create an informal tapestry hedge screen.

Picea mariana 'Nana' (dwarf black spruce) creates a compact teardrop-shaped shrub.

ORNAMENTAL ATTRIBUTES 'Nana' is sometimes called "blue nest"; it has blue-green needles and a very dense, slow growing globe-shaped form reaching 2 feet tall and nearly 3 feet wide. 'Wellspire' is an upright cultivar that can become tree sized, growing 6–15 feet tall and 3 feet wide.

Pinus banksiana cultivars
Dwarf Jack pine
Pinaceae (pine family)
Landscape group: dwarf conifers

Dwarf Jack pines are usually selections of genetic aberrations.

HOW TO GROW They are very tolerant of harsh conditions, often growing in pure sand and the poorest soils in full sun and windswept locations. See *Pinus banksiana* in "Evergreen Trees" for more details.

LANDSCAPE USE Dwarf Jack pine's unique growth makes it highly desired as a focal ornamental evergreen. It creates a windswept look where a full-sized tree wouldn't fit.

ORNAMENTAL ATTRIBUTES 'Al Johnson' is a ruggedly small cultivar that looks like it's growing in a harsh site. 'Angel' is a small version of the species, growing about 8 feet tall. 'Manomet' is a dense and wide, bushy cultivar that

Pinus banksiana 'Angel' (dwarf jack pine) holds its position between other conifers in a garden in Overland Park, Kansas.

grows around 3 feet tall and 6 feet wide. 'Schoodic' is a dense, ground-hugging prostrate form. 'Uncle Fogey' is a wild-growing undulating and weeping selection from Minnesota that creates a rugged, windswept look to a landscape and is found in many nurseries. Dwarf Jack pines display some of the most gnarly and unique forms of any evergreen and, unlike more dwarf conifers, most will eventually produce the ornamental cones.

Pinus resinosa 'Morel' (dwarf red pine) is a mushroom-shaped selection of the Minnesota state tree.

Pinus resinosa cultivars

Dwarf red pine

> Pinaceae (pine family)
> Landscape group: dwarf conifers

Dwarf red pines are virtually all propagated from witches' brooms or genetic aberrations of the tree.

HOW TO GROW They require well-drained soils and languish in wet clay soils. They do not tolerate extreme heat or drought. See *Pinus resinosa* in "Evergreen Trees" for more details.

LANDSCAPE USE Dwarf red pines are used as ornamental specimen evergreens in the landscape.

ORNAMENTAL ATTRIBUTES The long, dark green needles contrast well with other evergreens. 'Don Smith' is a flat-topped cultivar growing 4 feet in 10 years. 'Morel' is a mushroom-shaped, billowy mound taller than wide like its namesake fungus. Both these cultivars produce ornamental cones. 'Quinobequin' is an upright globe, growing 2–3 feet in 10 years.

Pinus strobus 'Curtis Dwarf' (dwarf eastern white pine) thrives in a garden in Overland Park, Kansas.

Pinus strobus cultivars

Dwarf eastern white pine

> Pinaceae (pine family)
> Landscape group: dwarf conifers

There are dozens of cultivars of eastern white pine. 'Nana' is a catchall name for several dwarf witches' brooms selections of eastern white pine that grow into compact, rounded plants. Dwarf white pine cultivars are widely available at nurseries.

HOW TO GROW Eastern white pine cultivars grow best in moist, well-drained soils. Plants struggle in wet clay soils and during extreme dry spells, but are overall the most widespread and adaptable of all the native pines. They make fine container plants. See *Pinus strobus* in "Evergreen Trees" for more details.

LANDSCAPE USE These dwarf cultivars are used in mass plantings as well as in borders and hedges. They lend themselves well to tapestry hedges with other evergreens.

ORNAMENTAL ATTRIBUTES 'Blue Shag' sports subtly bluish needles and grows into a taller mound 6 feet tall with a wider spread. 'Densa' is a compact form with more irregular character growing 4–5 feet tall. 'Reed's Point' has a dense form 3 feet tall and 4 feet wide with blue-green needles. 'Windswept' is from a witches' broom in Rockford, Illinois, and is as rugged-looking as its name suggests, growing just 3 feet in 10 years. The long, soft needles with contrasting bluish undersides make dwarf white pines perhaps the showiest dwarf conifers. Some dwarf selections produce the ornamental cones which add character to the plant.

This silhouette of wild *Pinus virginiana* (Virginia pine) in Ohio shows a witches' broom.

Pinus virginiana cultivars
Dwarf Virginia pine

> Pinaceae (pine family)
> Landscape group: dwarf conifers

Plantsmen have propagated few dwarf selections of Virginia pine though the species often sports witches' brooms.

HOW TO GROW Virginia pines thrive in a variety of poor soils and in summer heat. See *Pinus virginiana* in "Evergreen Trees" for more details.

LANDSCAPE USE These bushy dwarf pines work well in poor soil sites in the Lower Midwest, where they may be combined with red cedar—the two species often grow together in the wild. The short needles are "off" golden green in winter but contrast nicely with other plants in the winter landscape. I have not see cones produced on any dwarf Virginia pines.

ORNAMENTAL ATTRIBUTES 'Driscoll' forms a compact, rounded mound 3 feet tall and wide with more golden needles in winter.

Rhododendron maximum
Rosebay rhododendron

> Ericaceae (heath family)
> Landscape group: eastern heaths

HOW TO GROW Rosebay rhododendron is found only sparingly in the Appalachian Plateau of Ohio, where it can reach tree-sized proportions in sheltered ravines. It is very challenging to cultivate outside its native haunts in the Lower and Eastern Midwest, so is best used only where native. Well-drained, acidic soils are a must.

LANDSCAPE USE This species creates a dramatic large foundation plant for winter-shaded north- or east-facing facades or in sheltered narrow side yards between closely set homes or buildings.

ORNAMENTAL ATTRIBUTES Showy pinkish white flowers occur in late spring or early summer nestled at the base of new growth. The foliage is large and showy, fun to watch in winter as it rolls up in subfreezing weather. The colder the weather, the tighter the roll—as fine as narrow tubes when subzero temperatures are reached.

RELATED PLANTS Most successful rhododendrons grown in the Midwest are hybrids with the Catawba rhododendron (*Rhododendron catawbiense*) and Carolina rhododendron (*R. caroliniense*) from the high elevations of the southern Appalachian Mountains.

Rhododendron maximum (rosebay rhododendron) flowers light up a shady woodland in early summer.

Taxus canadensis
Canada yew

> Taxaceae (yew family)
> Landscape group: conifers

HOW TO GROW Canada yew is mainly found in the Upper Midwest from central Minnesota throughout the Driftless Area and southeastward as a relict in locations from east central Indiana to southeast Ohio, and around the Bluegrass Region in the Pottsville escarpment of Kentucky. Canada yew grows mainly on moist, north- or east-facing cliffs and steep embankments where deer won't devour it in winter. The plant is in decline and in trouble because of browsing by deer.

Masses of *Taxus canadensis* (Canada yew) can be seen at Malanaphy Springs, Iowa.

Thuja occidentalis 'Tiny Tim' is a classic mini-version of American arborvitae.

LANDSCAPE USE Canada yew is an open, sprawling shrub 3–6 feet tall and not like the dense yews commonly cultivated. Its form is probably why it's very rare in cultivation but it is worth preserving where native. I envision planting it in a moist, shady site beneath common witchhazel and paper birch for a sublime experience through the seasons.
ORNAMENTAL ATTRIBUTES Plants are dioecious. Female plants produce a single brown seed rolled in a fleshy aril that turns bright red when ripe in summer into fall. I have seen migrating warblers eat these in autumn.

Thuja occidentalis cultivars
Dwarf American arborvitae

Cupressaceae (cypress family)
Landscape group: dwarf conifers

No native evergreen has more dwarf selections than *Thuja occidentalis*. There are too many cultivars to list. Many are globe shaped, while others are upright. Some have foliage that stays greener in winter; others have golden new growth. A few dwarf cultivars have prickle-type juvenile foliage and don't look anything like the typical species. American arborvitae is naturally dwarfed into a shrub on many rock outcrops within its native range.
HOW TO GROW Dwarf American arborvitae cultivars prefer moist soils that are neutral to alkaline, where the roots are in well-drained soil with a constant supply of moisture. These conifers are not tolerant of excessive heat and drought. See *Thuja occidentalis* in "Evergreen Trees" for more details.

LANDSCAPE USE These cultivars are commonly used as foundation plantings, as low screens and hedges, or as striking ornamentals. They also make outstanding container plants.
ORNAMENTAL ATTRIBUTES The mossy splayed foliage mainly of fine-scaled leaves that adorn the twigs give it a luscious texture. Typical plants often discolor to a more golden green in winter, but many cultivars retain a lively green in winter. 'Golden Globe' retains yellowish needles, and more prickly leaved (actually, juvenile foliage) 'Rheingold' has orange-yellow needles in summer, turning copper orange in winter. 'Hetz Midget' is a reliable favorite with lush green classic arborvitae growth. It's a true genetic dwarf that originated from seed and produces cones. It grows just 3 feet tall and wide and requires no shearing. 'Tiny Tim' is even smaller with wonderful whorls of foliage.

Tsuga canadensis cultivars
Dwarf eastern hemlock

Pinaceae (pine family)
Landscape group: dwarf conifers

There are many dwarf prostrate, weeping and white-needled cultivars of eastern hemlock sold at midwestern nurseries—far too many to list.
HOW TO GROW Plant eastern hemlock cultivars in moist, well-drained soil as they are equally intolerant of wet feet or excessively dry sites. They need protection from hot summer winds and shade from the hottest afternoon sun.

Tsuga canadensis 'Jervis' (dwarf eastern hemlock) brightens up a winter trough garden.

Yucca filamentosa (Adam's needle) flowers gleam in the intense light of early summer.

See *Tsuga canadensis* in "Evergreen Trees" for more details.

LANDSCAPE USE Eastern hemlock cultivars are tolerant of shade and shearing. They can be used as a low hedge plant or in foundation plantings. Pendulous and variegated cultivars make outstanding focal ornamental shrubs.

ORNAMENTAL ATTRIBUTES The adorable small, deep green needles of hemlock make it an exquisite evergreen in all seasons. 'Cole's Prostrate' is one of the finest and unique of the many pendulous growing cultivars. 'Jervis' is a mini-sized plant growing less than an inch per year and is suitable for a container. 'Lewis' gives the look of a miniature eastern hemlock tree, growing just 4 feet in 10 years. 'Gentsch White' and 'Moon Frost' have whitened needle edges that really lighten up a dark, shady space, though the white edge of 'Gentsch White' discolors in winter.

Yucca spp.
Yuccas
Asparagaceae (asparagus family)
Landscape group: cacti and succulents

Yuccas are woody based with evergreen, swordlike foliage. They are indicative of southwestern landscapes in the minds of most gardeners. As evergreen shrubs, they make appropriate additions to gardens. Two species are native to the Midwest.

Yucca filamentosa / *Yucca flaccida* / *Yucca smalliana*
Adam's needle yucca
Asparagaceae (asparagus family)
Landscape group: cacti and succulents

Adam's needle yucca is a complex of three very similar or synonymous yuccas native to the southeastern United States but has been widely cultivated across the Midwest since settlement. It persists or naturalizes from cultivation so it is difficult to determine its native distribution. In many areas its pollinating yucca moths have spread along with it.

HOW TO GROW Adam's needle yucca grows in almost any upland soil from clay to sand, has a tenacious root that is nearly impossible to remove, and even survives in light shade though it doesn't flower in shade.

LANDSCAPE USE It can be used in the landscape like soapweed and Arkansas yuccas but has a much larger flower spike—reaching above the spiky basal leaves and forming a spectacular candelabrum of white pendant flowers 4–6 feet tall in early summer. These glowing flower spikes along with the fact that Adam's needle yucca was often planted in cemeteries have given them the name "ghosts in the graveyard."

Yucca glauca (soapweed yucca) flowers are greenish cream often brushed with rose.

Yucca arkansana (Arkansas yucca) flowers are similar in color to those of soapweed yucca but usually have blunt-tipped tepals.

ORNAMENTAL ATTRIBUTES Where yucca moths are present, Adam's needle yucca produces seedpods that hold well into winter.

RELATED PLANTS There are many cultivars of Adam's needle yuccas, a couple with more filamentous edged "hairy" leaves, and several with varying degrees of cream to yellow variegation.

Yucca glauca

Soapweed yucca

Asparagaceae (asparagus family)
Landscape group: cacti and succulents

HOW TO GROW Soapweed yucca is a Great Plains species growing eastward into the tallgrass prairie in South Dakota, Nebraska, Kansas, and disjunct eastward in the Loess Hills of western Iowa and northwestern Missouri. It grows in well-drained soils on steep loess, rocky, or gravelly substrates and requires full sun. It is one of the most heat and drought tolerant of all midwestern plants. Soapweed yucca is easy to grow from seed and is hardy throughout the Midwest.

LANDSCAPE USE This yucca makes a rugged evergreen for a harsh site where many evergreens would struggle and actually makes a striking companion to shrubby junipers. The sunny, windswept, and rocky end of my driveway is adorned with thriving yuccas.

ORNAMENTAL ATTRIBUTES The spiky 2-foot-long narrow foliage grows in a striking tuft and is grayish green all year. In early summer, the plant produces a 2- to 3-foot spike of greenish cream pendant flowers that are followed by up-facing seedpod fruits that ripen blackish and remain into winter. The flowers are pollinated by the yucca moth and nothing else, so no pods are produced where the moth is absent (unless you hand pollinate them). I do have yucca moths in my neighborhood so get to enjoy the fruits of their labor.

RELATED PLANTS Arkansas yucca (*Yucca arkansana*) is very similar to the soapweed yucca but native in the Ozark Highlands and southern Kansas southward into Texas. It is found in hot, sunny glades and grows well in gardens northward at least to USDA zone 5b.

Small Shrubs

Midwestern native shrubs defy fitting the mold of a traditional landscape. Many evolved in a landscape of fire with the constant battle between prairie and forest. To survive in that ecotone, many spread by underground rhizomes and produce thickets—a behavior not tolerated in most human-controlled landscapes.

Shrubs in the rose family (Rosaceae) are the most diverse, dominant, and important to the web of life group in the midwestern landscape. In that family, brambles (*Rubus*)—blackberries, raspberries, and dewberries—and wild roses (*Rosa*) are the largest groups of shrubs in our region.

ORNAMENTAL SHRUBS FOR TRADITIONAL LANDSCAPES

Amorpha canescens (leadplant)

Aronia melanocarpa (black chokeberry)

Ceanothus americanus (New Jersey tea)

Dasiphora fruticosa (potentilla)

Diervilla lonicera (northern bush-honeysuckle)

Hydrangea arborescens (wild hydrangea)

Hypericum kalmianum (Kalm's St. John's wort)

Rhododendron prinophyllum (roseshell azalea)

Rhus aromatica var. *arenaria* (low fragrant sumac)

Symphoricarpos albus (common snowberry)

Vaccinium angustifolium (lowbush blueberry)

Vaccinium corymbosum (highbush blueberry)

Viburnum dentatum (southern arrowwood)

Viburnum recognitum (smooth arrowwood)

Amelanchier stolonifera (low serviceberry) shines with upright clusters of white flowers for a brief time in early spring.

Amelanchier stolonifera
Low serviceberry

Rosaceae (rose family)

Amelanchier stolonifera is one of three botanical names for virtually identical serviceberries that grow as low, often thicket-forming shrubs. The other two names are *A. humilis* and *A. spicata*.

HOW TO GROW All three low serviceberries are native across the Upper Midwest from Ohio, Michigan, and northern Indiana westward to northeastern Kansas and points north. Missouri populations are not considered native. The native species grow in forest edges, openings in woodlands, and savannas in upland soils. They may be cultivated in most well-drained soils from sand to gravel and rock, as well as in good garden loam and clay-loam in full sun or partial shade. Low serviceberries are propagated by seed or by dividing suckers from a thicket.

LANDSCAPE USE Low serviceberries make an informal low hedge, shrub mass for woodland edge, and even a good plant for a foundation planting. They spread into a thicket but not aggressively. The fruits ripen in early summer and are sweet and delicious if you can beat songbirds to them. These shrubs are choice plants for an edible landscape, food forest, and bird garden.

ORNAMENTAL ATTRIBUTES Showy white, upright clusters of flowers appear in early spring as the leaves are expanding. The fruits begin green, become red, and ripen purplish but are quickly consumed by wildlife. Plants may have warm golden to orange and red fall color.

Amorpha canescens
Leadplant

Fabaceae (legume family)

HOW TO GROW Leadplant is native across the entire central and western Midwest as far eastward as northern and central Indiana and southwest Michigan. It is found on upland prairies and savannas including those with bedrock near the surface. It was common on hill prairies above lead mines of the Driftless Area, which is how it got its common name. The shrub grows in almost any well-drained soil in full sun or no more than morning shade. It is extremely heat and drought tolerant and is hardy across the entire Midwest. Leadplant blooms on new wood so it can be cut back in winter and will flower by midsummer.

LANDSCAPE USE Leadplant rarely grows more than 3 feet tall but can be maintained as a 30-inch-tall herbaceous

207

Amorpha canescens (leadplant) is suitable for a traditional landscape as depicted here at Powell Gardens, Missouri.

Amorpha nana (dwarf indigobush) produces beautiful flowers in early summer.

perennial. It's a tidy plant suitable for a traditional landscape but it is slow to grow and so only available at specialty nurseries in small sizes. Once established it is very long-lived and easy to maintain by pruning in late winter or early spring.

ORNAMENTAL ATTRIBUTES The early summer flowers are dark indigo with contrasting yellow-orange stamens and they are set in up-facing, multibranched spikes atop finely textured gray pinnately compound leaves. The flowers produce clusters of little legume pod fruits that adorn the plant in winter.

Amorpha nana
Dwarf indigobush

Fabaceae (legume family)

HOW TO GROW Dwarf indigobush is found in the western Midwest from the Dakotas and western Minnesota, southward to east central Iowa and north central Kansas where it grows on upland prairies of loamy soil. In a garden setting, it requires full sun and just about any well-drained soil. It is hardy throughout the Midwest. Because it blooms only from old wood, if you cut it back like leadplant it will not flower until the following growing season.

LANDSCAPE USE Dwarf indigobush is a short landscape shrub rarely more than 2 feet tall. It's another premier plant that is rarely or never found in landscapes because it takes so long to produce a landscape-sized plant.

ORNAMENTAL ATTRIBUTES In early summer, spikes of indigo purple flowers with contrasting golden stamens adorn this shrub. The pinnately compound leaves are green and finely textured. My botanist friend Mark Leoschke notes that the leaves have translucent glandular dots, which you can see by holding them between you and the sun. Tiny legume pod fruits adorn the plant into winter.

Aronia melanocarpa (black chokeberry) makes an attractive small shrub for a traditional landscape.

Aronia melanocarpa
Black chokeberry

Rosaceae (rose family)

Chokeberries are all the rage now with their dark, bitter fruit rich in antioxidants. The fruits are edible and don't make one choke but have a memorable aftertaste. They make excellent preserves reminiscent of cherries, and juice from the fruit is becoming popular in juice blends. Most varieties have been bred in Russia though several unique natural forms of chokeberry grow wild across the Midwest. Black and purple chokeberries are similar species treated separately in this book, though they are lumped into one species by many botanists.

HOW TO GROW Black chokeberry is found from eastern Minnesota and northeastern Iowa eastward across the Upper Midwest and sparingly southward in the Eastern Midwest. In the wild it grows on sandy soils, from dry sandstone cliffs to sand lenses around wetlands. This shrub is easy to transplant and can be cultivated in a wide range of soils from sand to clay. Plant black chokeberry in full sun to light shade. From my experience, the plant is not tolerant of conditions in the western Lower Midwest.

LANDSCAPE USE Black chokeberry grows 3–5 feet tall, is tidy and well behaved, and makes an ornamental shrub for a traditional garden.

ORNAMENTAL ATTRIBUTES The foliage and flower stems are smooth and glossy on most forms. The leaves turn rich saturated shades of red in the fall. Black chokeberry produces clusters of showy white flowers with pink stamens that are presented well as they are set above the leaves. The fruit is inky black when ripe, often beautifully polished or glossy in fall, and creates a striking contrast to the red fall color. The dried fruit remains on the plant all winter.

RELATED PLANTS The cultivar 'Professor Ed' was selected from Wisconsin and is a premier example of this species. Red chokeberry (*Aronia arbutifolia*) is found only on the eastern fringe of the Midwest in eastern Ohio. It has showy red fruit that persist into winter and is almost exclusively available as the cultivar 'Brilliantissima.'

Purple chokeberry (*Aronia prunifolia*) grows nearly through the core of Wisconsin and across lakeside Michigan, through the Chicago region, northern Indiana, and over most of Ohio. It is found mainly on acidic or sandy margins to wetlands, growing in sandy or peaty aerated soils, but not in standing water. Surprisingly, this translates into a shrub that is easy to grow in a wide range of soils from clay to sand. Purple chokeberry thrives in full sun to light shade and grows well across the entire Midwest. It is similar to black chokeberry, but with downy hairs on its leaves and flower stalks, grows larger (5–8 feet or more), has flowers with dark pink to red stamens, and produces more purplish black fruit. Most chokeberry cultivars sold for fruit production fall under this species. Some botanists consider purple chokeberry a hybrid between black and red chokeberry.

Ceanothus americanus
New Jersey tea, American ceanothus
Rhamnaceae (buckthorn family)

HOW TO GROW Look for New Jersey tea on upland prairies and savannas where it thrives in full sun to partial shade in rich, well-drained soils. It blooms on new wood and so is adapted to prairie fires or being cut off, blooming on new growth like an herbaceous perennial.

LANDSCAPE USE New Jersey tea makes a tidy low shrub 2–3 feet tall suitable for a traditional landscape. It just takes time to produce and so is available only in small sizes from native plant nurseries and never as a landscape-sized plant. Once established it is very long lived and can easily be maintained by late winter or early spring pruning.

ORNAMENTAL ATTRIBUTES The white clusters of midsummer bloom are very showy and attract a wide array of pollinating insects. The flowers produce seed capsule fruits that turn black when ripening and explode when fully ripe; capsule remnants may remain on the plant into winter.

209

Aronia prunifolia (purple chokeberry) fruits are purplish black and popular with commercial berry growers.

Frothy white clusters of flowers adorn *Ceanothus americanus* (New Jersey tea) on Harlem Hills Prairie in Rockford, Illinois.

Ceanothus herbaceus
Redroot, inland ceanothus

> Syn. *Ceanothus ovatus*
> Rhamnaceae (buckthorn family)

HOW TO GROW Look for redroot in upland prairies, especially those with bedrock near the surface including hill prairies, similar savanna habitats, and on sandy shorelines. Redroot blooms on year-old wood so if a plant gets burned or cut off, flowering will be sacrificed. Its botanical name is most inappropriate because it is a true shrub and not herbaceous.

LANDSCAPE USE Redroot makes a tidy low shrub 3 feet tall (rarely taller) suitable for a traditional landscape. Like its relative New Jersey tea, it just takes time to produce and so is available only in small sizes from native plant nurseries and never as a landscape-sized plant.

ORNAMENTAL ATTRIBUTES The white clusters of late spring bloom are very showy and attract a wide array of pollinating insects. The flowers produce seed capsule fruits that also turn black—ripening about the same time as New Jersey tea blooms.

An eastern tailed blue visits the flowers of *Ceanothus herbaceus* (redroot).

Dasiphora fruticosa
Potentilla, shrubby cinquefoil

> Syn. *Potentilla fruticosa*
> Rosaceae (rose family)

This widely cultivated shrub is known as potentilla by gardeners and nursery professionals, but botanists call it shrubby cinquefoil since *Potentilla* is the genus name of related wildflowers. It will take time for nonbotanists to accept the genus *Dasiphora*. Potentilla grows wild across much of the Northern Hemisphere but is native in the Midwest from central Minnesota to northeast Iowa, northeastern Illinois, central Indiana, and western and northern Ohio and points northward. It grows in incredibly diverse habitats from dry cliffs to wet fens, usually in sites where vegetation is short and this small shrub can compete. Potentilla is easy to grow in a garden in full sun or just partial shade, but suffers in rich heavy clay soils. It does best in the Upper and Eastern Midwest and is more fickle to grow in the western Lower Midwest as it does not like the heat and humidity of AHS heat zone 7. Nursery growers propagate potentilla cultivars by cuttings, but it may also be grown from seed.

LANDSCAPE USE Potentilla grows about 3 feet tall and is a compact, rounded shrub commonly found in most midwestern nurseries. It is suitable for a traditional landscape where it is often included in low shrub borders.

ORNAMENTAL ATTRIBUTES Potentilla has finely divided, gray-green foliage and is speckled with yellow flowers, heaviest in early summer. It produces some flowers into fall.

RELATED PLANTS Nurseries sell many cultivars of potentilla, which are cultivated for their more abundant flowering in various flower shades from bright to soft yellow, white, and even pink. Most of the cultivars are not from midwestern ecotypes of the shrub.

A shrubby cinquefoil, *Dasiphora fruticosa* Dakota Sunspot ('Fargo'), stands out with gray-green leaves and yellow flowers.

Diervilla lonicera
Northern bush-honeysuckle

Caprifoliaceae (honeysuckle family)

A relative of the garden-mainstay weigela, diervilla would be just as popular if its flowers were as large. Its common name also is a problem now that invasive exotic honeysuckles conjure up such negative connotations (and rightfully so).

HOW TO GROW Northern bush-honeysuckle grows wild in cool rocky woods across the Upper Midwest and sparingly southward but is absent from most areas once dominated by prairie. It is hardy throughout, even thriving in the heat of the Lower Midwest. Northern bush-honeysuckle is easy to transplant and thrives in almost any well-drained soil. It is drought tolerant once established and can grow in full sun northward, but grows best in light shade to partial (afternoon) shade.

LANDSCAPE USE Northern bush-honeysuckle suckers into a thicket but slowly and so is not a nuisance. It rarely grows more than 4 feet tall, makes a fine low shrub or hedge, and may be planted closely to create a sturdy mass planting. It has proven a sturdy choice for parking lot islands in the Upper Midwest.

ORNAMENTAL ATTRIBUTES The greenish yellow flowers are delightful upon close inspection. The foliage is clean and neat with some plants emerging coppery in spring. The plant sports nothing spectacular but is a garden-worthy sleeper more dynamic than many evergreen shrubs. Northern bush-honeysuckle's fall color can be shades of red from burgundy to scarlet.

Diervilla lonicera (northern bush-honeysuckle) flowers are glowing yellow upon close inspection.

Dirca palustris
Leatherwood

Thymelaeaceae (mezereum family)

Leatherwood has neat pliable stems and leathery bark apparently used as a leather substitute by Native Americans. It always reminds me of the popular succulent jade plant, which is a common houseplant, as the stems seem inordinately thick for its size.

HOW TO GROW Leatherwood is native across the Eastern Midwest and westward to central Minnesota, western Iowa, and the Ozark Highlands. It's found in mesic forests from floodplain terraces to sheltered north-facing slopes in the western Midwest. Shade is a must for this species but otherwise it will grow in moist, well-drained woodland soils. You may propagate leatherwoods from fresh seed sown immediately.

LANDSCAPE USE Leatherwood is a subtle but workhorse shrub for a woodland garden or shady traditional landscape. It grows slowly into a robust shrub 4–6 feet tall and occasionally taller with great age.

ORNAMENTAL ATTRIBUTES The dainty little yellow bell-shaped flowers are borne in the leaf axils in early spring. The berries ripen red and drop by late spring. The foliage turns rich butter yellow in the fall and the stems are smooth and brown, becoming lighter gray with maturity and standing out in the winter landscape.

Dirca palustris (leatherwood) has yellow bell-shaped flowers that appear in early spring.

The shorter flowers of *Dirca decipiens* (western leatherwood) provide nectar for a spring azure.

The lacecap flowers of *Hydrangea arborescens* (wild hydrangea) shine in the midsummer sun.

NOTES The largest leatherwoods I have ever encountered were 8 feet tall and wide and wrongly cut down by volunteer trail workers as the plant has simple leaves and looks like brush to the uninitiated. If they only knew the shrubs were probably as old as the mature trees in the forest and never would have overtaken the trail.

RELATED PLANTS Western leatherwood (*Dirca decipiens*) is a relatively newly described species that earned its botanical name by deceiving botanists. It was discovered in suburban Kansas City at Overland Park Arboretum where it grows widely on a north-facing forested slope above Wolf Creek. It has since been found in a couple of locations in the Ozark Highlands of Missouri and northwestern Arkansas. Western leatherwood is virtually identical to leatherwood but blooms several days later and is more tolerant of heat and summer dryness. It grows well in just about any well-drained woodland soil and also requires shade. Western leatherwood is an early spring nectar source and butterfly magnet.

LANDSCAPE USE Wild hydrangea is a showy-flowering shrub for moist shade but flowers more heavily with more light and will grow in full sun if moisture is available. It reaches 4–5 feet tall and its flowers attract many pollen-collecting insects, especially bumblebees and beetles.

ORNAMENTAL ATTRIBUTES Lightly fragrant, creamy white fertile flowers are surrounded by a few showy sterile flowers creating what's called a lacecap flower. The flowers appear in midsummer while few plants are in bloom. Fall color can be a lovely whitish yellow. Fertile flowers produce a flat head of tiny dry seedpod fruits that mature rich brown, while the fertile flowers age to beige, and together they hold on the plant well through winter.

RELATED PLANTS There are many cultivars of wild hydrangea whose flowers are comprised almost completely of showier sterile flowers called mopheads—note these provide almost nothing for pollinating insects. The most popular cultivar is the extremely showy 'Annabelle' originally found in Anna, Illinois. New cultivars have pink and even greenish sterile flowers.

Hydrangea arborescens
Wild hydrangea

Hydrangeaceae (hydrangea family)

HOW TO GROW Wild hydrangea is native across the Lower Midwest from the Ozark Highlands to the Ohio Appalachian Plateau and sparingly northward in Illinois, Indiana, and northeast Ohio. It has naturalized northward and is hardy throughout the Midwest. It grows in shaded moist rocky outcrops and slopes in mesic forest. It's easy to propagate from its dustlike seeds and also easy to root from cuttings.

Hypericum kalmianum
Kalm's St. John's wort

Clusiaceae (St. John's wort family)

HOW TO GROW Kalm's St. John's wort is endemic to the Great Lakes and found in sandy soils close to Lakes Erie, Huron, and Michigan and westward into the sand counties of central Wisconsin. It grows wild in sandy lakeshores, sand prairies, and sand savannas but is easy to cultivate in almost any well-drained soil in full sun to partial shade. Kalm's St. John's wort thrives across the entire Midwest including southward through the Lower Midwest. It self-sows sparingly and is very easy to propagate from its tiny seed or cuttings. Note that it is relatively short-lived.

LANDSCAPE USE It makes an adorable small, tidy shrub that is cultivated in traditional landscapes. It grows just 3–4 feet tall and makes a fitting low hedge, foundation plant, or shrub mass. Its flowers are rich in pollen and visited by frenzied pollen-collecting bees.

ORNAMENTAL ATTRIBUTES The foliage is small, narrow, and long, creating a fine texture, often bluish green and turning shades of yellow, orange, and red in fall. The flowers cover the plant in midsummer as a mass of yellow, with abundant bushy stamens. The seedpod fruits mature brown and adorn the plant through winter. The bark also is attractive and exfoliates to show mahogany tones underneath.

Hypericum kalmianum (Kalm's St. John's wort) makes a fitting shrub for a traditional landscape.

Hypericum prolificum (shrubby St. John's wort) forms a very sculpted shrub with age.

RELATED PLANTS Shrubby St. John's wort (*Hypericum prolificum*) is native across the Eastern Midwest and westward to eastern Missouri and the Ozark Highlands; it has naturalized northward. It grows wild in savannas, woodland edges, and disturbed ground where it self-sows abundantly. It becomes much taller than Kalm's St. John's wort, easily reaching 4–6 feet, but is also relatively short-lived. Because it self-sows and grows larger, it is less often seen in traditional landscapes but better suited to natural gardens in masses along a woodland edge or informal shrub border. The foliage, flowers, fruit, and bark are similar to Kalm's St. John's wort but the form of the plant may become more sculptural as it matures, creating some fantastic forms.

Rhododendron prinophyllum
Roseshell azalea

Syn. *Rhododendron roseum*
Ericaceae (heath family)

Azaleas remain one of the favorite garden shrubs of Americans. Although none are native throughout the Midwest, the ranges of several species skirt our region. Roseshell azalea is the hardiest with the largest native midwestern range.

HOW TO GROW Roseshell azalea is native across the southern Ozarks and in the Appalachian Highlands of Ohio in acidic

The clove-scented flowers of *Rhododendron prinophyllum* (roseshell azalea) are delightfully pink.

Rhododendron calendulaceum (flame azalea) blooms at the Missouri Botanical Garden in St. Louis.

The flowers of *Rhododendron periclymenoides* (pinxterbloom azalea) show spectacular long, curved stamens.

woodlands. Like most azaleas, it demands low pH soil, something that is absent from much of the Midwest, but soil can be readily amended in many areas with organic matter and sulfur-based soil acidifiers to make it azalea friendly. Roseshell azalea is hardy across the entire Midwest, but demands afternoon shade or light shade, doing especially well in sandy soil areas. It is not a good choice for alkaline prairie sites. Northeastern strains of this shrub do not survive in the heat of the western Lower Midwest where the local Ozarks strain is successful.

LANDSCAPE USE Roseshell azalea is a well-behaved medium-sized shrub 3–5 feet tall that works well in a woodland garden, shrub border, or near a favorite outdoor seating area.

ORNAMENTAL ATTRIBUTES One finds roseshell azalea in bloom not by sight, but by its clovelike alluring fragrance that permeates a wide area. The delightful, true pink flowers peak in midspring as the leaves emerge and attract lettered and Nessus sphinx moths for pollination. The shrub sports clean foliage that turns bronzy golden shades in fall. Roseshell azalea is the parent to several azaleas in Minnesota's hardy hybrid Northern Lights Series.

RELATED PLANTS Flame azalea (*Rhododendron calendulaceum*) produces yellow, orange, or red nonfragrant flowers in mid to late spring that are pollinated by hummingbirds. It is native only in the Appalachian Highlands of southern Ohio in the Midwest and southward through the Appalachian Mountains. It is cold hardy in USDA zone 5 but prefers cooler summers and is challenging to grow in AHS heat zone 7.

Pinxterbloom azalea (*Rhododendron periclymenoides*, syn. *R. nudiflorum*) is similar to roseshell azalea with mid to late spring nonfragrant, pink flowers of spectacularly long, curved stamens. It is native in eastern Ohio and around the Bluegrass Region of Kentucky with an outlier population in southern Illinois. It is cold hardy to USDA zone 5.

Swamp azalea (*Rhododendron viscosum*) has western populations that were formerly known as the Texas azalea (*R. oblongifolium*). These western swamp azaleas are native in the lower Ozark Highlands in Arkansas and Oklahoma (and southward to eastern Texas and western Louisiana) and not found in swamps but in upland oak-hickory and shortleaf pine forests. This plant has intensely fragrant, early summer white flowers that are pollinated by moths. It's a great plant for an evening garden and grows very well northward to at least the Kansas City region (USDA zone 6), being one of the best native azaleas at Powell Gardens.

Rhus aromatica var. *arenaria*
Low fragrant sumac

Anacardiaceae (cashew family)

This variety of fragrant sumac is not always botanically recognized but describes the low growing forms of the plant.

HOW TO GROW Low fragrant sumac is found from northeastern Illinois, northern Indiana, Michigan, and northern Ohio, most often on sandy or gravelly soils in open oak-hickory woodlands.

LANDSCAPE USE It makes a fine low shrub planted in mass or as a shrub spilling over a wall. It has been widely planted in parking lot islands across the Midwest.

ORNAMENTAL ATTRIBUTES Like the typical species, var. *arenaria* produces nectar-rich yellow flowers in spring before the leaves, and the male flowers are showier than the female. However, the fall color, when trees are planted in the Lower Midwest, is not as vibrant as that of the typical species and rarely is red.

RELATED PLANTS The popular cultivar 'Gro-Low' was selected from the Chicago region.

Ribes missouriense

Missouri gooseberry

Grossulariaceae (gooseberry family)

Several species of gooseberries and currants are native to the Midwest, but the two described in this book are the most garden worthy. All the species make nice additions to natural gardens and should be saved where they are native. Gooseberries and currants produce nectar-rich flowers and edible berries (prized for pies and preserves) that are also enjoyed by small mammals and birds. There are many cultivars of gooseberries and currants grown for fruit production but most of these are selections from nonnative species.

HOW TO GROW This is the most widespread species of gooseberry in the Midwest but is absent from much of the Eastern Midwest. It grows wild mainly in dry upland woodlands and is most abundant in woods that were previously grazed by cattle. Missouri gooseberry is extremely easy to cultivate in almost any well-drained soil in full sun to shade. Wherever a stem touches the ground, it roots into a new plant and that's the easiest way to propagate the Missouri gooseberry. It is generally not available in nurseries.

LANDSCAPE USE Missouri gooseberry is a sleeper of a shrub but an outstanding choice for a natural woodland garden. It can be planted as a low barrier hedge as it usually grows 3–5 feet tall, becomes a thicket over time, and has thorny stems. The shrub is a great choice for a food forest or edible landscape.

ORNAMENTAL ATTRIBUTES The longer I live with this plant, the more special it has become. It is one of the first shrubs to leaf out in spring. The pale cream flowers in midspring are elegant shooting stars of reflexed petals behind a pointed, beaklike center. Bumblebees are the pollinator and often grab my attention that it is in bloom before actually seeing the flowers. The maple leaf–shaped foliage can become sparse by late summer but late fall color is stunning with shades of bronzy orange, pinkish rose, or yellow. The fruit is edible when green and ribbed with lighter longitudinal stripes; at this stage it makes the best pies, but becomes a bit sweeter after it fully ripens and turns purplish black. How does one describe the special flavor of a gooseberry? Try one.

RELATED PLANTS Swamp gooseberry (*Ribes hirtellum*) replaces Missouri gooseberry to the north and east of its range and so is native from central Minnesota eastward across Michigan to the northeastern half of Ohio. It grows in more moist to wet soils, usually around wetlands but also in moist forested slopes.

Rhus aromatica var. *arenaria* (low fragrant sumac) grows on pure sand at the Indiana Dunes National Lakeshore.

Ribes missouriense (Missouri gooseberry) leaves turn various shades from gold to rose in late autumn.

Ribes odorata
Clove currant

Grossulariaceae (gooseberry family)

Whenever I smell the pleasurable scent of this shrub's bloom, I am transported back to childhood walking home from school through old neighborhoods where this shrub was once prized. It lost favor in modern times but has regained popularity recently, mainly because of the native plant and edible landscape movement.

HOW TO GROW Clove currant has naturalized throughout the Midwest so its native range is a bit blurry but it is usually found in dry, rocky habitats on the edges of cliffs, glades, and in prairies. Its actual native range may just skirt the west edge of the Midwest. It is easy to cultivate in any well-drained soil and grows well in full sun to partial shade but blooms best in more sun.

LANDSCAPE USE This was a popular lawn shrub in historic landscapes and is of smaller stature, usually growing 3–5 feet tall. I have seen some strains that sucker while others do not. Plant it where you can enjoy its heavenly fragrance. Clove currant has a rather open and informal stature and so is not a great choice for a formal landscape. It makes a fitting component in a perennial border where later-blooming perennials mask it in late summer as it may drop many leaves prematurely or go dormant early. The nectar-rich flowers attract many butterflies and other pollinators. The fruits are delicious for people and wildlife, making it a choice small shrub for an edible landscape or wildlife-themed garden.

ORNAMENTAL ATTRIBUTES The clusters of yellow, tubular flowers, which are produced in early to midspring, are gorgeous to look at and perfume a wide space. This shrub leafs out very early too, providing a good backdrop to the flowers. The maple leaf–shaped leaves often drop over the summer, some persisting and showing burgundy red to wine purple fall color. The fruit are flavorful black currants.

RELATED PLANTS The cultivar 'Crandall' can produce more fruit than the species, but all currants do that when two or more noncloned plants are grown for cross-pollination.

American black currant (*Ribes americana*) is native across the Upper Midwest with a few populations southward to Kentucky in the Eastern Midwest. It grows mainly in wetlands in full sun to partial shade and propagates wherever its branch tips touch the ground. The spring flowers are cream-colored, less showy, and not fragrant compared to clove currant, but they produce delicious black currants in mid to late summer.

Ribes odorata (clove currant) is a fragrant mass of yellow at Shaw Nature Reserve in Franklin County, Missouri.

Ribes americana (American black currant) is adorned with subtle creamy yellow flowers in pendant clusters.

Rosa spp.
Wild roses
Rosaceae (rose family)

Wild roses are very similar species of shrubs and often challenging if not impossible to identify. They all have single pink flowers followed by red fruits known as hips that color up in fall and remain on the plants into winter. Most are best utilized for natural landscapes as they sucker into thickets and are considered too invasive for traditional landscapes. They are iconic plants of the Midwest whose flowers are always a welcome sight in early to midsummer. None of the native species has repeat flowering like landscape roses but they have less disease prone foliage, more colorful fall foliage, and more abundant fruit hips for the fall and winter landscape. Wild rose is the state flower of Iowa and North Dakota.

Rosa arkansana
Arkansas, low prairies wild rose
Rosaceae (rose family)

HOW TO GROW Arkansas wild rose is native to upland prairies from northeastern Oklahoma north to Canada, eastward to most of Illinois, and sparingly as far east as Michigan. It grows in full sun to partial shade in almost any well-drained soil. It can be propagated from seed or by division.

LANDSCAPE USE Arkansas wild rose suckers into a low open thicket of wispy stems rarely more than 2 feet tall and densely covered by fine, straight prickles so it is best grown where it can intermingle with other plants including shorter native grasses such as little bluestem and sideoats grama and sturdy perennials such as baptisias and silphiums. It's an aggressive nuisance in a perennial border but a perfect component in a prairie or meadow planting where it has to compete with other plants.

ORNAMENTAL ATTRIBUTES The typical pink wild rose flowers are borne in clusters of two to four in early summer, but flowers can be variably colored and sometimes hot pink, white, and even striped petals. The fruit is a typical red rose hip though more smooth with persistent sepals, and the fall color can be a nice blend from yellow to red.

RELATED PLANTS Carolina wild rose or pasture wild rose (*Rosa carolina*) looks and grows very similarly to Arkansas wild rose. It is native across much of Eastern North America including all of the Eastern and Lower Midwest northwestward to Wisconsin and Iowa. It is more of a successional

Rosa arkansana (Arkansas wild rose) produces classic, single pink wild rose flowers.

Rosa carolina (Carolina wild rose) blooms at Searls Prairie in Rockford, Illinois.

species found on the edge of woodlands, hedgerows, meadows, and pastures. Carolina wild rose needs the same cultural requirements as Arkansas wild rose and has the same use in the landscape as it also runs. Its flowers are usually borne singly and are pink or rarely white.

217

Rosa palustris (swamp wild rose) flowers in the wilds of Fernwood, Michigan.

Rosa blanda (smooth wild rose) hips are pear-shaped and orange before they ripen red.

Rosa blanda
Smooth wild rose
Rosaceae (rose family)

This was the wild rose most common where I grew up in northeast Iowa and we just knew it as the state flower. It's by far the least thorny of the wild roses.

HOW TO GROW Smooth wild rose is found in upland savannas, open woodlands, and hedgerows. It grows in any moist, well-drained soil in full sun to partial shade. It can be divided from a clump, readily rooted from cuttings, or grown from seed.

LANDSCAPE USE Smooth wild rose grows 4–5 feet tall and suckers into a thicket. It makes a good shrub for massing on the edge of a natural landscape, creating great nesting and cover habitat for wildlife. It also is a good erosion control shrub for an embankment where it will help hold soil by its suckering nature and sturdy root system.

ORNAMENTAL ATTRIBUTES Smooth wild rose has typical early summer, single pink wild rose flowers. Its twigs are often smooth with few or no thorns. The orange, pear-shaped rose hips ripen glossy red in the fall, and age ruddy red as they hang on the plant into winter.

Rosa palustris
Swamp wild rose
Rosaceae (rose family)

I know the swamp wild rose from the wild in premier natural areas in Illinois, Indiana, and Michigan and not from disturbed habitat like many of the other wild roses.

HOW TO GROW It grows in wetlands, mainly in sandy marshes, shrub swamps, or tree swamps where it may grow in standing water. It will grow in good garden soil but prefers a moist to wet soil with a lower pH in full sun or partial shade. It is easy to propagate from seed.

LANDSCAPE USE Swamp wild rose usually grows in a clump in the wild and forms a 5- to 6-foot-tall upright vase-shaped shrub with pendant branches. It does sucker, but it's a fine shrub for a rain, pondside, or wetland garden. The thorny, arching stems create habitat for nesting songbirds.

ORNAMENTAL ATTRIBUTES The shrub produces typical pink wild rose flowers in midsummer (later than most wild roses) and rose hips in fall that hang on the shrub through winter.

Rosa setigera
Prairie wild rose, climbing wild rose, Illinois wild rose
Rosaceae (rose family)

HOW TO GROW Prairie wild rose grows in upland savannas, the edges of woodlands, and hedgerows where it thrives in moist, well-drained soil in full sun or partial shade. It may be pruned like any climbing rose as its stems, known as "canes," usually grow their first year, flower and fruit on their second year, and decline or die on their third year. By pruning out the third-year dying canes, you can keep the shrub more vigorous and suitable for a traditional landscape. Without pruning the shrub becomes a thicket with a core of dead, thorny stems. Prairie wild rose is resistant to rose rosette disease and more resistant to blackspot foliar issues than most landscape roses. It's easy to grow from seed.

LANDSCAPE USE Prairie wild rose can replace any climbing rose in a traditional landscape. It makes a large 5-foot-high haystack of arching stems and can climb to 8 feet or more when rambling over other plants or when trained on a trellis. It does not sucker and its stems do not root where they touch soil. If not pruned annually it is best reserved for a natural landscape where it makes prime wildlife habitat and is a favorite shrub for nesting songbirds—a great replacement for nonnative and invasive multiflora rose. It makes a stunning combination with American elderberry and indigobush that bloom at the same time.

ORNAMENTAL ATTRIBUTES Around the summer solstice clusters of flowers open rich pink but fade to light pink creating a two-toned effect when in full bloom. The hips are the most beautiful of any wild rose; they are more abundant and round (the calyx does not persist on its end) and are showy in the winter landscape.

Rosa woodsii
Prickly wild rose
Rosaceae (rose family)

HOW TO GROW Prickly wild rose is very closely related to smooth wild rose and may be its western counterpart

growing in Minnesota and the Dakotas and west of the tallgrass prairie in Nebraska and Kansas. It grows in prairie, open woodland, woodland edge, and other successional habitat where it thrives in well-drained soils in full sun or light shade. It can be propagated from seed or by division.

LANDSCAPE USE Prickly wild rose suckers into a dense thicket, forming a solid shrub border, especially workable as a thorny barrier in a natural landscape. We planted it at Powell Gardens to discourage visitors from taking a short cut from our parking lot and it has worked for over 15 years. It is too aggressive for a traditional shrub border or at the back of a perennial border. It usually grows 4–5 feet tall.

ORNAMENTAL ATTRIBUTES Flowers are produced in late spring and are pink but smaller than average. Typical red rose hips ripen in the fall and remain on the plant into winter.

Rubus spp.
Brambles—blackberries and raspberries
Rosaceae (rose family)

With the popularity of the edible landscape movement, there is a new interest in cultivating our delicious native brambles. They are divided into four main groups: blackberries, black raspberries, red raspberries, and dewberries—all with edible berries. (Note that dewberries lack support in their stems so trail along the ground or sprawl over other plants; they are described in the groundcover chapter.) The first-year stems of all native brambles grow solely with foliage (primocanes) while flowers and fruits are produced on

Rosa setigera (prairie wild rose) is a spectacular sight during its summer solstice bloom.

Rosa woodsii (prickly wild rose) blooms at Powell Gardens, Missouri.

Like all blackberry species, *Rubus ostryifolius* produces beautiful white flowers in late spring.

second-year stems (floricanes) and the stem dies following fruiting.

Commercial berry growers will warn of the diseases found on the native species (various rusts and galls) and recommend not having them near the cultivated versions. In my natural landscape the diseases are present but there are always plenty of plants flowering and fruiting well. Most naturalists agree that the native bramble species seem to have the best flavor despite their thorns or smaller size. None of the bramble species are ideal formal garden plants, but are ideal in food forests, and add incredible beauty and prime wildlife food and habitat to a natural garden or along the edge of a traditional garden.

Rubus allegheniensis / Rubus ostryifolius
Blackberries

Rosaceae (rose family)

Blackberries are a mind-boggling array of similar species. Allegheny blackberry (*Rubus allegheniensis*) is perhaps the most widespread across the Midwest, but each region of the Midwest has its own suite of species. In my area of the Lower Midwest, *R. ostryifolius* is prevalent. Blackberries represent a very complex group of species and I'll admit they pretty much all look alike and I can't tell them apart without carefully keying out their botanical features. Welby Smith's *Trees and Shrubs of Minnesota* depicts this complexity well.

Cultivated blackberries are hybrids of American species; most of the breeding work has been done in Arkansas so that many of these "improved" selections are not fully hardy in the Upper Midwest even though there are wild types across the entire Midwest. Most of the new selections are thornless as blackberries have wicked "cat's claw" thorns. A few new blackberry cultivars actually bloom and fruit on first-year canes.

HOW TO GROW Blackberries are easy to cultivate in most soil types in full sun or partial shade. Be sure you really want the wild species because of their wicked thorns and because they spread the most aggressively (by underground rhizomes) of the brambles. In late winter, wearing leather gloves, I remove unwanted, wayward plants; these divisions are the easiest way to propagate blackberries.

LANDSCAPE USE Blackberries are best relegated to a natural landscape where they provide exceptional nesting and protective cover habitat for wildlife in open woodland or along the edge of forests. The flowers are nectar- and pollen-rich for a plethora of pollinators, while humans and wildlife eat the edible fruits alike.

Rubus allegheniensis (Allegheny blackberry) produces an abundance of fruits that initially turn red then ripen black.

ORNAMENTAL ATTRIBUTES Blackberries are classic four-season ornamental plants: they display showy white flowers in late spring followed by delicious berries that change from green to red to black when ripe in summer. The plants produce riotous fall colors in a range of reds, possibly the showiest fall color of any native shrub, and the winter stems are often ornamental reddish tones.

Rubus occidentalis (black raspberry) fruit turns red and then ripens black.

Rubus occidentalis
Black raspberry
Rosaceae (rose family)

Black raspberries reflect one species, *Rubus occidentalis*, which includes three popular cultivars grown for fruit production because of their disease resistance—'Black Hawk', 'Bristol', and 'Jewel'.

HOW TO GROW Black raspberry is found across the entire Midwest where it is a denizen of open woods, woodland edges, and hedgerows. Plant it in full sun or part shade in average to dry soil. Unlike other brambles, it does not spread by underground rhizomes, but via its arching canes that root wherever they touch the ground. In late winter, edit out wayward plants; severing and transplanting the rooted cane tips is the best way to propagate the plant.

LANDSCAPE USE Like all brambles, black raspberry is best in a natural setting unless it is controlled and maintained for fruit production.

ORNAMENTAL ATTRIBUTES The winter stems are a stunning violet unlike those of any other plant. The flowers in late spring are not showy but very nectar rich and one of the best for pollinators including hummingbirds, swallowtail butterflies, and bees. The fruit gradually turn red then ripen black when they are edible and delicious. This shrub is simply gorgeous when fruiting.

Rubus odoratus (purple-flowering raspberry) makes a lovely addition to an edible landscape.

Rubus odoratus
Purple-flowering raspberry
Rosaceae (rose family)

Red raspberries are from a group of species, though only two are particularly garden worthy: American red raspberry (*Rubus strigosus*) is cultivated for its classic tasty and colorful fruit, while purple-flowering raspberry (*R. odoratus*) is grown primarily for its gorgeous flowers.

HOW TO GROW Purple-flowering raspberry is found across the eastern fringe of the Midwest with isolated outliers westward into Illinois. It's native in moist open forests and forest edges. It is easy to grow in moist semishaded sites, where it is shaded from the scorching southwest sun in the western Midwest.

LANDSCAPE USE Purple-flowering raspberry is best utilized on the edge of woods or in a more natural landscape where this thicket-forming plant can spread about.

ORNAMENTAL ATTRIBUTES This species is the most ornamental of the brambles with large, showy purple flowers, beautiful maplelike foliage, and simply gorgeous winter stems of exfoliating light cinnamon bark. Its red berries are thinner fleshy caps than American red raspberry, but they are

Rubus strigosus (American red raspberry) has a stem that looks furry in the winter.

very tasty too, with a zippier raspberry flavor. I have read gardeners' accounts that their purple-flowering raspberries never produce fruits, but all plants grown from several sources at Powell Gardens, near Kansas City, produce fruit.

Rubus strigosus
American red raspberry

Syn. *Rubus idaeus* var. *strigosus*
Rosaceae (rose family)

Our native American red raspberry, whether just a variety of a species found across the Northern Hemisphere or its own species, is a moot point for anyone wanting to grow native plants. Most of the cultivars of red raspberries grown for backyard or commercial berry production are from the nonnative version.

HOW TO GROW American red raspberry is found in open woods, forest edges, and successional areas across the entire Upper Midwest with a few localized populations southward. Here it is easily cultivated in a variety of soils in full sun to part shade, but it requires more water and shade southward in warmer areas. This shrub spreads vigorously from underground rhizomes so it needs to be controlled. It can be trained in a row—like cultivated raspberries where post-fruiting and wayward canes can be readily removed. Propagate red raspberry by these divisions.

LANDSCAPE USE American red raspberry is best as a naturalized plant on the edge of woods where it is companion to other shrubs or large perennials. Without competition, it quickly becomes a solid mass of canes. It can be formally grown in a traditional edible landscape but requires extra care of removing dead canes.

ORNAMENTAL ATTRIBUTES The flowers adorn second-year (floricane) stems and are not showy but are still nectar rich and attract many insects. The foliage is quite special with a strikingly whiter underside. The stems are bristly, looking furry brown in winter. The red fruits borne atop the stems are colorful and delicious, but variable in flavor.

Salix spp.
Willows
Salicaceae (willow family)

Several small shrub willows are found across the Midwest, mainly in the Upper Midwest and usually associated with wet prairies, sedge meadows, and fens. My fondest willowy memories come from botanizing those wetlands. Willows

are seldom grown by nurseries and are not seen in traditional landscapes though they offer so much utility and beauty. Let's start a new trend: gardening with native willows. I dream of a midwestern parking lot edged not with Japanese spireas, alpine currant, and Knockout roses, but with leadplant, New Jersey tea, and native shrub willows. Even the mow, blow, and go crew could maintain it.

HOW TO GROW Willow seed germinates immediately and only on moist, exposed soil so shrubs are often found growing in wetlands though they are adapted to almost any soil from moist to wet. If a willow shrub becomes too twiggy or wayward growing, it may be cut back and treated like an herbaceous perennial. The best time to cut them back is right after flowering in spring. Willows require full sun or only partial shade and may be propagated by cuttings or by seed.

LANDSCAPE USE Shrub willows can be used anywhere a small shrub is needed in the landscape. They support an amazing array of caterpillars and nesting songbirds will thank you for planting them as caterpillars offer a wealth of protein for birds to feed their nestlings. Willow flowers (both male and female) also produce nectar and pollen early in the season that sustains a wealth of bees and other pollinators.

ORNAMENTAL ATTRIBUTES The male flowers of catkins are yellow with pollen; most emerge with silky hairs and are known as pussy willows. Female flowers contain nectar and produce the plant's seed capsule fruits that look neat when they burst with the cottony tufted seeds; the cotton acts to carry the seed in the wind.

Salix candida (sage willow) is native in the Midwest from eastern South Dakota to northern and east central Iowa, northern Illinois, northern Indiana, and northern Ohio and points north. It is often found in fens, growing just 3 feet tall. Its silvery foliage is lovely through the growing season. A cultivar 'Silver Fox' from Canada is newly available.

The silvery foliage of *Salix candida* (sage willow) is a standout in the wild or in a garden all growing season.

The autumn fruits of *Salix serissima* (autumn willow) look like cotton balls hanging on the shrub.

Salix pedicellaris (bog willow), seen here growing wild at Crossman Prairie in Iowa, makes a very refined, small shrub.

Unique foliage and dwarf habit distinguish *Salix humilis* var. *tristis* (low prairie willow) from other shrub willows.

Salix humilis var. *tristis* (low prairie willow) is like a mini-version of prairie willow (*Salix humilis* var. *humilis*—see the Small Tree and Large Shrub chapter) growing just 3–4 tall. It's found across much of the Upper Midwest growing in prairies and sedge meadows. Its leaves are noticeably grayish beneath creating a nice contrast all through the growing season.

Salix pedicellaris (bog willow) is native from North Dakota southeastward through the northeastern half of Iowa, northern Illinois, northern Indiana, and northern Ohio and points north. It is often found in fens and bogs, grows 3–4 feet tall, and has cute little leaves with rounded tips. The fine-textured leaves have untoothed (entire) edges and are only about 1½ inches long.

Salix serissima (autumn willow) is native from eastern North Dakota to southeastern Minnesota, northeastern Illinois, northeastern Indiana, and northeastern Ohio and points north. It grows in fens and bogs and though it may be a large shrub or small tree, it's usually a small shrub in the Midwest, less than 6 feet tall. Its seed capsule fruits do not open until autumn, but then show cottony tufts in the late fall landscape.

Spiraea alba
Meadowsweet spirea

Rosaceae (rose family)

HOW TO GROW Meadowsweet spirea is native mainly across the Upper Midwest from the eastern Dakotas southward across Iowa to northern Missouri and eastward across northern and central Illinois, Indiana, and Ohio. It grows in moist to wet soils mainly in mesic to wet prairies of higher pH and also grows well in good garden soil. It flowers best in full sun but also grows well in partial shade. It blooms on new wood and so it can be treated like an herbaceous perennial. Meadowsweet spirea spreads by underground rhizomes to form a thicket and is readily propagated by division.

Spiraea alba (meadowsweet spirea) creates a mass display in front of the Peggy Notebaert Nature Museum in Chicago.

Spiraea tomentosa (hardhack spirea) blooms on the Hoosier Prairie in northwest Indiana.

LANDSCAPE USE Meadowsweet makes a fine rain garden or wet swale mass planting, especially good at controlling erosion with its extensive root system. It makes a good companion with aggressive grasses like prairie cordgrass or bluejoint and is a beautiful plant to include in a wet prairie reconstruction.

ORNAMENTAL ATTRIBUTES Pyramidal cones of white, frothy flowers appear in midsummer. These ripen into fruit clusters comprised of seed capsules that become rosy red and ripen brown, lasting through the winter.

Spiraea tomentosa
Hardhack spirea, steeplebush

Rosaceae (rose family)

HOW TO GROW Hardhack spirea is native in scattered locals across the Eastern Midwest, westward across northern Illinois and Wisconsin to central Minnesota. It is found in wet prairies, shrub swamps, and open forest swamp habitats where it grows in moist to wet soils that are usually sandy based and more acidic with a lower pH. Hardhack spirea will grow in good garden soil but can suffer chlorotic yellowed foliage in alkaline soils. It blooms best in full sun but will grow in partial shade. Because it blooms on new wood like meadowsweet, it can be treated like a perennial. It also spreads by underground rhizomes to form a thicket. The term "hardhack" refers to plows cutting through its hard-to-hack rhizomes.

LANDSCAPE USE Hardhack spirea makes a good thicket-forming shrub for a rain or sandy wet swale mass planting.

Symphoricarpos albus (common snowberry) always captures attention with its white fruits in the winter landscape.

With its hardhack extensive roots, it controls erosion exceedingly well. It works well in a sandy wet prairie or wetland restoration with lower pH soils.

ORNAMENTAL ATTRIBUTES The flower clusters are very similar in shape and form to meadowsweet, but are rosy pink. The fruit capsules mature brown and adorn the winter landscape.

Symphoricarpos albus
Common snowberry

Caprifoliaceae (honeysuckle family)

HOW TO GROW Common snowberry is native sparingly across the Upper Midwest in mesic woodlands from North Dakota eastward to northeast Iowa, northern Illinois, and Ohio. It grows to about 3 feet tall and spreads lightly into a clump. It should be cultivated in moist, well-drained soils in full sun to shade. It's not a good choice for warm areas of AHS heat zone 7 of the western Lower Midwest. It is difficult to grow from seed, but it is easy to divide out plants from a clump.

LANDSCAPE USE Common snowberry makes an appealing shrub mass for a moist, shady site and works well in mossy, shady sites where few other shrubs will grow.

ORNAMENTAL ATTRIBUTES The ½-inch round white berries that ripen in late summer and remain on the plant into winter are simply stunning.

Symphoricarpos orbiculatus
Indiancurrant coralberry, "buckbrush"

Caprifoliaceae (honeysuckle family)

HOW TO GROW Indiancurrant coralberry is native mainly across the Lower Midwest and sparingly northward, naturalizing well north of its native range into Minnesota, Wisconsin, Michigan, and Ontario. It's common in disturbed dry upland woods, successional areas, and hedgerows. It's hardy throughout the Midwest and grows in moist to dry, well-drained soils in full sun to shade. It fruits heavier in full sun. It's easiest to propagate from dividing out plants from a shrub mass.

LANDSCAPE USE This shrub grows about 4 feet tall but spreads widely by ground-hugging stolons to create a thicket—usable as a tall groundcover. It works well for bank stabilization or as a shrub mass under ornamental and shade trees.

ORNAMENTAL ATTRIBUTES The flowers in mid to late summer are not showy but are visited by bumblebees and produce

Symphoricarpos occidentalis (western snowberry) flowers stand out amid prairie grasses at Blue Mound State Park, Minnesota.

Symphoricarpos orbiculatus (Indiancurrant coralberry) stands out in fall with an abundance of colorful fruits.

colorful reddish to pinkish-rose berries that ripen in fall and linger on the plant after leaf drop.

RELATED PLANTS Western snowberry (*Symphoricarpos occidentalis*) is similar to Indiancurrant coralberry but with showier pink budded, white flowers followed by white berries. It's found mainly across the Upper Midwest in prairies and open habitat.

Blueberries are beautiful on *Vaccinium angustifolium* (lowbush blueberry) and delicious to eat.

Vaccinium spp.
Blueberries
Ericaceae (heath family)

Blueberries are currently extremely popular shrubs as their fruit is well known for its health benefits and because the edible landscape movement is in full swing. Virtually all plants cultivated for fruit production or landscape use are select cultivars of highbush or lowbush blueberries and their hybrids.

HOW TO GROW Blueberries are native across two-thirds of the Midwest but limited to sandy, igneous or cherty acidic soils often on the edges of wetlands or in oak-hickory and pine-oak savannas and barrens. If blueberries are cultivated in the Corn Belt or other rich prairie soils, the soil must be amended with acidic organic matter, raised for better drainage, and treated with soil acidifiers. Blueberries flower and fruit best in full sun but also do well in partial shade, surviving with little flowering and fruiting in dense shade. They require two different plants or cultivars for cross-pollination and good fruit set.

LANDSCAPE USE Blueberries make great ornamental and edible landscape shrubs, often grown in raised beds to easily prepare their required soil and for easy picking.

ORNAMENTAL ATTRIBUTES Blueberries are highly ornamental with white pendant urn-shaped flowers in midspring, colorful delicious blue fruit in summer, and gorgeous fall colors that are mainly red but can be a blend from yellow to orange and red to burgundy purplish red.

Vaccinium angustifolium (lowbush blueberry) produces an abundance of white, bell-shaped flowers.

Vaccinium angustifolium (lowbush blueberry) is native across the Upper Midwest from central Minnesota and northeast Iowa eastward to northern Ohio. It usually grows just 18 inches tall and spreads slowly into a thicket. Its fruits are often marketed as wild blueberries as the species is seldom cultivated commercially for fruit production, though wild stands are harvested. Its smaller blueberries are considered more flavorful than most highbush blueberries. There are many hybrids between highbush and lowbush blueberries made for improved hardiness, larger berries, and more productive fruiting. Most of the hybrid cultivars are from Minnesota and have "north" or northern places in their names. These highbush-lowbush or "half-high" hybrids have proven garden worthy southward into the Kansas City region.

Vaccinium corymbosum (highbush blueberry) is native from relict stands in central Wisconsin and northern Illinois eastward across Lower Michigan and northern Indiana to northeastern Ohio and southward on the Appalachian escarpment to the eastern edge of the Bluegrass Region. It is

Vaccinium corymbosum (highbush blueberry) can have as beautiful fall color as any ornamental shrub.

226

Viburnum acerifolium (maple-leaved viburnum) is a standout in fall attire at Warren Woods State Park, Michigan.

227

Vaccinium pallidum (dryland blueberry) often shows a unique pinkish fall color.

Vaccinium stamineum (deerberry) flowers are starrier than the bell-shaped flowers of the other blueberries.

a nonsuckering shrub that usually grows 4–6 feet tall in the Midwest but it can become a large shrub 8 feet tall and over 10 feet wide. Many cultivars of highbush blueberry have been cloned for their superior fruiting traits and widely cultivated for fruit production, especially in western Michigan. Their fruits are the typical blueberries found in grocery stores.

Vaccinium pallidum (dryland blueberry) is found throughout the Ozarks and Appalachian Highlands and sparingly in between, northward into northern Illinois and Lower Michigan. Its small fruit is less desirable as a quality edible berry and is often colloquially called huckleberry.

Vaccinium stamineum (deerberry) is native in the Ozark Highlands eastward to southern Indiana's Shawnee Hills and in southern and eastern Ohio into Ontario. It is rarely found in nurseries and usually grows only 3 feet tall in the Midwest though it's a large shrub in the American South. The pendant white flowers open up more starlike so are quite showy though the berries are yellowish to purplish and not palatable.

Viburnum acerifolium
Maple-leaved viburnum

Adoxaceae (moschatel family)

HOW TO GROW Maple-leaved viburnum is found across most of the Eastern Midwest and westward across northeastern Illinois and most of Wisconsin, recently discovered in

the northeastern corner of Iowa. It's found in mesic forests, usually in acidic soils, and is very shade tolerant. It will
grow in moist garden soils in partial shade to full shade.
Maple-leaved viburnum is very challenging to propagate
and is why this shrub is so rarely encountered in cultivation.
LANDSCAPE USE Maple-leaved viburnum grows just 4 to
rarely 6 feet tall and so is a prime low shrub for a shady site.
It's a good choice for a densely shaded site where few other
shrubs will grow well.

ORNAMENTAL ATTRIBUTES Maple-leaved viburnum is aptly
named for its three-lobed (maplelike) leaves and is a real
standout because of its normally pink fall color like no other
shrub—though fall color may vary from yellowish to wine
purple. The flowers are flat clusters of cute but tiny white
blossoms. The beadlike fruits change from green to red,
then ripen blue-black set upon contrasting reddish stems.

Viburnum dentatum (southern arrowwood) produces an abundance of
blue fruit in late summer.

Viburnum dentatum
Southern arrowwood

Adoxaceae (moschatel family)

There are four species of viburnums known as "arrowwoods" with rounded, toothed (dentate) leaves that are
very similar. These can be used quite interchangeably in the
landscape. They get their name from the long and straight
shoots that sprout from the crown of a mature shrub.

HOW TO GROW Southern arrowwood is native from northern Arkansas, northeastern Missouri, across Illinois, central and southern Indiana to Ohio and points east and south.
Its leaves have 10–20 pairs of teeth with undersides covered in fine, downy hairs. Native to forested slopes, forest
edges, and woodland openings, this species grows well in
almost any moist, well-drained soil in full sun to light shade
but flowers and fruits best in full sun. It may be propagated
from cuttings or by a tedious long process from seed.

LANDSCAPE USE Southern arrowwood viburnum is often a
medium-sized shrub in the 5- to 8-foot-tall range, but occasionally can be a large shrub. It makes perfect informal
hedges, shrub masses at the back of a perennial border, and
forms great wildlife habitat and bird garden shrubs. The
stinky flowers attract many flies, beetles, and some bees
and butterflies. The round fruits ripen in late summer and
are a feast for migrant birds including the eastern kingbird,
which mainly eats insects before migration.

ORNAMENTAL ATTRIBUTES In late spring, malodorous flat
clusters of white flowers with a haze of yellow anthers create an overall creamy look. The foliage is clean and often
neatly pleated, turning rich shades from burnt orange to
reds and even burgundy tones in fall—sometimes pale

Viburnum dentatum (southern arrowwood) can produce distinctive pale yellow to pink fall color.

Viburnum rafinesquianum (Rafinesque's viburnum) has wonderfully pearly white flower buds.

yellow to pinkish yellow in heavy shade. The berries ripen blue in late summer but seldom persist after leaf drop.

RELATED PLANTS Many cultivars of southern arrowwood are available at most nurseries, whereas locally native forms of the species may be quite difficult to procure.

Kentucky or softleaf viburnum (*Viburnum molle*) is native from the Ozark Highlands northward to southeastern Iowa and western Illinois, and eastward to Indiana, Ohio, and across Kentucky into Tennessee. Its leaves are more rounded and heart-shaped with 20 pairs of teeth compared to other arrowwoods, and the flowers are similar, but the fruits are more flattened, also ripening blue-black. Its gray stems exfoliate revealing brown inner bark, making it unique in the winter landscape. Cultural requirements, propagation methods, and landscape use are identical to that of the other arrowwoods.

Smooth or northern arrowwood (*Viburnum recognitum*) is native mainly in northeastern North America but is found as far west as Wisconsin and the Ozark Highlands. Its leaves are similar to southern arrowwood but with smooth undersides. Otherwise, it shares the same habitat as southern arrowwood and has the same uses in the landscape.

Viburnum rafinesquianum
Rafinesque's viburnum, downy arrowwood

Adoxaceae (moschatel family)

HOW TO GROW Rafinesque's viburnum is native across much of the Upper Midwest and southward into the Ozark Highlands of Missouri, Arkansas, and Oklahoma. It grows in moist to dry upland forests, often on slopes and cliffs. It requires moist garden soils in full sun to light shade. Propagate Rafinesque's viburnum by cuttings.

LANDSCAPE USE This shrub usually grows about 5 feet tall, occasionally taller when reaching for light. It makes a lovely but subtle shrub for a natural woodland or wildlife-themed garden.

ORNAMENTAL ATTRIBUTES The flat clusters of pearly white buds open to tiny, darling white flowers, larger and not as malodorous as the other arrowwood viburnums. The leaves are toothed, usually with around 10 pairs of teeth, and they turn dull red to burgundy shades in the fall. The fruits are larger and flattened, ripening dark bluish black, and seldom persist after leaf drop.

Vines

Vines are an important component of a healthy, functional, and beautiful landscape. This will make arborists and foresters shudder as uncontrolled vines can smother young trees. Many native vines certainly can be rampant so I will overuse "garden thug" in this chapter, as that's a well-known gardener's term. With careful placement and diligent pruning, however, vines can be adequately domesticated. The rewards are great including exceptional seasonal fragrance and color, a plethora of beneficial insects including butterflies and hummingbird-like sphinx moths, improved birdlife in the garden, and cloaks of green plants in locations that would otherwise be impossible.

Vines come in three groups:

WOODY VINES: those with woody stems that can become quite massive

PERENNIAL VINES: those that are perennial and die back to the ground each winter but return the following spring

ANNUAL VINES: those that are annual and die with a fall freeze, only to return by reseeding the following growing season.

Two significant groups of vines in the Midwest are grapes and clematis. Wild grapes (*Vitis*), raccoon grape (*Ampelopsis*), and the creepers (*Parthenocissus*) are related vines in the grape family (Vitaceae). They comprise the largest component and biomass of midwestern native vines. The grape vine group is underutilized for their shading capabilities on outdoor structures, homes, and businesses reducing cooling costs and saving energy. They are host to more species of insects (their foliage also hosts more species of large sphinx moths than any other group of plants) and provide

sustenance for wildlife more than any other group of vines.

True (wild) grapes are edible and have been grown by humans for food and drink since times unmemorable. Viticulture has become an important agricultural practice across the Midwest. Though most cultivated grapes originated in Europe, the most productive grapes in the Midwest are hybrids with native species that are better adapted to our continental climate. Wild grapes have a promising future, as does the whole grape family.

Clematises belong to the buttercup family (Ranunculaceae) and are beloved vines grown worldwide for their beautiful flowers and fruit. Some are woody but most are nearly herbaceous and dieback to near ground level each winter. All clematises are extensively hybridized around the world for larger and more colorful flowers

One little-discussed value of native vines is that if you inherit a nonnative tree in a landscape, you can plant appropriate vine species to climb up it and enhance its value to nature around you. Try planting a Virginia creeper (very

The colorful fruit of *Ampelopsis cordata* (raccoon-grape) is usually hidden among the late-summer foliage.

VINES FOR A TRADITIONAL LANDSCAPE
Celastrus scandens Autumn Revolution ('Bailumn')
    (American bittersweet)
Clematis crispa (curly leather flower)
Clematis versicolor (pale leather flower)
Dioscorea villosa (wild yam)
Lonicera flava (yellow honeysuckle)
Lonicera reticulata (rock honeysuckle)
Parthenocissus inserta (grape creeper)
Parthenocissus quinquefolia (Virginia creeper)
Passiflora lutea (yellow passionvine)
Smilax lasioneura (carrionvine)
Wisteria frutescens (Kentucky wisteria)

high value) on a ginkgo or zelkova (trees of almost no value). Some nonnative trees like Norway maples have too dense a shade while London plane trees will slough off vines as they shed their bark.

Ampelopsis cordata
Raccoon-grape

Vitaceae (grape family)
Landscape group: woody vines

This relative of true grapes and invasive exotic porcelain berry (*Ampelopsis brevipedunculata*) is never described as a landscape plant. It is just as rambunctious as grapes, growing huge and engulfing everything in its path as well as self-sowing abundantly in disturbed soils. In harsh sites of poor and droughty soils, it can be quite a spectacular vine where given room.

HOW TO GROW The vine is common across the Lower Midwest in floodplain forests as far north as Omaha along the Missouri River and southeast Iowa along the Mississippi River. It also thrives along hedgerows and all successional habitats. It's hardy across the Lower Midwest and into USDA zone 5. Raccoon-grape is tolerant of a wide range of soils and easily grown from seed. It is never available from nurseries.

LANDSCAPE USE It should be considered for green infrastructure uses of naturally cooling buildings as it will grow in urban soils and thrive in heat and drought. In places where it is native, it should be kept, as it is such a phenomenal plant for insects and wildlife.

ORNAMENTAL ATTRIBUTES The trunks become large and beautiful, corrugated by thick bark. The flowers are little noticed but visited by a wealth of small insect pollinators. The fruits (inedible) in early fall are a gorgeous range of colors as they ripen, from fuchsia pink to aqua and navy blue and relished by songbirds and raccoons. Fall color is a consistent and sunny yellow.

NOTES I have read so many accounts that advise to avoid this plant. I inherited it in my past and existing landscapes where I grew to love the plant growing on the edges of the garden (on droughty, upland soils). It is most vigorous and invasive in deep, rich soils of our major river floodplains.

The flowers of *Apios americana* (groundnut) are not pendant, though otherwise the plant looks remarkably like a wisteria.

Apios americana
Groundnut

Fabaceae (legume family)
Landscape group: perennial vines

HOW TO GROW Groundnut is native across the entire Midwest and virtually all of eastern North America. It requires full sun or partial shade in rich, moist soils. Groundnut vines grow 8 to rarely 12 feet.

LANDSCAPE USE This is a rambunctious vine for a naturalistic setting (too invasive for a traditional garden) suitable as a soil-stabilizing plant on an embankment. It can safely weave through established prairie plantings and not smother them. It's naturalized on a pedestrian bridge abutment at Powell Gardens where it climbs up into the railings, less prolifically than a wisteria would. The tubers and beans are edible so this is a good addition to a food forest.

ORNAMENTAL ATTRIBUTES The lush foliage looks almost identical to a wisteria but with fewer leaflets. The late-summer flowers are very unique and ornamental in a short cluster; individual pealike flowers are a two-toned pale pink and rusty burgundy color.

Campsis radicans
Trumpetvine

Bignoniaceae (bignonia family)
Landscape group: woody vines

A notorious garden thug, trumpetvine can be spectacular when grown in the right place where its suckering habits and rampant growth can be readily maintained. It's in the same family as catalpa trees.

HOW TO GROW Trumpetvine's native haunts are floodplain forests in the Lower Midwest, but it has naturalized northward. It climbs trees by rootlets that form along its stems and trunk. Trumpetvine roots everywhere the stems touch the ground and spreads by underground rhizomes as well. It's adaptable to all soil types except those that are excessively dry and is hardy to -25°F so needs a sheltered site in colder parts of the Upper Midwest. It requires full or part sun and will be weak and not flower in dense shade.

LANDSCAPE USE Plants are slow to establish but don't be fooled, the old adage sleep, creep, and leap is a sure thing with this vine so beware where you plant it. Planting it against a home is a very bad choice as it can ruin siding and sprout up all around the foundation. I have seen it spectacular in a garden, planted where it will grow up a tree, pole, or post, and where hardscape or a lawnmower can keep the suckers at bay. It can sometimes be trained to look almost like a small tropical tree when grown on a low post. The plant is considered invasive in locations beyond its native haunts.

Campsis radicans (trumpetvine) can look like a small tree when trained on a garden post—as seen here at Reiman Gardens in Ames, Iowa.

The winter fruit of *Celastrus scandens* (American bittersweet) holds its color well into winter.

ORNAMENTAL ATTRIBUTES The large, glorious flowers bloom over a long period through the intense summer sunshine and are usually orange but vary from yellow (on rare occasion) to almost red. They're especially brilliant when backlit by sunshine. Hummingbirds pollinate them by day and sphinx moths by night, producing pendant woody capsules that split down the middle and add a bit of décor to the plant in winter. The foliage is lush and almost tropicalesque but can cause a skin rash in sensitive people. It's host to the plebian sphinx moth.

RELATED PLANTS Atomic Red ('Stromboli') trumpetvine is a dark red budded, orange-red flowering cultivar hardy in USDA zone 4. The yellow-flowering cultivar 'Flava' is less invasive and rarely spreads by suckers.

Celastrus scandens
American bittersweet

Celastraceae (staff tree family)
Landscape group: woody vines

This vine is prized for its orange-hulled, vermillion fruit that brighten up the midwestern landscape as it becomes bare in autumn. The plant would be more popular in gardens were it not dioecious: vines are either male or female. Bittersweet is in the same family as wahoo.

HOW TO GROW Bittersweet is found in dry woods and woodland edges and hedgerows throughout the Midwest. It is easy to grow in almost any well-drained soil in full sun to part shade. Plant several seedlings together to ensure a male and female vine for fruiting. The vine can be slow growing after transplanting but is extremely drought tolerant once established.

LANDSCAPE USE American bittersweet is best grown on a fence or among established shrubs or small trees in a woodland edge. It rarely grows more than 30 feet and only when it is trying to reach light. It can add ornamental appeal to an established small tree, growing up into the crown and flowering (small, green and little noticed) and fruiting, showing its great color after the host tree drops its leaves. At Powell Gardens, a female vine self-sowed in exotic *Heptacodium miconioides* (seven sons) and it is a stunning composition that I never would have dreamed up. I have read that this vine is an invasive self-sower in gardens but that has not been the case from my own experience and could be based on the invasive exotic Oriental bittersweet (*Celastrus orbiculatus*). Yes, it occasionally self-sows but usually takes time to establish.

ORNAMENTAL ATTRIBUTES The best display is the fruit on female vines, and a male must be present to pollinate the flowers. The fall color is also consistently an outstanding light yellow.

RELATED PLANTS A new selection Autumn Revolution ('Bailumn') from Minnesota has mainly perfect flowers (both male and female parts) so is self-fertilizing (termed self-fruitful by fruit growers).

Clematis—Herbaceous Vines
Leather flowers
Ranunculaceae (buttercup family)
Landscape group: perennial vines

Four native clematises are essentially herbaceous, woody only at the base, so naturally die back to or near the ground every winter. They also climb by twining their leaf petioles around support but are not nearly as invasive as virgin's bower, a woody clematis vine.

HOW TO GROW Herbaceous clematises are slow to establish but expect them to grow about 6 feet in a single season, thriving in a variety of well-drained soils from moist to slightly dry, flowering best in full sun but tolerant of light shade. They are a challenge to grow from seed and difficult to root from cuttings but occasionally self-sow where conditions are ideal. Because they are nearly herbaceous, the vines can be cut to the ground in late winter or early spring and they will bloom on new wood. All are hardy throughout the Midwest and can cross-pollinate between species in a garden setting, creating your own hybrids. Some plants in cultivation appear to be of hybrid origin.

LANDSCAPE USE These clematises are much more garden friendly than virgin's bower and can be planted on trellises at the back of a border or against a home or arbor. They can be allowed to ramble over or through larger perennials and shrubs without smothering them.

ORNAMENTAL ATTRIBUTES The flowers are hanging bells comprised of four leathery sepals with self-fertile perfect flowers, so a single plant will produce the lovely fruiting heads after blooming. All are long blooming from midsummer until fall with peak bloom in late summer and often sporting flowers and seed heads at the same time. Bumblebees pollinate them.

Clematis crispa (curly leather flower, swamp leather flower, blue jasmine) has native haunts confined to mesic and mesic floodplain woodlands and edges in the lowestmost Mississippi Valley of the Midwest. It's a long-blooming species flowering from midsummer into fall even in the Upper Midwest with peak bloom in late summer. The flowers have the most dramatically curled "crisped" edged sepals of any leather flower and vary from dark to light blue-purple (flower color appears darker in full sun). This species is not as tolerant of dry conditions as the other leather flowers.

Clematis pitcheri (pitcher's leather flower) is found in woodland edges and open woods and is more widespread across the Lower Midwest from southern Indiana to southeastern Nebraska. Its flowers vary from uniformly steely blue-purple to bicolored blue-purple with creamier sepal ends but always looking like upside-down urn. My new vine

Clematis crispa (curly leather flower) has dramatically curled sepals.

Clematis pitcheri (pitcher leather flower) rambles compatibly through perennial plantings.

Clematis viorna (vase leather flower) blooms lightly, but over an extended period of time.

The silky plumed fruit of *Clematis versicolor* (pale leather flower) captures light when backlit.

Clematis occidentalis (purple clematis) blooms along a trail in Van Peenen Park, Decorah, Iowa.

from a Missouri source taught me this, and then I looked on the website of Lady Bird Johnson Wildflower Center, which includes images that show this variation. Although the flowers are not showy from afar, they are beautiful on close inspection. The fruits are less showy than those of other herbaceous clematis, with shorter silkless-awned achenes. This species is the most vigorous of our leather flowers growing as much as 10 feet.

Clematis versicolor (pale leather flower) is found in woodland and woodland edges from Kentucky westward to the lower Ozarks of Missouri. It's more tolerant of dry conditions as it often grows on rocky glades. Probably the most ornamental of the leather flowers, it always grabs attention while in full bloom in mid to late summer and while fruiting and blooming in late summer into fall. Flowers are rosy purple to rosy lavender fading to creamy greenish on the recurved tips of the sepals for a nice two-toned look. The fruits are the showiest of the herbaceous clematis displaying a luminous bouffant of silky-awned achenes

Clematis viorna (vase leather flower) is found on rocky woodland edges across the Lower Midwest from Ohio to southern Missouri. This clematis produces dramatic individual fruit, forming showy, silky plumes. The sepals of the flower are thick and ribbed, rosy purple to lavender on the outside curling back at the tips to reveal their creamy, woolly edged inside.

Clematis occidentalis var. *occidentalis*
Purple clematis

Syn. *Clematis verticillaris*
Ranunculaceae (buttercup family)
Landscape group: woody vines

HOW TO GROW This rare woody clematis grows wild in the Midwest solely in the Driftless Area of Minnesota, Wisconsin, Iowa, and Illinois where it is confined to cool, moist talus slopes and rocky woods. In the cooler Upper Midwest, plant it in a location that is not exposed to the hot southern or western sun. Purple clematis blooms on second-year stems, with new growth from near the base each season that will flower the following season. Prune out stems that have flowered in late winter.
LANDSCAPE USE It makes a fine small vine for a rock garden, trellis, or to ramble through spring wildflowers and shrubs.
ORNAMENTAL ATTRIBUTES The 3-inch, four-petaled flowers are produced in midspring and are a lovely blue-purple. The flowers are self-fertilizing and are followed through summer by showy silky seed heads similar to virgin's bower.

Blooming *Clematis virginiana* (virgin's bower) cloaks a large decorative post at the Minnesota Landscape Arboretum.

Clematis virginiana
Virgin's bower

Ranunculaceae (buttercup family)
Landscape group: woody vines

Don't confuse native virgin's bower with the highly invasive and nonnative sweet autumn clematis (*Clematis terniflora*), which has similar and wonderfully fragrant flowers but threatens natural communities being even more invasive and growing larger than our native species.

HOW TO GROW Virgin's bower is common in second-growth and edge habitats across the entire Midwest except for the extreme west and northwest and hardy in all zones. It's an easy-to-cultivate and rambunctious woody vine that readily grows 8–15 feet and can be quite invasive in a garden setting. It is tolerant of a wide range of well-drained soils from moist to dry and in full sun to light or partial shade. Plants are dioecious (either male or female) and climb by winding their leaf petioles around supports. Virgin's bower self-sows readily and can root wherever the stems touch the ground. It is easily grown from seed and rooted divisions. You may prune back to stem buds in early spring, but cutting this vine to the ground will sacrifice flowers.

LANDSCAPE USE This vine is best grown in natural landscapes along the edge of woodlands or along hedgerows where it creates marvelous habitat. It also can cover a large arbor where it can be controlled around the edges. Virgin's bower can be used as a groundcover on steep embankments.

ORNAMENTAL ATTRIBUTES It is striking during its midsummer bloom as a blanket of white flowers (male flowers more showy than female) with a corresponding cloud of little pollinating insects. Female vines produce beautiful silky fruiting of plumed achenes that hold into early winter. The autumn seed heads are the greatest ornamental assets of the plant.

Cucurbita foetidissima
Missouri gourd, buffalo gourd

Cucurbitaceae (cucumber family)
Landscape group: perennial vines

HOW TO GROW Missouri gourd is a southwestern native plant that has naturalized north and eastward into Iowa, Kentucky, Michigan, and Ohio. It's probably native in the Flint Hills and Osage Plains and is usually found in disturbed soils.

LANDSCAPE USE This is a very unique vine for a dry site with poor soil where it could be allowed to sprawl as a groundcover or climb a fence or arbor. The seeds are listed as edible if boiled or roasted and the flowers also are listed as edible though bitter.

ORNAMENTAL ATTRIBUTES The foliage is coarse and gray-green while the orange-yellow flowers are large and showy (similar to a squash blossom) and are either male or female. The fruit look like mini watermelons and are quite beautiful in fall.

Dioscorea villosa
Wild yam

Dioscoreaceae (yam family)
Landscape group: perennial vines

HOW TO GROW Wild yam is dioecious vine native across the Eastern and Lower Midwest and in the Upper Midwest as far west as central Minnesota and southeastern Nebraska. It's found in moist woods, including mesic floodplains, and in hedgerows. It will grow in full sun to part or light shade, and is very drought tolerant but prefers moist, well-drained soils. Plants can be grown from seed or division of tubers.

LANDSCAPE USE This elegant vine can ramble through established plantings or weave through a fence or trellis. It grows at least 6 and as much as 15 feet on an arbor. Note that

Cucurbita foetidissima (Missouri gourd) produces stunningly golden flowers partly shaded by gray-green leaves.

The elegant leaves of *Dioscorea villosa* (wild yam) make it a great foliage plant from spring through fall.

The flowers of *Echinocystis lobata* (wild cucumber) are held upright above the plant.

although wild yam is a related to cultivated yams, the wild yam tubers are *not* edible.

ORNAMENTAL ATTRIBUTES This vine has beautiful foliage through the growing season with heart-shaped leaves with neatly precise veining. The leaves turn a golden yellow in autumn. Flowers are showiest on male vines in creamy, frothy clusters. The winged seeds of female vines are light green in summer and ripen brown, hanging delightfully into winter.

Echinocystis lobata
Wild cucumber

> Cucurbitaceae (cucumber family)
> Landscape group: annual vines

HOW TO GROW Wild cucumber is native across the Upper Midwest and sparingly southward in all midwestern states. It is readily grown from seed, prefers rich, moist soil, and can grow a whopping 15 feet in ideal conditions. I have been unsuccessful in growing it in my droughty clay soil. It flowers most heavily in full sun and will not grow in dense shade.

LANDSCAPE USE This is an attractive annual for a trellis or natural garden where it can scramble over existing shrubs. The fruits are not edible.

ORNAMENTAL ATTRIBUTES The summer flowers are greenish or yellowish white and male flowers quite showy when produced in abundance. The foliage is lobed and interesting and the fall fruit look like a prickly 2-inch gherkin—ornamental as it dries and lingers on the plant into winter.

Humulus lupulus
Native hops

> Urticaceae (nettle family)
> Landscape group: perennial vines

HOW TO GROW Hops are found native across almost the entire Midwest and most of the Northern Hemisphere. It's hard to judge native varieties (var. *lupuloides* and var. *pubescens*) from escaped cultivated forms. Hops grow easily from seed and cuttings.

LANDSCAPE USE This overly rampant vine is not suitable for a traditional landscape, and the true native varieties do not produce the desirable flavored bitters traditionally used in brewing beer. They're being used in breeding traditional hop cultivars more suitable for cultivation in the Midwest and perhaps a specialty microbrewery may find their flavor suitable. Shoots are edible and tasty too when properly

Humulus lupulus (native hops) produces long spears of growth that make a delectable vegetable.

Ipomoea pandurata (potato vine) produces large, white flowers like morning glory.

prepared. Native hops can be used for a sturdy groundcover along steep banks and may be left where wild.

ORNAMENTAL ATTRIBUTES The female vines produce the unique hop fruit that are actually quite showy in late summer as they turn light lime-green, and then mature dry brown and hang into winter.

Ipomoea pandurata
Potato vine

> Convolvulaceae (morning glory family)
> Landscape group: perennial vines

HOW TO GROW Potato vine is native across the Lower Midwest and northward into southern Michigan, southeastern Iowa, and southeastern Nebraska. It grows in full sun to part shade in well-drained soils, usually found wild in edge habitats. It can be grown from seed.

LANDSCAPE USE This perennial morning glory relative is best suited to a natural landscape or food forest. The vine grows 20 feet or more but is never smothering except in rich, moist soils, often sprawling along the ground or through larger perennials and shrubs. Its tuberous root can become huge giving it the alternate name "man-of-the-earth." Native Americans utilized the potato vine's root as a food source.

ORNAMENTAL ATTRIBUTES The late-summer flowers are large and showy white with a purplish center—visited by butterflies. The foliage is host to several tortoise beetles and the marvelous pink-spotted hawkmoth.

Isotrema tomentosa
Woolly pipevine

> Syn. *Aristolochia tomentosa*
> Aristolochiaceae (pipevine family)
> Landscape group: woody vines

My love for butterflies introduced me to this vine and led me to plant it in my garden. It is one of only two species of native midwestern plants that host the beautifully spectacular pipevine swallowtail.

HOW TO GROW Native pipevine is found across the Lower Midwest in low woods along streams and rivers but is cold hardy through the entire Midwest (USDA zone 4). It's easy to grow in any woodland or organic soil, readily transplants, and is easy to propagate from root divisions or grown from seed.

LANDSCAPE USE This vine is really invasive, so be sure you plant it where you can embrace its large size and suckering root sprouts. It is nigh impossible to remove once established, but is a phenomenal screening plant and covers many a large porch. The eastern native big-leaf pipevine (*Aristolochia macrophylla*) was the preferred species on Victorian landscapes and does not sucker and is widely cultivated across the Midwest.

ORNAMENTAL ATTRIBUTES The large, heart-shaped leaves are aptly named with a coating of woolly hairs that make this plant far more heat and drought tolerant than the eastern species. The summertime flowers are very unique and quite a conversation piece. If you plant two different-sourced

The flower of *Isotrema tomentosa* (woolly pipevine) looks quite bizarre.

The berries and foliage of *Lonicera reticulata* (rock honeysuckle) are a real standout in early autumn.

plants, you can improve pollination and enjoy the simply beautiful podlike capsule. These mature in autumn and dry out into black 12-sided capsules full of flaky seeds.

NOTES Note that if you plant it, they will come—pipevine swallowtails, that is. Female butterflies will lay groups of striking orange-brown eggs on fresh growth. Eggs hatch into little armies of toxic black, red-tipped caterpillars, which devour the new growth. The caterpillars can keep the plant's growth in check, often completely defoliating young plants.

Lonicera spp.
Honeysuckle vines

Caprifoliaceae (honeysuckle family)
Landscape group: woody vines

The invasive exotic Japanese honeysuckle (*Lonicera japonica*) vine along with the exotic bush honeysuckles have given our native honeysuckles in the genus *Lonicera* a challenging stigma to break through. None of the native species behave in such a smothering way and they fit nicely into any garden. Native honeysuckles climb by twining around their support. They're much more ornamental plants when cultivated and spared wild competition. Our honeysuckles' best show is their fruit in fall. Unbeknownst to most gardeners, they are not self-fertilizing so two noncloned plants or two different clones of these species are needed to produce fruit. Bluebirds in my garden relish these berries. There are only two species, rock and glaucous honeysuckle, found across nearly the entire Midwest in rocky woodlands though they can be quite local in some areas.

Lonicera reticulata
Rock honeysuckle, grape honeysuckle

Syn. *Lonicera prolifera*
Caprifoliaceae (honeysuckle family)
Landscape group: woody vines

I learned this plant a long time ago and admired its clusters of golden flowers set in a bowl of its leaves. Leaves are paired along the stem but near the end of stems they are fused into what looks like a single round leaf that the stem appears to pierce (called perfoliate). These perfoliate leaves can be exquisitely glaucous (waxy coated) as to look silvered or bluish. Flowers and fruit set on the end perfoliate leaves.

HOW TO GROW This vine is easy to grow in about any well-drained soil and is extremely drought tolerant. It does not overtake its surroundings or self-sow.

LANDSCAPE USE Rock honeysuckle is an ideal vine to climb up a small trellis or garden sculpture. It can even be planted without support where it can take on a haystacklike growth making you think it is a shrub. The vine usually grows 4 feet tall in the wild, but will grow 6 feet or more in cultivation.

ORNAMENTAL ATTRIBUTES The plant becomes a fountain of branches of the unique perfoliate leaves "a shish kabob" of 3-inch round leaves looking like a eucalyptus. Each end branch is studded with creamy yellow flowers that age to orange in late spring, followed by orange-red berries in fall. Why this great plant is not better known is puzzling. The flowers are nectar rich and visited by many pollinators from hummingbirds to various bees. Birds relish the fruit and the foliage hosts several unique caterpillars, most noticeably

241

Lonicera dioica (glaucous honeysuckle) is a delicate vine with deep, almost maroon-colored flowers.

Lonicera flava (yellow honeysuckle) has lightly fragrant golden blooms.

Lonicera sempervirens (trumpet honeysuckle) flowers are a classic design to be pollinated by hummingbirds.

the hornworm of the bumblebee clearwing, a day-flying sphinx moth that mimics a bumblebee.

RELATED PLANTS There is a ghostly glaucous selection of this vine called Kintzley's Ghost from Iowa.

Glaucous honeysuckle (*Lonicera dioica*) is more diminutive and vinelike than rock honeysuckle with foliage that is more elongated so the stems look like they pierce a "bow tie" of perfoliate leaves. The flowers are set between the end perfoliate leaves and are orange, fading to maroon and followed by orange berries if there is another vine for cross-pollination.

There are several other native vine-honeysuckles that have more northern or southern affinities. The northern species are more demanding and challenging to grow while the more southern are easy but not fully hardy in the Upper Midwest. The best of the southern species is the yellow honeysuckle (*Lonicera flava*) with slightly fragrant golden-yellow flowers and a twining vine to 8 feet. Yellow honeysuckle is native in the Ozarks with a few isolated locations in Illinois, Kentucky, and Ohio.

Trumpet honeysuckle (*Lonicera sempervirens*) is a more southeastern species that has now naturalized over much of the Lower and Eastern Midwest. It has the showiest tubular flowers of brilliant scarlet (sometimes yellow) that are pollinated by hummingbirds.

Menispermum canadense
Canada moonseed

Menispermaceae (moonseed family)
Landscape group: woody vines

I learned about this widespread midwestern vine at an early age because it grew on the west trellis of my Grandma's back porch as a sunscreen. I only wish I had asked how it got there: Did it self-sow? Or was it dug from the farm's woods?

HOW TO GROW This species is found in woodlands moist to dry across the entire Midwest and is hardy throughout. It's easy to grow in any woodland or humus-rich soil. Although it is often considered invasive as it spreads by rhizomes and sprouts about, it never really overtakes anything. Once you plant it, it's hard to remove. Canada moonseed is almost never sold but easy to transplant by rhizomatous divisions.

LANDSCAPE USE This tough vine can be used as a rugged screen on a trellis or allowed to ramble through native shrubs and small trees in a woodland garden (as I inherited it as a native in my current woodland garden). It can also be used as a groundcover in rich soils.

Menispermum canadense (Canada moonseed) makes a fine summer foliage vine.

Parthenocissus quinquefolia (Virginia creeper) climbs the stone facade near the entryway of a home.

ORNAMENTAL ATTRIBUTES This plant really is a sleeper. It has sturdy maplelike leaves and subtle creamy flowers hanging like mini chandeliers, which are followed by dark blue-black berries in fall on into winter. Inside these berries are the namesake seeds.

Parthenocissus quinquefolia
Virginia creeper

> Vitaceae (grape family)
> Landscape group: woody vines

This gorgeous native vine deftly demonstrates how exotic plants are favored over natives. Its Japanese counterpart is called Boston ivy (*Parthenocissus tricuspidata*), the plant of Ivy League schools and beloved Wrigley Field in Chicago. As we learn the importance of natives and how Boston ivy is becoming invasive, it's time to embrace and grow our even more ornamental Virginia creeper, which is native across nearly the entire Midwest.

HOW TO GROW Virginia creeper is easy to cultivate in any well-drained soil in full sun or shade. The plant is available from many nurseries. It grows huge over time, readily 20 feet to over 50 feet.

LANDSCAPE USE This beautiful vine climbs by tendrils with adhesive disks at their ends that literally stick themselves to whatever they are climbing on. It makes a fine vine to cover almost anything with a curtain of living plant where no trellis is applicable. It is at its best when used like it grows in nature: up a tree trunk and into the major limbs of a shade tree. It also can be used as a groundcover on steep banks and other woodland settings.

ORNAMENTAL ATTRIBUTES This magnificent cloak of green will grow almost anywhere, and the foliage consistently

Parthenocissus inserta (grape creeper) makes a fine groundcover as seen wild at the Minnesota Zoo.

turns fiery red in early fall as the inedible blue-black fruit ripen. The early fall color may draw attention to birds who seek its ripe fruit as important fuel for migration. It is host to many of the same unique moths that are hosted by its cousin the grapes, so is one of the best plants for providing food for nesting as well as migrating birds. The flowers are little noticed but very rich in nectar for pollinators including bees and hummingbirds.

RELATED PLANTS Two selections are available: Star Showers ('Monham') has unique variegated foliage and 'Yellow Wall' has foliage that turns yellow in the fall.

Grape creeper (*Parthenocissus inserta*) is first cousin to the Virginia creeper and nearly identical but found more often north and westward. It does not grow as large and it climbs by grapelike tendrils that wrap around or cling to its support. This vine often sprawls over the ground where foliage may be slightly larger with a more abrupt and toothed end to the leaf. Another underutilized native, it's nearly impossible to find in nurseries.

The flowers of *Passiflora incarnata* (maypop passionvine) are incredibly intricate.

Passiflora lutea (yellow passionvine) has wonderful markings on its leaves, which are showier than its small flowers.

Passiflora incarnata
Maypop, passionvine
Passifloraceae (passion flower family)
Landscape group: perennial vines

HOW TO GROW The maypop is native across the Lower Midwest but is hardy northward throughout the Midwest if planted next to a warm foundation. It grows best in full sun in moist, well-drained soils. It can be grown from seed, cuttings, or divisions.

LANDSCAPE USE It's a garden thug in traditional landscapes as it can spread by underground rhizomes and come up everywhere. In the Upper Midwest winter, cold limits its spread and it can be an OK vine for a trellis against a home. It's best in a natural garden or food forest where it can ramble through established perennials and shrubs. It's a good butterfly garden plant as it's the host plant of the variegated and Gulf fritillary, the latter rarely found north of the southernmost Midwest.

ORNAMENTAL ATTRIBUTES The flowers are intricately stunning, purplish white with an amazing structure of wavy segments above the petals. The lobed foliage is lovely all season and fruit (where cross-pollinated) matures yellow and is an edible large berry.

RELATED PLANTS Yellow passionvine (*Passiflora lutea*) also is native across the Lower Midwest and has small, yellowish flowers that are not particularly showy. It produces small inky fruit that are edible though not too tasty—wild turkeys and songbirds devour them. The maple-shaped leaves are often naturally veined with lighter markings and quite showy. Yellow passionvine is not nearly as invasive as the maypop and makes an interesting foliage vine for a trellis.

Smilax hispida
Hispid greenbrier
Smilacaceae (greenbrier family)
Landscape group: woody vines

Gardeners usually consider this widespread vine another pest or garden thug but once you get to know it, it becomes a visceral component of a midwestern woodland. It hosts two of the most fabulous caterpillars in my landscape: the harvester—our only carnivorous butterfly—and the curvelined owlet—a moth whose caterpillar looks like fresh bird poop dripping in the middle. How's that for a plant to hook children on nature? This vine is very spiny and tenacious— that's how it gets its winner-of-the-botanical-death-match title and is the proverbial briar patch. It thus creates fine habitat for wildlife to find shelter and nesting.

HOW TO GROW No nursery sells this plant, but it is native in almost any hedgerow or woodland across the entire Midwest. It is extremely drought tolerant and does readily self-sow and the seedlings are a bit challenging to pull.

LANDSCAPE USE This vine is tardily deciduous and often green into November when the rest of a deciduous forest is at rest. It's best in a naturalistic setting on the edge of a woods or climbing up a woodland tree or large shrub. It grows 15–30 feet or higher, climbing by tendrils. New basal shoots on established plants grow faster than any plant I know. These tender new shoots are edible and delicious and that's what attracts the woolly aphids that the harvester caterpillars dine on.

ORNAMENTAL ATTRIBUTES Fine clusters of green flowers are borne in rounded umbels in late spring after the leaves flush. The fruit undergo an amazing transformation in late

The black berries of *Smilax* (greenbrier) hang on the vine into winter.

The fall color of *Toxicodendron radicans* (climbing poison-ivy) is usually brilliant scarlet.

summer as they change from green to polished black. No other plant has such a gorgeous, unique green-black hue. The fruit hang on the vine into winter and are relished by birds.

Smilax lasioneura
Carrionvine

> Syn. *Smilax herbacea*
> Smilacaceae (greenbrier family)
> Landscape group: perennial vines

This vine is so entertaining and colorful in the fall that it has me scratching my head why it is not grown by native plant nurseries, though it is available as seed. It does not self-sow abundantly like its woody cousins the greenbriers.

HOW TO GROW Carrionvine is found across the entire Midwest except for eastern Ohio and is hardy everywhere. It usually grows in partial shade—found wild in open woodlands, savanna, and fencerows—in well-drained, upland soils that are moist to summer dry. It can be easily grown from seed. It usually is inherited in a landscape as a wild or bird-planted vine.

LANDSCAPE USE Carrionvine is a welcome sight on the edge of a woodland or planted along a fence. It grows 6–8 feet tall, but as a single stem that climbs by tendrils.

ORNAMENTAL ATTRIBUTES The vine is entertaining as soon as its spear emerges from the soil in spring and grows as fast as any plant. It produces green clusters of flowers in late spring that mature to blue-black clusters of fruits in fall as the leaves turn yellow. Carrionvine often goes unnoticed until it's in fruit, becoming showy in fall and into the early winter landscape.

Smilax lasioneura (carrionvine) produces ornamental clusters of blue-black berries that are striking in fall and into winter.

Toxicodendron radicans
Climbing poison-ivy

> Anacardiaceae (sumac family)

Say what? Are you kidding me? I am sensitive to the contact dermatitis this plant causes, but it is one of the most important vines for wildlife.

HOW TO GROW It is not a vine you want to plant, but if it is native in a wild portion of a property, away from human contact, then an open mind should allow it to stay. You could say its removal would reduce seedlings, which are spread by bird droppings, but it is so widespread in most of its range that removal has little effect. I reuse plastic bags to remove seedlings. The flowers are pale green and little noticed.

LANDSCAPE USES The vine climbs by rootlets, growing 20 feet to occasionally well over 50 feet in height up large, floodplain trees. If nothing to climb up is available, then the vine spreads over the ground and roots everywhere.

ORNAMENTAL ATTRIBUTES Female vines produce berries that mature white and hang off the vine into winter. The berries are relished by an astounding diversity of birds. The leaves are as, per the saying, "leaves of three, let it be," but turn beautiful shades of red like its relatives the sumacs.

RELATED PLANTS This has finally been botanically separated from the shrubby western poison-ivy (*Toxicodendron rydbergii*) native farther north and west.

Vitis aestivalis (summer grape) smothers a clump of young sassafras.

Vitis spp.
Wild grapes

Vitaceae (grape family)
Landscape group: woody vines

At least one species of wild grape can be found across the entire Midwest but as many as seven species are native in the Lower Midwest. Wild grapes are thought of as garden thugs, generally not available from any nursery, native focused or otherwise. My home, ensconced in a second-growth woods, has three wild species and I cannot imagine the home garden and landscape without them. I forgive them as thugs and weed out their seedlings from formal beds and trim them from various shrubs and trees they otherwise would overtake. It is worth the effort. All native grapes have edible fruit that are delicious eaten fresh, made into juice or wine, or for preserves. Japanese beetles feed heavily on grape foliage and can be a serious pest (and on related *Ampelopsis* and *Parthenocissus*).

HOW TO GROW Wild grapes are easy to grow and tolerate a wide variety of soils from moist to dry. They need full sun to thrive and under such conditions they can take over any object and smother surrounding plants. They all can grow 35–50 feet and sometimes more. They climb by tendrils that can wrap around any small object or cling to rough bark or even brick. You can train them on a trellis just like a domesticated grape cultivar grown for fruit production, and yes, grow them for their fruit. They can take extreme pruning and develop large trunks as they age. Riverbank grape is the rootstock of many a cultivated grape as our native grapes are resistant to phylloxera (a root aphid) which when accidentally brought to Europe, wiped out the European grapes until understocks were changed to our native species.

LANDSCAPE USE The trick is to plant wild grapes where they will have to fight for light with existing established trees. Make sure if you grow them on any object or shrub that you

The male flowers of *Vitis cinerea* (gray grape) perfume the midsummer air above fully mature leaves.

The leaf underside of *Vitis labrusca* (fox grape) is quite striking.

Like all wild grape fruit, that of *Vitis riparia* (riverbank grape) is beautiful, delicious, and attract wildlife.

247

prune them judiciously or have space around them to keep them in bounds. Pruned female vines are important components to edible landscapes and food forests. I feel these magnificent native plants are still not used to our advantage as they create an unbeatable green curtain for even large, urban buildings, helping with summer cooling—a practice that is gaining popularity abroad and sure to arrive here. Simply plant a male where fruit would be inappropriate. I planted one on the west side of my parents' house for just that, covering a west window and cooling the house, but it must be heavily pruned every year. Grapes are extremely sensitive to certain herbicides like 2,4-Dichlorophenoxy-acetic acid (abbreviated 2,4-D) so I like them as canaries in the mineshaft monitoring neighborhood pesticide misuse.

ORNAMENTAL ATTRIBUTES I cherish the fragrance of their bloom, the colors of their emerging leaves, the wealth of insects they are host to, and the grand finale in the fall: the purplish black edible fruit on female vines. The vine's papery strips of bark are a favorite material birds use for lining their nests. Cut vines make great wreaths and other garden décor.

The delicate springtime male flowers of *Vitis riparia* (riverbank grape) fill the air with a pleasant aroma.

Vitis vulpina (frost grape) leaves are similar to those of riverbank grape but without the fringe of hairs around the leaf edge.

Vitis aestivalis (summer grape) is the second most wide-spread grape growing across the entire Eastern and Lower Midwest and northward up the Mississippi River to southeastern Minnesota and up the Missouri River to Omaha. Summer grape prefers upland habitat and is intolerant of wet soils. The flowers bloom around the summer solstice, their sweet fragrance something to look forward to. The early foliage can have distinctive U-shaped sinuses and is quite lovely, being two-toned from a dark green surface to a lighter whitish underside of cottony and rusty hairs. This grape is parent to the 'Norton' wine grape, the state grape of Missouri.

Vitis cinerea (gray grape, winter grape) is found mainly across the Lower Midwest native northward to the southwest corner of Iowa, into northern Illinois, and north central Indiana. No grape has more beautiful foliage than this species: young foliage is pink, unfurling to silvered hairy foliage in spring, and finally maturing to gray-green with inset leaf veins. The underside of the foliage is also lighter with an almost cottony surface. It also blooms around the solstice with exquisitely fragrant flowers.

Vitis labrusca (fox grape) is more restricted to the Eastern Midwest and prefers acidic soil. It's naturalized westward into Missouri from cultivated plants. It bears the largest fruit of our native species and has a foxy aroma, hence the common name. This grape is the most widely cultivated of our native grapes with several cultivars for fruit production and it is parent to the popular 'Concord' grape.

Vitis riparia (riverbank grape) is the only grape that grows wild across the entire Midwest. It is well named as it can be inundated by floodwaters for periods but still thrives in well-drained soils and is found growing in almost any woodland or hedgerow. The fleecy green flowers go unnoticed to the eye but are extremely fragrant and call attention when they bloom in late spring. Foliage emerges pink in spring and early leaves can have V-shaped sinuses but most leaves are rather simple with three main lobes.

Vitis vulpina (frost grape) is native across the Lower Midwest and northward to southwestern Michigan, northern Illinois, and eastern Nebraska. It's easy to overlook. The first time I found this plant was because I tasted the fruit, which does have a musky, foxy aroma: its botanical name *vulpina* means fox. Far less vigorous than riverbank grape, it otherwise is very similar, but with leaves that are rounded or with just two slight side lobes. Petioles are smooth rather than hairy on riverbank grape.

This wild-collected (grown from cuttings) *Wisteria frutescens* (Kentucky wisteria) shows exceptionally large flower panicles.

Wisteria frutescens
Kentucky wisteria

> Syn. *Wisteria macrostachya*
> Fabaceae (legume family)
> Landscape group: woody vines

Fragrant, pendant strands of purple flowers draping springtime leafless vines are what come to mind when one mentions wisteria. Well that would be the Asian native species which are quick-growing massive vines that can engulf anything they are planted on including mature shade trees. Our native species has similar flowers, though less showy and sans fragrance, but produced on a vine that never grows that big or problematic.

HOW TO GROW This vine is native mainly in floodplain forests of the Lower Ohio and Mississippi Rivers with outlying populations beyond in similar habitat. It is hardy throughout the entire Midwest (USDA zone 4). It is easy to grow in any good garden soil and is easily propagated from cuttings or seed. Seedlings can be prolific in disturbed soil.

LANDSCAPE USE Kentucky wisteria is a beautiful vine for a pergola, arbor, porch, or other structure suitable for a vine. It can be planted on larger trees and will climb up them but not overtake them like the Asian species. It grows 12–15 feet, occasionally larger.

ORNAMENTAL ATTRIBUTES This vine blooms in late spring so is never subject to flower-bud freeze damage like its Asian counterparts. The lavender-blue or rarely white flowers are produced after the plant leafs out so are less showy than the Asian species. Some individual plants can have large flowers on 12-inch-long racemes, and several cultivars are now readily available. The foliage is pinnately compound. It is a host to the silver-spotted skipper butterfly whose caterpillar wraps the foliage around itself as a shelter for feeding and pupating. The fruit is a hanging pea pod (legume) that matures to a polished mahogany and can hang through the winter. In late winter during periods of low humidity, the pods audibly snap open flinging the seeds a good distance and leaving empty and open twisted pods.

RELATED PLANTS The Kentucky and more southeastern American wisteria were lumped into a single species though they are quite different at their extremes. The midwestern forms have the largest flowers and leaves. Midwestern cultivars include: 'Aunt Dee', 'Blue Moon', and Summer Cascade ('Betty Matthews')—all from Minnesota.

249

Prairie Perennials

Perennials are herbaceous plants whose various root structures survive the winter below ground so they are protected from the harsh realities and weather extremes of a midwestern winter. They return each spring, flowering and fruiting in the growing season, then dying back to their base each winter. Perennials with attractive inflorescences (flower spikes or flower heads) or ornamental infructescences (fruiting structures) are included in this book. Many perennials are very long lived and grow incredibly deep root systems. Almost all are hardy throughout the Midwest so only those that have specific hardiness requirements are noted under How to Grow.

Perennials are the most diverse group of plant types in the Midwest so they are assigned into three groups—Prairie, Woodland, or Wetland—based on their primary original habitats and the conditions under which they best grow in a garden or landscape. Each group is treated in a separate chapter. It's gratifying to know that the Midwest "heart of North America" is where many of the world's most popular temperate climate perennials originate.

Prairie and meadow perennials, the subject of this chapter, originally grew native to upland prairies and thrive in full sun in moist to dry soils. Many of these plants also grow well in the partial shade of savannas and will linger as woodland becomes more closed and shaded. They languish and never flower when grown in full shade. Upland prairie plants grow best in well-drained soils that may be seasonally wet to dry. Perennial forbs and grasses were the dominant plants of the midwestern prairies and were a major factor in the development of the prairie's deep organic soils. They create an amazing biomass below ground via their roots, which over millennia of death and decay created the Midwest's premier black soil defined by this organic matter.

Surprisingly, prairies often look different each year: weather, management regimes, and critters all play a role in the performance of various plant species each season.

Adequate rainfall creates robust growth while a drought stunts many species, a fall fire promotes forbs while an early spring fire promotes grasses, and a boom year in vole numbers may spell doom for the corms of tuberous plants like blazingstars the voles find delicious.

The most important plant families in upland prairie are grasses, composites, legumes, carrots, and mints. No other group of perennials reflects the midwestern spirit of place better than the grass family (Poaceae). Keep in mind that grasses are highly flammable and were once the main fuel of prairie wildfires. They grow exuberantly following being burned off in spring, but maintaining them by cutting them back in late winter or early spring is more feasible in a traditional or community landscape.

Grasses can be grouped into warm- versus cool-season types. Warm-season grasses are slow to emerge in spring, greening up in late spring, growing best in the heat of summer, flowering in late summer into early fall, and turning warm colors in autumn. Cool-season grasses grow

Andropogon gerardii (big bluestem) and *Sorghastrum nutans* (Indian grass) are two of the four main tall grasses of the midwestern prairie.

best in cool weather and emerge in early spring, flower by early summer, and have ripened seed fruits by late summer. Cool-season grasses often have green basal leaves that regrow in the fall and persist through the winter.

Grasses also can be grouped by how they spread. Those that stay in clumps are known as bunch type grasses, while those that spread by underground rhizomes are called turf type grasses and their profiles are in the groundcover chapter.

The roots of some perennial grasses have been documented to grow down 15 feet. It is the aboveground stems and blades billowing in the wind that define a midwestern prairie experience. They were the main original fodder for the American bison, which grazed on grasses and much less so on forbs. Only a handful of managed bison herds have been reintroduced to the larger remnants of prairie in the Midwest.

Composites, members of the aster family (Asteraceae), are plants whose inflorescences are comprised of multiple tiny flowers massed together in a head. Many of these have ray flowers around the edge of their heads that look like petals with the rest of the flowers in the center known as disc flowers. Composites produce copious amounts of nectar and pollen to entice pollinators and provide abundant seed fruits that are often rich in oil and nutritious food for wildlife.

Asters are some of the best-known composites. Are you an aster master? That's the name "prairie people" in Rockford, Illinois, gave those who knew or could identify all the native asters. The asters can be an intimidating bunch to identify. Be sure and look beyond their flowers to their stems, leaves, habit, and habitat and then it's not that hard to identify most of them. Many wildflower guides depict only the blossom, but its difficult to discern asters just by their flowers.

Asters have been split from the genus *Aster* into several new genera that old-time botanists may have a hard time getting used to. *Symphyotrichum*, *Eurybia*, and *Ionactis* just don't rhyme with anything. (My botanist friend Mark Leoschke calls this the aster disaster.)

All asters are easy to grow from seed, several of them a bit rambunctious in their self-sowing. Frost aster (*Symphyotrichum pilosum*) is a ubiquitous beautiful, though weedy, species that appears in all disturbed ground with abundant white flowers in mid fall—it is rather short lived so should be welcomed as a good cover crop. Aster's seed fruits create a tawny froth atop plants when ripe in fall—individually cute with their tufts of hairs (pappus) that help them blow in the wind. Aster seeds are highly sought by songbirds and small mammals and aster foliage is host to the pearl crescent butterfly that overwinters as caterpillars in the fallen leaves

around the plant. Included here are only the best aster species suited to a garden.

Legumes enrich the prairie soils with available nitrogen, one of three essential plant elements required for plant growth. Nitrogen is abundant in air, but in a form that plants cannot utilize. The nodules on roots of most legumes contain bacteria that transform and fix this nitrogen into a usable form for plants. These bacteria, called inocula, often aid in the germination of a specific legume's seeds and can be purchased and incorporated with seed planting. The legume family (Fabaceae) is also known as the pea or bean family.

Plants in the carrot family (Apiaceae) are perennials whose flowers are usually in flat-topped clusters called umbels. The flowers attract a diverse group of insects and are premier pollinator plants. Carrot family plants listed here have acrid foliage and stems, shunned by many herbivores and making them ideal deer-resistant plants.

Members of the mint family (Lamiaceae) are superior insectaries. These plants have an abundance of nectar-rich flowers produced over a relatively long period. The pungent aromas of their foliage and stems make them unpalatable to many herbivores, so they are the ultimate deer-resistant plants.

The diversity of a high-quality prairie remnant may be nothing short of astounding, and only the most appealing plants for cultivation are included here. Gardeners wanting to fully restore or recreate a prairie will have to delve well beyond the plants in this book, though these plant profiles are a great starting point.

PERENNIALS FOR A TRADITIONAL PERENNIAL BORDER IN FULL SUN

Asclepias tuberosa (butterfly milkweed)
Baptisia australis varieties (blue wild indigo)
Coreopsis grandiflora (showy coreopsis)
Coreopsis lanceolata (lanceleaf coreopsis)
Echinacea purpurea (purple cornflower)
Liatris pycnostachya (prairie blazingstar)
Liatris spicata (marsh blazingstar)
Rudbeckia missouriense (Missouri black-eyed Susan)
Schizachyrium scoparium (little bluestem)
Solidago speciosa (showy goldenrod)
Sporobolus heterolepis (prairie dropseed)
Symphyotrichum leave (smooth aster)
Symphyotrichum novaeangliae (New England aster)
Symphyotrichum oblongifolium (aromatic aster)

Andropogon gerardii
Big bluestem
Poaceae (grass family)

Big bluestem is a midwestern icon: one of the tall grasses of the tallgrass prairie.

HOW TO GROW This grass is native across the entire Midwest and most of North America from the Rockies to the Atlantic. It is hardy throughout the Midwest, but regional forms should be grown to ensure plants are appropriate for each site's growing season as southern plants may not flower in northern regions. Big bluestem grows largest in full sun and deep rich soils but does fine in partial shade with weaker flowering and is shorter in poorer, drier soils. It's easiest to grow from seed.

LANDSCAPE USE Big bluestem is most suitable to natural landscapes and restorations because it readily self-sows into tall, dense masses. Tall means 3 feet in poor soils in dry years and up to 8 feet in rich soils in moist years. It is best grown with other plants to give it some competition, which will also reduce seedlings. In prairie plantings I recommend low seeding or planting rates so that smaller plants have a chance to establish as this grass will otherwise outcompete them.

ORNAMENTAL ATTRIBUTES Elegant tall flowering stems are topped with a three (or more) parted inflorescence that looks like a turkey's foot—turkey foot is another name for the grass. The basal tuft of foliage is 18–36 inches tall, often highlighted with steel blue and sometimes wine purple. Fall color is variable but always warm tones: from gold and orange-russet to reddish.

The tall culms of *Andropogon gerardii* (big bluestem) display a blend of colors in early fall.

Asclepias syriaca
Common milkweed
Apocynaceae (dogbane family)

Milkweeds (*Asclepias* spp.) are premier insectary plants that attract hordes of butterflies and other pollinators. Most are rather short-lived perennials, except for the butterflyweed, which may live for decades. Milkweeds are the sole host plant for the monarch—along with a few other closely related species in the old milkweed family (Asclepiadaceae), which is now lumped with the dogbane family.

HOW TO GROW Common milkweed is native to central and northeastern North America including the entire Midwest. It's most often found in disturbed open ground from meadows to roadsides but has prairie and savanna origins and requires full sun to only partial shade. It is difficult to

The fragrant flowers of *Asclepias syriaca* (common milkweed) look like pink popcorn balls.

transplant and can be aggressive once established, but it's easy to pull unwanted spears of new plants, which won't return in that location.

LANDSCAPE USE Common milkweed is best in natural landscapes where it grows around 3 feet tall and can be allowed to run underground into a thicket of plants. In good soils without competition, established plants may grow 5 or 6 feet tall.

ORNAMENTAL ATTRIBUTES The 4-inch-wide spheres of flowers are distinctively and pleasantly fragrant. Each flower is bicolored—light pink with rosy pink petals—and individual flowers age tarnished orange-pink. The flowers occur in the leaf axils (where the leaf meets the stem) on the upper portion of the plant. Leaves are broad and flat displaying a coarse texture. The fall seedpod fruits are warty and fat, bursting with fluffy seeds, and remain though the winter.

RELATED PLANTS Blunt-leaved milkweed (*Asclepias amplexicaulis*) is native across most of the eastern and central United States and is found in all of the Midwest except the Dakotas. It is similar to common milkweed but grows 2–3 feet tall with stylish, curvaceous, undulating leaf margins. It grows in poorer, dry soils—especially sandy soils. The flowers are in looser umbels looking like an explosion of pink with greenish-reflexed petals.

Sullivant's or prairie milkweed (*Asclepias sullivantii*) is native from North Dakota and Oklahoma eastward to Michigan and Ohio. It also is similar to common milkweed but has smooth leaves with veins of pink. Its flowers are larger, fewer in a cluster, and also deeper rosy pink than common milkweed. Sullivant's milkweed grows 2–3 feet tall and also does not sucker or run and stays in a tighter clump or creates a dense patch with a huge rhizome.

Asclepias tuberosa
Butterfly milkweed
Apocynaceae (dogbane family)

Is there any wildflower more energizing, stimulating the mind with its warm vibrant color? The activity of all its pollinators adds to its buzz as a masterpiece of art and science.

HOW TO GROW Butterfly milkweed is native throughout the entire Midwest and beyond: across eastern North America and into the southern Rocky Mountains. It grows in mesic to dry prairies, savannas, glades, and other open lands, flourishing in average to well-drained soil that may be rocky, gravelly, or sandy in full sun. Some strains grow in pure clay such as those native on Powell Gardens' prairie remnants (at least one mail-order nursery sells such a strain), but most strains demand fast-draining soils.

LANDSCAPE USE This milkweed makes an ideal perennial for a traditional perennial border and grows 24–30 inches tall. It thrives in harsh sites as well, making an excellent plant in a rock garden or other hot, sunny location with good drainage.

ORNAMENTAL ATTRIBUTES Butterfly milkweed produces the most vibrant orange flowers of any native plant. Occasionally flowers are vermillion red, golden orange, or even

Asclepias sullivantii (Sullivant's milkweed) in a roadside remnant in Illinois.

The curvaceous leaves beneath open, sphere-shaped flower clusters identify this as *Asclepias amplexicaulis* (blunt-leaved milkweed).

256

A vibrant mass of *Asclepias tuberosa* (butterfly milkweed) thrives at Matthaei Botanical Gardens in Ann Arbor, Michigan.

An extensive mass of *Asclepias verticillata* (whorled milkweed) perfumes the air at Nachusa Grasslands, Illinois.

golden yellow. They occur in early to midsummer with an occasional flower later in summer. The plant is leafier than most milkweeds. Leaves are narrow, strappy, and fine textured. Narrow pod-shaped fruits hold well into winter, crowning the stems.

Asclepias verticillata
Whorled milkweed
Apocynaceae (dogbane family)

HOW TO GROW Whorled milkweed is native across eastern North America east of the Rocky Mountains and is found throughout the Midwest. It grows in upland prairies, glades (or other sunny rocky outcrops), meadows, and other disturbed open land. It thrives in almost any well-drained soil in full sun. Colonies of plants may be divided.

LANDSCAPE USE This milkweed is best in a natural landscape, growing between shorter prairie grasses. It reaches 2 feet

Astragalus canadensis (Canada milkvetch) flowers stand tall at Schaefer Prairie, Minnesota.

tall and spreads into a mass via underground rhizomes. The masses are neither smothering nor aggressive and mingle well with other plants.

ORNAMENTAL ATTRIBUTES The white flowers are intensely fragrant and perfume a wide area. The leaves are narrow and almost threadlike for a very fine texture. This thin erect fruit pods crown the plant in fall and remain into winter.

RELATED PLANTS There are several other prairie milkweed species, most rarely cultivated.

Astragalus canadensis
Canada milkvetch
Fabaceae (legume family)

The mere mention of this plant triggers a fond memory of sitting on a bench in the solace of the meadow/prairie section of the Eloise Butler Wildflower Garden and patiently watching the varied and various pollinator visitors gain sustenance from this plant in bloom. It's a premier pollinator plant.

HOW TO GROW Canada milkvetch is native across the entire Midwest though more localized in the Eastern Midwest. It grows in mesic soils that may be organic black to sandy in prairies and savannas. In a garden setting it grows well in moist, well-drained soils in full sun to partial shade.

LANDSCAPE USE This somewhat sprawling perennial stays in a clump and grows 2–31/2 feet tall. It is best in a natural setting with companion grasses or sturdy perennials to keep it from flopping. It provides abundant nectar to a plethora of insects from bumblebees to honeybees and small butterflies like the eastern tailed-blue and gray hairstreak. The foliage is edible browse to many herbivores so needs protection where rabbits and deer are prevalent.

ORNAMENTAL ATTRIBUTES The inflorescences bear very showy, long-pointed spires of creamy flowers. The foliage also is very delightful with finely textured compound leaves whose vertical leaflets look like the teeth on a coarse comb.

Astragalus crassicarpus
Prairie plum, ground plum
Fabaceae (legume family)

Prairie plum should be called prairie snow pea: its unripe, inflated pea pods can be eaten raw with the same crunchy texture of a snow pea.

HOW TO GROW This species is a native of the Great Plains growing eastward to the Mississippi River and crossing into Wisconsin in the Twin Cities region and into Illinois in the St. Louis region. It is found in dry prairies in well-drained rocky or gravelly soils mainly on hill prairies and glades eastward. It struggles in good garden soil, as it requires a lean rock garden soil mix of equal parts of local gravel, compost, and topsoil. It is relatively easy to grow from seed.

LANDSCAPE USE Prairie plum is a decumbent perennial with ground-hugging stems, flowers, and foliage so is best sited in a rock garden, on top of a rock wall, or in raised bed. It shows best with gravel mulch and diminutive grasses, such as grama grasses, and spring wildflowers, such as blue-eyed grass, prairie violets, and violet wood sorrel. It's one of the earliest premier nectar sources visited by a diversity of pollinators; its ground-hugging nature ensures solar warmth and shelter from the spring winds. It should be included in insectaries gardens for early nectar and as a component in an edible landscape.

ORNAMENTAL ATTRIBUTES Plants emerge early and flower by midspring. The flowers are rosy or purplish pink on var. *crassicarpus*, yellowish on var. *trichocalyx*, which is native in the eastern Ozarks. The pea pods are somewhat showy but also ground hugging and the whole plant is ephemeral and goes dormant by midsummer.

The ground-hugging blossoms of *Astragalus crassicarpus* (prairie plum) are some of the first bright flowers on the prairie in spring.

Baptisia spp.
Wild indigos
Fabaceae (legume family)

Wild indigos are magnificent, long-lived perennials, many whose presence and stature are as significant as much-loved and traditional, though nonnative, peonies. Wild indigos are gaining in garden popularity and all the native species have now been hybridized into a menagerie of flower colors (just like peonies) from weird dark to reddish chocolate brownish purple to shades of yellow, from pale to bright sky blue to deep blue, and so on. The hybrids I've observed in the garden are still visited by their traditional pollinator—the bumblebee—and are also host to the wild indigo duskywing (a spread-wing skipper butterfly) and seem to produce viable seed. The new hybrids are bred to be tidy and full foliaged when compared to the native species.

The native midwestern wild indigos include three widespread and one minor species. They are legumes and nitrogen fixers, with deep extensive roots that are impossible to transplant. They are part of the reason prairie soils are so rich, so consider locally native wild indigo over hybrids, especially in natural landscapes and prairie restorations. Wild indigos emerge in spring looking like large asparagus spears. They often are listed as deer resistant but deer can crop the fresh new growth and flower buds, and the plant will not flower for the season. Deer don't like the mature foliage. A Great Plains moth, the genista broom moth, also has spread from its namesake host plant to wild indigo where armies of its caterpillar will defoliate the plant in later summer. Wild indigos become tumbleweed; the whole plant can snap off by late winter and blow with their cargo of rattling inflated pea pods (legumes) to disperse their seeds. Note that native weevils can be hard on seed production.

Baptisia alba var. *macrophylla*
White wild indigo
Syn. *Baptisia leucantha, B. lactea*
Fabaceae (legume family)

I grumble when I think about the new name that lumps the white-flowering wild indigos into a single species—*Baptisia leucantha* described the midwestern native type which grows like an elegant ballerina on tiptoes, its long stem carries its tutu of foliage above most of the surrounding prairie plants with a graceful spire of white flowers projecting skyward another 2 feet. Most gardeners do not appreciate this

The flowering stem of *Baptisia alba* var. *macrophylla* (white wild indigo) stands as elegantly as the finest ballerina.

Typical *Baptisia australis* (blue wild indigo) makes a shrublike perennial suitable for a foundation planting.

Baptisia australis var. *minor* (dwarf blue wild indigo) makes a compact leafy perennial with stunning spires of flowers.

lankiness and they prefer the fuller, bushier form, formerly named *B. pendula*, which is not native to the Midwest.

HOW TO GROW White wild indigo is found wild through most of the Midwest, absent only from the far northwest with a spottier range eastward through Ohio. It grows in mesic to dry upland prairies and savannas and does well in almost any well-drained soil. It can be grown from seed easily.

LANDSCAPE USE White wild indigo is an integral component of a prairie restoration. The midwestern native variety (*Baptisia alba* var. *macrophylla*) grows 4–6 feet tall (tip of the flower stalk) and looks best at the back of a traditional perennial border. It's superb in companion with shorter prairie grasses.

ORNAMENTAL ATTRIBUTES The foliage is the most blue-green of the wild indigos and contrasts nicely with other plants. The early summer flowers are milky white in tall spires showing well above the foliage and followed by a showy spire of pea pod fruits that ripen from green to black and remain on the plant as an ornamental spire through most of winter. The pea pod splits open at some point in winter so that the plant audibly rattles as the two parts clap together—the seeds gradually lose their moorings and fall out.

Baptisia australis
Blue wild indigo

Fabaceae (legume family)

Blue wild indigo is the most popular of the wild indigos in gardening, sold at most nurseries across the Midwest and well represented in perennial gardens of traditional landscapes. In the wild it is not that widespread—found native mainly across the Lower Midwest. The Ohio River region is where the traditional variety of this perennial is found wild (var. *australis*) in Kentucky, Indiana, and Ohio. Its most widespread variety (var. *minor*) is found in the Flint Hills, western Osage Plains, and Ozark Highlands, possibly native northward rarely (and questionably) into Nebraska and Iowa and eastward to western Kentucky. Variety *minor*, sometimes called dwarf blue wild indigo, is a much smaller and tidier plant growing half the size of variety *australis*.

HOW TO GROW Blue wild indigos grow in mesic to dry upland prairies and savannas, often in rocky soil. They do well in most well-drained soils in full sun or partial shade and are easy to grow from seed.

LANDSCAPE USE Most plants sold at typical nurseries are the eastern (variety *australis*) form that grows quite large into a robust perennial 3–4 feet tall and wide. Plants native west of the Mississippi and purchased at native plant nurseries are the variety *minor* and grow just 18 inches to not much more than 2 feet tall. Obviously, a gardener needs to know which form they have and use it appropriately in a traditional perennial border, in a foundation planting, or plant the variety indicative of their region in a prairie restoration.

ORNAMENTAL ATTRIBUTES In late spring the stunningly beautiful indigo blue flowers appear in spikes above the foliage. The foliage is handsome and tidy turning an interesting black in the fall. The pea pods form green and ripen to black in autumn, adorning the plant through winter.

RELATED PLANTS Yellow wild indigo (*Baptisia tinctoria*) is mainly an East Coast species but does grow native westward into Ontario, northeastern Ohio, southeastern Michigan, and northwestern Indiana with a few scattered and isolated rare populations westward into northeastern Illinois, Wisconsin, Minnesota, and Iowa. It is found in poor, sandier, often acidic soils usually in sand prairies and sand savannas. The flowers are pale to pure yellow and occur in early summer in loose clusters around the bushy plant. The plant reaches just 2–3 feet tall and wide.

In Indiana, yellow wild indigo has hybridized with white wild indigo to produce a creamy flowered plant: Deam's wild indigo (*Baptisia ×deamii*) named after Indiana's legendary botanist.

Baptisia bracteata

Cream wild indigo

Syn. *Baptisia leucophaea*
Fabaceae (legume family)

HOW TO GROW Cream wild indigo is native from the Flint Hills and Ozark Highlands northward to southeastern South Dakota, southern Minnesota, and Wisconsin, and eastward across Illinois to western Kentucky, northeastern Indiana, and barely into Michigan. It grows in mesic

The pendant flowers of *Baptisia bracteata* (cream wild indigo) surround its shell of gray-green foliage.

Bouteloua curtipendula (side-oats grama) crowns a hillock at Harlem Hills Prairie in Rockford, Illinois.

to dry upland prairies and savannas which translates into surviving in most well-drained garden soils (sandy, gravelly, or black organic) in full sun to partial shade. It can be fickle to establish. Cream wild indigo likes companion plants such as sideoats grama, little bluestem, and prairie dropseed. It is best propagated from seed or transplanted as seedling plugs.

LANDSCAPE USE This wild indigo makes a fine edge perennial in a traditional perennial border as long as companion plants are used. It shows nicely with shorter perennials and grasses in rock gardens. It's a spring necessity plant for a prairie reconstruction, best established as a seedling plug.

ORNAMENTAL ATTRIBUTES Cream wild indigo is the first of the native species to flower in midspring with gorgeous creamy yellow flower clusters splaying outward from a low dome of gray-green foliage. They remind me of tortoises when blooming on a burned prairie. The foliage is handsome all season, and the pea pods form low, disguised in the foliage of companion prairie plants.

Bouteloua curtipendula
Side-oats grama

Poaceae (grass family)

HOW TO GROW Side-oats grama is a Great Plains and south-western warm-season grass but native eastward across the entire Midwest. It is localized in the Eastern Midwest to drier microhabitats. It grows in dry upland prairies and performs well in gardens in well-drained soil in full sun. It is easy to grow from seed.

LANDSCAPE USE This shorter grass rarely reaches more than 30 inches tall. It is clump forming and does not self-sow wildly so makes a fine addition to a traditional garden. It looks good as a single plant or planted in masses. It mixes well with little bluestem and prairie dropseed but can be smothered by big bluestem, Indian grass, and switchgrass.

ORNAMENTAL ATTRIBUTES Side-oats grama is named for its inflorescence consisting of spikelets that line one side of the stem. A single clump of grass can produce a full crown of these inflorescences. It's most beautiful when in bloom as the anthers are red, making each spikelet look like a firecracker. The plant looks great in fall as it turns straw colored and holds well into winter.

Bouteloua gracilis
Blue grama

Poaceae (grass family)

HOW TO GROW Blue grama is another Great Plains and southwestern warm-season grass. It is native on the western edge of the Midwest from the Dakotas and western Minnesota, Nebraska, western Iowa and Kansas, and westernmost Missouri. It has naturalized eastward very sparingly into Wisconsin, Illinois, and Michigan. Blue Grama is a component of the mid- and short-grass prairies so is found into the Midwest only in dry upland prairies. It does well in gardens across the Midwest but requires well-drained soil and a hot dry site in full sun. It's easy to grow from seed or by dividing a clump.

LANDSCAPE USE This very tidy and compact grass grows 2–3 feet tall. It is a delightful grass to edge a sunny perennial border, for a rock garden, above stonewalls, or in scorching "hell strips" between roads and sidewalks.

ORNAMENTAL ATTRIBUTES The plant is most beautiful in flower with inflorescences of comb-shaped spikelets that look like eyebrows (a nickname is eyebrow grass). The grass turns rich orange-brown in fall and stays a warm brown into winter.

Bouteloua gracilis (blue grama) stands above *Sedum* Autumn Joy ('Herbstfreude') in a traditional perennial border.

RELATED PLANTS There is a heavy-flowering, taller selection with spikelets that mature light tan called 'Blonde Ambition.'

Hairy grama (*Bouteloua hirsuta*) is very similar to blue grama but smaller, growing only 8–10 inches tall. It is found in similar dry upland prairies, especially gravel and hill prairies across the Upper Midwest west of Lake Michigan, west and southward in the Great Plains. Its diminutive size makes it vulnerable to be smothered by larger plants, so it is best as a very refined rock garden plant or even in a container.

Callirhoe bushii
Bush's poppy-mallow

Malvaceae (mallow family)

Five species of *Callirhoe* (poppy mallows) are native to the Midwest. Purple poppy mallow and light poppy mallow are Great Plains species, Bush's poppy mallow and standing poppy mallow are Ozark Highland endemics. These four species make good garden subjects while the fifth species, clustered poppy mallow (*Callirhoe triangulata*), is a sand prairie specialist challenging to grow in a garden. Poppy-mallows are often called winecups because their up-facing, five overlapping petaled flowers form bright fuchsia-pink cups.

Callirhoe bushii (Bush's poppy-mallow) forms an upright bushy perennial of stunning fuchsia flowers at Powell Gardens.

Masses of *Callirhoe involucrata* (purple poppy-mallow) are stunning while in bloom.

A sweep of blooming *Coreopsis lanceolata* (lanceleaf coreopsis) is visited by great spangled fritillaries.

262

HOW TO GROW Bush's poppy-mallow has a small native range in the western Ozark Highlands and is naturalized elsewhere. It grows in well-drained prairies, open woodlands, and glades and demands well-drained garden soil in full sun or partial shade. Bush's poppy-mallow is easy to propagate from seed.

LANDSCAPE USE This poppy-mallow grows 18–30 inches tall, more upright than purple poppy-mallow though it can flop without adjacent plants as support. It self-sows in ideal garden conditions, so is best in a natural landscape where it makes a vibrant display.

ORNAMENTAL ATTRIBUTES The classic bright fuchsia-pink flowers are the largest of the winecups and produced in abundance through midsummer.

RELATED PLANTS Standing poppy-mallow (*Callirhoe digitata*) is another more upright poppy-mallow that also is a western Ozark Highlands endemic, naturalized as far away as northeastern Illinois. It's classic winecup flowers stand 3 feet tall atop naked, wiry stems—quite an exaltation of flowers in the bright summer solstice sunlight. This poppy-mallow may be the one best suited to a traditional perennial border in well-drained soil.

If you enjoy the bright fuchsia flower colors of poppy-mallows and have pure sand, plant halberd-leaved poppy-mallow (*Callirhoe triangulata*).

Callirhoe involucrata
Purple poppy-mallow

Malvaceae (mallow family)

HOW TO GROW Purple poppy mallow is mainly native to the central and southern Great Plains. It is found eastward to western Iowa, Missouri, and Arkansas though it has naturalized widely across the Midwest. It grows in dry upland prairies and glades and is easy to cultivate in lean, well-drained soils in full sun. The plant is wide-spreading and should be trimmed back to its base in late summer as the plant goes dormant. Propagate this poppy-mallow by cuttings or seed.

LANDSCAPE USE This low, wide-spreading plant is under 12 inches tall but may spread 4 feet across, which makes it a good choice for a rock garden and to spill over a wall. Its

Great Plains origin makes it a good choice for very hot, dry, and rocky sites where other perennials may languish. It self-sows in sites where it thrives.

ORNAMENTAL ATTRIBUTES The plant produces an abundance of classic winecup flowers in early through midsummer and sparingly later. The maplelike divided foliage is handsome early in the season through flowering.

RELATED PLANTS Light poppy-mallow (*Callirhoe alcaeoides*) is similar but with tinier pale pink or white flowers and a smaller habit. It is native to dry upland prairies and glades in the eastern half of Nebraska, Kansas, and Oklahoma eastward into Missouri and Arkansas and is naturalized elsewhere. Its cultivar 'Logan Calhoun' has white flowers and is the main form of the species in cultivation.

Coreopsis lanceolata
Lanceleaf coreopsis

Asteraceae (aster family)

HOW TO GROW Lanceleaf coreopsis is native from the Ozark Highlands northeastward to Lake Michigan and Central Wisconsin sand plains. It grows native almost exclusively in sand savannas and sand prairies. Unlike many plants from such an environment, it is easy to cultivate in almost any well-drained garden soil. It requires full sun or partial shade and is hardy throughout the Midwest. It is easy to grow from seed.

LANDSCAPE USE Lanceleaf coreopsis makes a tidy perennial usually no more than 2 feet tall when in midsummer flower. It self-sows into masses, but only lightly, so is suitable for a traditional garden. It is a phenomenal plant for hot, dry "hell strips" and other sandy or rocky sites and requires no extra irrigation to survive in such settings.

ORNAMENTAL ATTRIBUTES The 2- to 3-inch early summer flowers are brilliant and saturated golden yellow like no other. These flowers are formed in a dome over the foliage but for only about two weeks. The basal leaves grow only about 8 inches long and remain lush through the season. Fruiting heads turn brownish black when ripe and are sought by songbirds.

RELATED PLANTS Showy coreopsis (*Coreopsis grandiflora*) is very similar to lanceleaf coreopsis but found in rocky prairies and has smaller, less golden flowers and does not persist in a garden setting as easily as lanceleaf coreopsis. Both lanceleaf and showy coreopsis species are hybridized into many cultivars with extra ray flowers and a longer flowering period. I don't observe pollinating insects visiting the hybrids, but I have seen them seeking the disc flowers of the true species.

Coreopsis grandiflora (showy coreopsis) produces stunning yellow flowers in midsummer.

Coreopsis palmata (stiff coreopsis) spreads into spectacular masses of yellow flowers in midsummer, seen here at Paintbrush Prairie, Missouri.

Coreopsis palmata
Stiff coreopsis

Asteraceae (aster family)

HOW TO GROW Stiff coreopsis is native throughout most of the Midwest, but in the Eastern Midwest is only found in northwestern Indiana and southwest Michigan. It grows in dry to mesic prairies and savannas. Because it cannot compete with lush and larger-growing plants, it is more often found with shorter grasses. It's easy to cultivate in any well-drained garden soil but runs via rhizomes into a larger colony—not obnoxiously so if given companion plants like little bluestem, sideoats grama, or prairie dropseed. It thrives in full sun or partial shade and is easy to grow from divisions or seed.

LANDSCAPE USE Stiff coreopsis grows around 2 feet tall and spreads into a mass so needs to be planted with other shorter grasses (mentioned above) and companion forbs like pale purple coneflower and butterfly milkweed and the low shrub leadplant. It's integral to most prairie reconstructions where its nectar-rich flowers attract many pollinators and its fall and winter seeds are utilized by songbirds.

ORNAMENTAL ATTRIBUTES The flowers are bright true yellow and up-facing above the leafy stems in early summer. The leaves are rather unusual, slim and three-parted creating a unique fine texture to a planting.

Coreopsis tripteris
Tall coreopsis
Asteraceae (aster family)

HOW TO GROW Tall coreopsis's range is centered on the Ohio River valley but ranges westward throughout the Ozarks and much of Missouri and northward into southern Iowa, Michigan, and Ontario. It's a common component of tall-grass prairie plantings in much of the Midwest so has naturalized northward and is hardy throughout the Midwest. Tall coreopsis thrives in moist, rich soil in savannas and prairie openings and grows vivaciously in well-drained good garden soil in full sun or partial shade. In poorer, drier soil it is less aggressive and shorter. It is easy to grow from seed or divide from a clump. It can be cut back to bloom later and reduce its height.

LANDSCAPE USE Tall coreopsis is aptly named and easily grows 5–8 feet in open clumps. It makes an architectural statement with sturdy vertical stems of neat foliage. Without competition, it can be fairly aggressive in rich soil and self-sow abundantly. It makes a fine open screen or back-of-the-border plant. It is at its best in a natural landscape mixed with native grasses and other forbs.

ORNAMENTAL ATTRIBUTES The sturdy tall stems are rhythmically tiered with pairs of three-parted strappy foliage. The up-facing, true yellow ray flowers surround a dark center of dark red-brown disc flowers. The flowers are set atop the stems in flat clusters and produce blackened seed heads that hold through winter. The plant can turn shades of orange to burnt red in autumn.

RELATED PLANTS A cultivar 'Lightning Flash' has bright butter yellow new foliage and grows shorter and floppier, greening up by midsummer. My own select seedling of wild plants with yellow spring foliage is a sturdy mature height of just 3 feet.

Dalea purpurea
Purple prairie-clover
Fabaceae (legume family)

Prairie-clovers (*Dalea* spp., syn. *Petalostemon* spp.) are related to the nonnative clovers (*Trifolium* spp.) but are not sprawling plants; rather, they are upright perennials that stay in a clump. They produce very delicate, fine-looking

Coreopsis tripteris (tall coreopsis) produces a flat-topped cluster of flowers atop tall, leafy stems.

The flowers of *Dalea purpurea* (purple prairie-clover) open from the base in an ethereal tutu of purple petals.

A mass of *Dalea candida* (white prairie-clover) blooms at Gensburg-Markham Prairie, Illinois.

leaves that are much more tolerant of sun, heat, and drought than their three-part-leaved brethren. The flowers are sublime and always remind me of the great Impressionist painter Degas's ballerinas. The flower head is shaped like a narrow thimble and the flowers open from the base in an ethereal tutu of petals with protruding orange stamens. The whorl of flowers blooms sequentially up the thimble, which produces an attractive fruit mass of seeds that holds into winter. The fruiting heads gradually disintegrate as the seeds are dispersed.

HOW TO GROW Purple prairie-clover is native as an isolated rare plant in the Eastern Midwest, but is widespread and common westward. It is found in mesic to dry upland prairies and savannas and grows well in almost any well-drained garden soil in full sun to partial shade. It is easy to grow from seed, though germinates better with inocula.

LANDSCAPE USE This is a fitting plant for a traditional perennial border though it is best in companion with shorter grasses. It thrives in extremely rugged sites like "hell strips" and rock gardens. It makes an important and beautiful plant in a prairie reconstruction.

ORNAMENTAL ATTRIBUTES The finely textured leaves are comprised of five thin, almost linear leaflets, and the flowers are exquisite fuchsia-pink with shockingly contrasting orange stamens.

RELATED PLANTS White prairie-clover (*Dalea candida*) is very similar looking with a very similar range, but found only in northwestern Indiana and western Kentucky in the Eastern Midwest (absent from Michigan and Ohio). The leaves are slightly larger and the flowers are white with orange stamens.

Delphinium spp.
Delphiniums, larkspurs
Ranunculaceae (buttercup family)

Gardeners usually think of tall perennials in shades of blue when it comes to delphiniums—and that would be the non-native hybrids at their best in the gardens of England or the Pacific Northwest. The Midwest has its own suite of native species in similar colors but with more refined flowering and adaptability to wind and hot summers. The native species are also called "larkspurs" though I think of those as the annual, nonnative species (*Consolida* spp.) in a garden.

Prairie and Carolina delphiniums or larkspurs, once considered two separate species, have been lumped into a single species by botanists and are now considered subspecies. I never would have thought that but Powell Gardens lies where the two subspecies meet—and I have experienced the native plants on the site's dry prairie ridge where they are a beautiful blend of the two with variable white and light blue flowers.

HOW TO GROW Prairie delphinium (*Delphinium carolinianum* subsp. *virescens*) is native in the western Midwest from northwestern Wisconsin southwestward to western Missouri and Oklahoma and westward throughout much of the Great Plains. Here it grows in dry upland prairies. Carolina delphinium (*D. carolinianum* subsp. *carolinianum*) is native from west central Illinois southward through the Ozark Highlands into Arkansas and Louisiana, scattered eastward as far as South Carolina. It grows on rocky prairie, savanna, and woodland glades. Both do well in almost any well-drained soil in full sun to partial shade and they can be grown readily from seed.

265

The light, lavender-blue flowers identify this as a Carolina delphinium (*Delphinium carolinianum* subsp. *carolinianum*).

White flowers distinguish prairie delphinium (*Delphinium carolinianum* subsp. *virescens*) from Carolina delphinium.

Echinacea pallida (pale purple coneflower) stands tall in a prairie border at Powell Gardens, Missouri.

LANDSCAPE USE Prairie and Carolina delphiniums are fairly ephemeral so they emerge in spring with divided basal leaves, bloom on 2- to 3-foot-tall spikes in early summer that quickly set seed, and the whole plant disappears by the end of summer. They are best as companion plantings in natural landscapes or in perennial borders with shorter grasses including little bluestem, sideoats grama, and prairie dropseed.

ORNAMENTAL ATTRIBUTES The flowers are welcome spires above the prairie—white-flowering in the prairie subspecies and light sky blue in the Carolina subspecies. The distinctive leaves are palmate and deeply divided.

Great spangled fritillaries vie for nectaring space atop *Echinacea angustifolia* (narrow-leaf purple coneflower).

Echinacea pallida

Pale purple coneflower

Asteraceae (aster family)

HOW TO GROW Pale purple coneflower is native through the mid-Midwest from the Flint Hills, Osage Plains and Ozark Highlands northward across central and Eastern Iowa, nearly all of Illinois and into southern Wisconsin, rarely eastward into northwestern Indiana, and across Michigan to Ontario. I grew up near its northernmost population on a unique private prairie remnant. It is found in dry upland prairies including hill, gravel, or otherwise rocky prairies. It grows in lean, well-drained garden soils but cannot compete with robust tall plants in rich, mesic prairies. It does best in full sun and is hardy throughout the Midwest. Pale purple coneflower is easy to propagate from seed, but extracting seeds from its sharp cones can draw blood.

LANDSCAPE USE This refined perennial is suitable for a traditional landscape perennial border where it does exceedingly well in droughty locales with poor soil. I have heard of plants failing in a traditional landscape because it is fertilized and irrigated. Pale purple coneflower flowers may reach 30–36 inches tall and are set above foliage that is seldom over 1 foot tall. Flowers are nectar rich and attract a wide range of pollinators including a specific species of mining bee (*Andrena helianthiformis*) that utilizes only pale and narrow-leaf purple coneflowers.

ORNAMENTAL ATTRIBUTES The early summer flowers are sublime—drooping straplike ray flower "petals" are soft, creamy pink surrounding an orange spiny cone of disc flowers, each dotted with white pollen on top. Few populations of pale purple coneflower are variable, but those at Harlem Hills Prairie near Rockford, Illinois, show a range of colors from rose to white. The sharply spiny seed heads turn almost black in winter and hold well all the way until spring.

RELATED PLANTS Purple and pale purple coneflowers never or almost never grow together in the wild and their bloom times barely overlap. When the two species are cultivated together, they can hybridize, producing intermediate plants.

Glade purple coneflower (*Echinacea simulata*) is very similar to pale purple coneflower but is found more locally on prairie glades—thin soils over bedrock. Glade purple coneflower's flowers have yellow pollen rather than white but are otherwise quite similar.

Narrow-leaf purple coneflower (*Echinacea angustifolia*) is a Great Plains species whose range reaches the western Midwest as far eastward as the Flint Hills, western Iowa, and western Minnesota. It demands perfect drainage and tolerates extreme heat and drought.

Echinacea paradoxa
Yellow purple coneflower
Asteraceae (aster family)

I like to call this plant by its common name because its botanical name aptly describes this paradox in the coneflower clan. Its yellow floral pigments are the reason for the outbreak of coneflower colors beyond purple. Have hybridizers gone too far?

HOW TO GROW This Ozark Highlands endemic grows on glades and thin-soiled prairies over bedrock. It demands good drainage and does best in full sun and scorching heat though it's hardy throughout the Midwest. It's easy to propagate from seed.

LANDSCAPE USE Yellow purple coneflower is well behaved and suitable as a traditional landscape perennial because it doesn't run or self-sow. The leaves may grow about 1 foot tall with flowers atop nearly bare stems reaching to 3 feet.

ORNAMENTAL ATTRIBUTES The flower head is large and bright comprised of true yellow, 3-inch ray flower "petals" that droop like a skirt around its core 2-inch cone of rich reddish-brown disc flowers. The dark, almost black fruiting head holds well into winter.

Echinacea purpurea
Purple coneflower
Asteraceae (aster family)

HOW TO GROW Purple coneflower is a popular perennial across the entire Midwest but was originally native mainly in the Lower Midwest and not primarily in prairies but in open woodlands and savannas. It has been planted in most prairie reconstructions and naturalizes almost anywhere in the Midwest. It grows well in moist to summer-dry upland soils in full sun to partial shade. It's easy to propagate from seed.

LANDSCAPE USE Very popular in landscaping, purple coneflower is often massed in perennial borders. It may self-sow into dense plantings and allowing it to do so is a good idea because it is fairly short-lived. Purple coneflower is stunning mixed with other grasses and forbs and is at its best in a semishady natural garden. The flowers are very nectar rich, attracting many pollinators including butterflies. American goldfinches seek its seeds as soon as they are formed. It's a preferred host plant for the silvery checkerspot whose young caterpillars feast in little armies and may skeletonize the plant.

ORNAMENTAL ATTRIBUTES The showy rosy-pink ray flowers surround an orangish cone of disc flowers, blooming from

Echinacea paradoxa (yellow purple coneflower) grows abundantly with *E. pallida* (pale purple coneflower) at Paintbrush Prairie, Missouri.

Echinacea purpurea (purple coneflower), grown from native Missouri plants, flowers prolifically in the Perennial Garden at Powell Gardens, Missouri.

midsummer and sporadically into fall. The fruiting heads are spiny, ripen almost black, and are quite showy in winter, more showy when capped with snowfall.

RELATED PLANTS There has been an explosion of cultivars and hybrids of purple coneflowers. Many of the hybrids are double and do not produce nectar for pollinators.

The seed heads of *Elymus canadensis* (Canada wild rye) are reminiscent of cereal grains.

Elymus canadensis

Canada wild rye

Poaceae (grass family)

HOW TO GROW Canada wild rye is native over much of temperate North America from the Rocky Mountains to New England and is found throughout the Midwest. This cool-season grass thrives with disturbance so is found sparingly in prairies but more often in old-field meadows, roadsides, and other open areas. It thrives in well-drained soils that are at least moist in spring. Plants are readily grown from seed.

LANDSCAPE USE Canada wild rye makes a suitable cover crop in a prairie reconstruction where it matures in a couple of seasons and suppresses weeds. It is rather short-lived and declines as more long-lived plants establish. Disturbance, usually by animals, creates a niche for it to reappear. It makes a distinctive ornamental grass for any sunny natural landscape.

ORNAMENTAL ATTRIBUTES The nodding, bushy seed heads are reminiscent of its cereal grain relatives and form by midsummer, then mature a blond or sandy tan. I was chided it matched my moustache (in color and texture) when I was a younger man. The showy seed heads hold through winter when they exquisitely capture and display snow.

Eryngium yuccifolium (rattlesnake master) in bloom steals the show from marsh blazingstars at Gensburg-Markham Prairie, Illinois.

Eryngium yuccifolium

Rattlesnake master

Apiaceae (carrot family)

HOW TO GROW Rattlesnake master is native across most of the Midwest, northward to southeast Nebraska, southeastern Minnesota, and southwest Michigan. It grows wild on mesic prairies and savannas and grows well in good garden soil in full sun or partial shade. It is easy to grow from seed.

LANDSCAPE USE This very distinctive perennial that matures about 4 feet tall and makes a striking addition to a perennial border. It can self-sow abundantly when cultivated without mulch or competing plants. Rattlesnake master makes great mass plantings and serves as an integral component in a prairie restoration where it holds its own well

Euphorbia corollata (flowering spurge) flowers are comprised of petaloid bracts that last and last like the red bracts of its relative the poinsettia.

in competition with other prairie grasses and forbs. It is an incredible insectaries plant attracting many pollinators including small butterflies.

ORNAMENTAL ATTRIBUTES The sturdy, lightly forked stems are crowned with spiny white, 11/2-inch-domed flower clusters. The basal foliage is spiky and bluish green—reminiscent of a yucca. The mature stems hold well into winter with striking brown seed heads.

Euphorbia corollata
Flowering spurge

Euphorbiaceae (euphorbia family)

HOW TO GROW Flowering spurge is native over most of the Midwest, absent only from western Minnesota and North Dakota. It's found wild in prairies and meadows in mesic to dry soils. It grows in almost any well-drained soil from sand to gravel, loam, and clay in full sun. Because it develops a deep, almost woody root, it is best propagated from seed.

LANDSCAPE USE The plant grows 3 feet tall and makes a heat- and drought-tolerant substitute for the popular perennial baby's breath. It is suitable for a traditional perennial border but is at its best in a restoration or natural garden.

ORNAMENTAL ATTRIBUTES The abundant, tiny white flowers have five, long-lasting petaloid bracts produced in an airy, flattened mass over the top of the plant in summer. The foliage often turns lovely shades of red in the fall.

Gentiana alba
Cream gentian

Syn. *Gentiana flavida*
Gentianaceae (gentian family)

HOW TO GROW Cream gentian is native to Ontario and all midwestern states except the Dakotas but it's most prevalent from the Ozark Highlands northward to the Driftless Area. It grows wild in mesic to dry prairie and savannas and is readily cultivated in moist soils in full sun or partial shade. It is slow and challenging to grow from seed. Bottle and cream gentians may hybridize when grown together in a garden.

LANDSCAPE USE I'm not sure why cream gentian is not found in traditional perennial borders as I find it perfectly suited for such. It grows 2–3 feet tall and fits well in any natural prairie and partially sunny woodland garden.

ORNAMENTAL ATTRIBUTES The late summer, milky white flowers (usually earlier than bottle gentian) adorn the top of the plant and are up-facing but open-ended with pale green

Gentiana alba (cream gentian) flowers show intricate green venations.

Gentiana andrewsii (bottle gentian), growing through vinca at my boyhood home was the first bottle gentian I grew.

striations. Florists would find it a divine subject to work with—and bottle gentian too if someone would grow them for cut flowers. Cream gentian's smooth green leaves are paired along the stem, alternating at 90-degree placements for a dapper look while not in bloom.

Gentiana andrewsii
Bottle gentian, closed gentian

Gentianaceae (gentian family)

Why would a flower have such an odd, bottle-shaped flower? A patient wait near this plant will reveal large bumblebees as its pollinators—strong enough to open its twisted "closed" and fringed top, they crawl inside the flower. The treat for pollination must be a good one from the outside look and sound of what's going on inside. Some pollinators cheat and chew into the flower.

HOW TO GROW This species is native across most of the Eastern and Upper Midwest, sporadically southward barely into Kentucky and scattered throughout Missouri. It's absent only from Kansas, Oklahoma, and Arkansas but will grow across the entire Midwest. Bottle gentian is found in mesic to wet prairies, fens, sedge meadows, and savannas. It is probably the easiest gentian to cultivate in rich, moist soils—good garden soil is just fine. It is sturdiest in full sun but will grow (being more decumbent) in partial to light shade.

LANDSCAPE USE Bottle gentian is best grown with other plants in a perennial border or natural garden as stems splay outward when they mature and are rather sprawling in shade. Good companion plants include tussock sedge, soft rush, and blueflag or zigzag irises. Bottle gentian can grow as tall as 3 feet but is usually lower because of its decumbent nature.

ORNAMENTAL ATTRIBUTES The flowers are nothing short of heavenly royal blue, a memorable experience to all who see them, aging tinted wine red or purplish. They bloom at the end of the growing season, usually in September to October so are a welcome last floral hurrah in any garden. The foliage is paired and rich green, turning gold with ruddy highlights after flowering.

The fruits of *Geum triflorum* (prairie smoke) give the plant its common name (when viewed from a distance).

Geum triflorum
Prairie smoke

Rosaceae (rose family)

This is Eloise Butler's three-flowered avens of the prairies and it is a must to experience while in fruit on a windswept prairie. There's wildness in its nature: fruits with long silky-haired awns that flow like a sumptuous head of hair—a perfect masterpiece that captures the spirit of its windswept home. It's too beautiful to just be about seed dispersal. When I was a young man, prairie smoke was one of my most beloved floral memories of annual late spring visits to Hayden Prairie.

HOW TO GROW Prairie smoke is native across the Upper Midwest from west central Michigan westward to northern Illinois, Wisconsin, northeast Iowa, Minnesota, and the Dakotas. It is found on alkaline glades and prairies that are moist to dry. It is hardy across the entire Midwest but requires good drainage and air circulation in the Lower Midwest. Plant it in full sun or only partial shade. It is readily propagated from seed.

LANDSCAPE USE This showy spring perennial is suitable for a rock garden, raised perennial border, or challenging "hell strip" site. Its basal foliage cannot be overshaded by robust perennials or it will die out though makes a fine companion to shorter grasses and other diminutive plants. It's challenging to put in a prairie restoration, best planted as a plug or bare-root plant in the company of shorter grasses. The fruiting plumes grow just 1 foot above the foliage.

ORNAMENTAL ATTRIBUTES The pendant spring flowers are maroon red in clusters of three—not particularly showy but worth watching as large bumblebees pollinate them. Each pollinated flower produces an exquisite fruit that looks like smoke over the prairie when in mass and observed from a distance. Another common name based on the fruiting is "old man whiskers," but I have never once gotten that impression from prairie smoke. The basal foliage is attractive, being deeply lobed and evergreen.

Helianthus spp.
Wild sunflowers

Asteraceae (aster family)

Perennial wild sunflowers native on midwestern prairies are iconic wildflowers that symbolize the spirit of our place almost like no other. Yes, they are well named as their flowers capture the color of our bright summer sunlight from dawn to dusk. Most perennial sunflower species spread aggressively by rhizomes and are best suited to natural landscapes or left in the wild while a few species stay in a clump and are more suited to a garden setting.

HOW TO GROW Wild sunflowers are aggressive in gardens because of their adaptations to surviving on the prairie in competition with robust plants. Sunflowers contain allelopathic chemicals that deter the growth of neighboring plants; they also have vigorous rhizomes to spread vegetatively. When planted in good garden soil without competition, they run amok smothering anything in their path while becoming a nuisance to control. They can self-sow abundantly as well. Each species has a slightly different niche thriving in soils moist to dry. All can be easily grown from divisions or seed.

LANDSCAPE USE The aggressive wild sunflowers are best when planted in a natural landscape or planted in a prairie reconstruction only after more conservative grasses and forbs are well established. They can be planted with the big prairie grasses and in harsher sites to help control them. Wild sunflowers are an important component of our flora with a wide array of insects that are tied to them as a host plant and source of abundant nectar and pollen. Songbirds and other wildlife seek wild the nutritious seed fruits of sunflowers.

ORNAMENTAL ATTRIBUTES The yellow to golden flowers are smaller than the classic annual sunflower but produced in showy clusters above the stem in late summer to early fall depending on the species. The flowers are a prairie icon and fit the aesthetics of our place. Most flowers have a very distinctive scent of pure cocoa.

Helianthus grosseserratus (sawtooth sunflower) grows wild in moist to wet prairie soils across nearly the entire Midwest. It forms tall (6–8 feet or more) dense stands in

Helianthus mollis (downy sunflower) creates a lovely mass in the meadow planting at Powell Gardens, Missouri.

Helianthus pauciflorus (stiff sunflower) stands tall at Flora Prairie, Illinois.

disturbed soils, fencerows, and roadsides. I look forward to its bloom in early fall when it is one of the showiest roadside flowers—a brilliant blaze of golden yellow. The long leaves have a lovely serrated edge. This species is an aggressive grower.

Helianthus mollis (downy sunflower) is native from the Flint Hills eastward through most of Missouri, Illinois, western Indiana, and sparingly into Michigan and Ohio. It grows wild on dry upland sandy or rocky prairies where it can create extensive masses. It has glorious earlier flowers in late summer that are true yellow and its foliage is

Helianthus grosseserratus (sawtooth sunflower) gleams beneath a bright blue sky.

rounded and handsome gray-green because of its covering of whitish hairs. It, too, is an aggressive species.

Helianthus pauciflorus (stiff sunflower, showy sunflower, syn. *H. rigidus*) is native to all midwestern states except Ohio and grows in dry upland prairies where it runs into an open patch when growing in competition with other native grasses and forbs. It also blooms in late summer, the sparse up-facing yellow flowers with a rich brown center on stout tall stems and just a few pairs of small, rigid leaves. The flowers produce an intense pure cocoa aroma and the species is aggressive growing.

Helianthus occidentalis
Western sunflower
Asteraceae (aster family)

One of three less-aggressive sunflowers and a bit more suitable in gardens.

HOW TO GROW Western sunflower is native to the Ozark Highlands and Driftless Region eastward to Michigan and sparingly in Kentucky and Ohio. It grows wild on dry prairies and savannas in poorer or thin upland soils where there is bare ground and short grass. Western sunflower's short stature of mainly basal leaves makes its spread by underground rhizomes not so invasive. Even in a garden setting, this smaller stature makes it easier to control as its size prevents it from smothering most plants.

LANDSCAPE USE Western sunflower's flowering stems grow no more than 3 feet tall above low foliage, which suits it to planting in a poor, dry site in scorching sun including "hell strips" to rock gardens. It should be included in prairie reconstructions on poor soils but will be overwhelmed by other plants in rich soils.

ORNAMENTAL ATTRIBUTES Western sunflower's wide basal leaves are beautiful rich green from spring through fall. Its 2-inch blooms open earlier than most wild sunflowers in late summer and the flowers adorn tall, thin stems that are nearly leafless.

Helianthus maximiliani
Maximilian sunflower
Asteraceae (aster family)

One of three less-aggressive sunflowers and a bit more suitable in gardens.

HOW TO GROW Maximilian sunflower is native mainly to upland prairies of the Great Plains on the western edge of the Midwest from the Dakotas and western Minnesota southward to Kansas. It has naturalized eastward throughout the Midwest so its exact original range is blurry. Maximilian sunflower grows well in any well-drained soil in full sun.

LANDSCAPE USE Maximilian sunflower is a bit seedy to be used in traditional perennial borders though it does stay in a clump. It grows 5–6 feet tall in dry soil but will be taller in good garden soil. It is a perfect sunflower for a natural landscape or prairie restoration with the larger grasses as competition.

ORNAMENTAL ATTRIBUTES The early fall flowers are a rich, true yellow and produced along much of the top of the stem, making it a very showy species. The leaves also are very handsome: narrow and folded, arched downward, and are a unique gray-green.

Helianthus occidentalis (western sunflower) blooms on tall stems at Nachusa Grasslands, Illinois.

Helianthus maximiliani (Maximilian sunflower) produces tall spires of classic yellow sunflowers.

Helianthus salicifolius (willow-leaved sunflower) is beloved by gardeners for its finely textured foliage.

Heliopsis helianthoides (false sunflower) blooms for a long period through summer and into fall.

Helianthus salicifolius
Willow-leaved sunflower
Asteraceae (aster family)

One of three less-aggressive sunflowers and a bit more suitable in gardens.

HOW TO GROW Willow-leaved sunflower is endemic to the Flint Hills and western Osage Plains in Kansas, Missouri, and Oklahoma where it is found on thin soils over bedrock. It is a rare clump-forming perennial sunflower that grows in any well-drained soil—overly robust in good garden soil where it will often reach a gangly 8 feet or more in height. It can be cut back before late summer to reduce its flowering height. Willow-leaved sunflower requires heat so is best grown in the Lower Midwest but is hardy northward through USDA zone 5.

LANDSCAPE USE willow-leaved sunflower is suitable for a traditional perennial border when cut back. It is an ideal plant for poor, rocky sites in scorching heat where it will be smaller and 5–6 feet tall when blooming in early fall.

ORNAMENTAL ATTRIBUTES The foliage is very narrow and straplike creating a very fine texture that's beloved early in the growing season before the plant becomes too leggy—cutting it back extends the handsome foliage season. The flowers are an outstanding bold yellow like no other sunflower.

Heliopsis helianthoides
False sunflower
Asteraceae (aster family)

HOW TO GROW False sunflower is native across most of temperate North America including the entire Midwest. It is found in dry to mesic disturbed prairies, savannas, and woodland edges. It grows in almost any garden soil in full sun or partial shade. It is easy to grow from seed or cuttings.

LANDSCAPE USE False sunflower is already popular in traditional perennial borders where it makes a long blooming, sturdy taller perennial. Most cultivated plants are cultivars with additional or two-toned ray petals considered showier.

ORNAMENTAL ATTRIBUTES The golden, orange-yellow, midsummer flowers of false sunflower shock me every season by how long lasting and impervious to heat and drought they are. Midwestern selections of wild plants have names like 'Tuscan Sun' and I have to scratch my head why Tuscany trumps the glory of our place.

NOTES False sunflower is occasionally host to red aphids that never appear to do much damage and are not present every year. If they appear bothersome, they should be washed off with nothing harsher than mild dish soap.

Heuchera richardsonii
Alumroot
Saxifragaceae (saxifrage family)

Heucheras (pronounced HOY-ker-ahs from its German namesake) are currently very popular perennials widely planted across the Midwest, especially the new hybrids with an amazing array of colorful leaves from gold to purple with variable silver and burgundy markings. The vibrant pink-flowering coralbells from the southwestern mountains also are beloved. Alumroot has plain green leaves but is hands down the hardiest species growing on prairies throughout the northern Great Plains. It has just recently become utilized in hybridization of hardier heucheras.

HOW TO GROW Alumroot is native from the Canadian Great Plains southeastward to southwestern Michigan, northwestern Indiana, south central Illinois, southern Missouri, and eastern Oklahoma. It's found on dry to mesic upland

273

Heuchera richardsonii (prairie alumroot) flowers are subtly beautiful in a light shade of green.

Koeleria macrantha (Junegrass) at Sugar River Forest Preserve outside Rockford, Illinois, flowers and fruits earlier than most grasses.

prairies, savannas, and open woodland. It grows in almost any well-drained soil in full sun to light shade and can readily be propagated from seed.

LANDSCAPE USE This handsome perennial can be planted in masses near the edge of a perennial border, in rock gardens, or in semishady woodland gardens. The leaves seldom reach 12 inches in height with stems of airy flowers rising 24–30 inches.

ORNAMENTAL ATTRIBUTES The evergreen leaves are handsome from spring through winter, sometimes with purplish undersides. The flowers need close inspection to appreciate them as they are light olive green, asymmetrical with a longer topside, and sport cute orange protruding stamens—they may not be pink like coralbells but hummingbirds nectar from them just the same.

Koeleria macrantha
Junegrass

> Syn. *Koeleria cristata*
> Poaceae (grass family)

HOW TO GROW Junegrass is found over most of the Midwest, though rare or absent from much of the eastern Lower Midwest. It is most common in dry upland sand and gravel prairies and in savannas. Because it is a cool-season grass,

it greens up and blooms earlier than most of the region's native grasses. It requires well-drained soil (rotting in clay) in a garden in full sun or partial shade. It's easy to grow from seed or by dividing a clump.

LANDSCAPE USE This clump-forming grass is suitable for a traditional landscape but requires such well-drained soil it is best grown in rock gardens or raised sand beds. The plant is reminiscent of the popular (nonnative) 'Karl Foerster' ornamental reedgrass (*Calamagrostis ×acutiflora*), but smaller, growing just 2–3 feet tall.

ORNAMENTAL ATTRIBUTES Junegrass blooms in May or June with upright plumes of flowers that are most striking when backlit, a sparkling golden pale green that tightens up to a spear shape after flowering and ripens tan with seeds. The basal tuft of leaf blades is light to bluish green.

Liatris spp.
Blazingstars
Asteraceae (aster family)

Blazingstars are a quintessential midwestern wildflower though they're found through much of North America. I group the various blazingstar species found across the Midwest into three categories: those with dense flower heads giving a wandlike appearance to the 3- to 5-foot flowering stem (*Liatris pycnostachya*, *L. spicata*); those with flowers in looser, buttonlike clusters along the stem that grow more than 2 feet tall (*L. aspera*, *L. ligulistylis*, *L. scariosa*); the more diminutive species, 1–2 feet tall with fewer or otherwise bottlebrush-looking flowers (*L. cylindrica*, *L. punctata*). Blazingstars are one of the few perennials that bloom from the top of the stem sequentially to the bottom.

Most blazingstar species form an underground corm that is a favorite food of voles, and many species are short lived. Losing plants to critters and a short lifespan translates into the reality that gardeners should let their blazingstars self-sow or they may lose them. It is amazing how blazingstar displays may vary year-to-year on some native prairies as well as in garden settings.

Blazingstars are amazingly rich in nectar and attract a wide array of pollinators, especially butterflies. You can have blazingstars blooming in your garden from midsummer into fall if you plant appropriate early to late-flowering species. All produce seed fruits that make an excellent food source for birds and other wildlife.

Migrating monarchs can't resist the nectar from *Liatris ligulistylis* (meadow blazingstar) on a roadside prairie remnant near Decorah, Iowa.

Liatris ligulistylis

Meadow blazingstar, Rocky Mountain blazingstar

Asteraceae (aster family)

Liatris ligulistylis and the two other buttonlike-flowering blazingstars bloom in late summer into fall when monarchs are migrating and can be among the best source of nectar for fattening up the butterflies. I've witnessed all three species adorned with flocks of monarchs, a sight I hope everyone and future generations will have.

HOW TO GROW Meadow blazingstar is native to the central and northern Rockies, but has a widespread range across the central Midwest from the eastern Dakotas and Wisconsin southward sporadically into Illinois and Missouri. I know it from mesic upland prairie remnants including roadsides.

LANDSCAPE USE The button-type blazingstars are best in a more natural landscape or planted in a mass of the smaller prairie grasses such as little bluestem, prairie dropseed, and sideoats grama. Meadow blazingstar establishes well in prairie reconstructions and is usually in peak bloom around Labor Day.

ORNAMENTAL ATTRIBUTES The lavender-purple flowers are part of the floral color scheme indicative of the prairie, though all blazingstar species occasionally produce a white-flowered form. All produce whitened tufts of fruit that also are quite attractive in the fall but they quickly disintegrate and disperse the seeds.

Nothing says autumn on the prairie like the buttonlike purple flowers of *Liatris aspera* (rough blazingstar).

Liatris punctata (dotted blazingstar) blooms below side-oats grama along a roadside in the Flint Hills of Kansas.

Liatris scariosa (savanna blazingstar) flower heads may be the largest of any midwestern blazingstar.

RELATED PLANTS Rough blazingstar (*Liatris aspera*) grows wild across the entire Midwest but is more localized in most of Indiana, Ohio, and Kentucky. It can be found on drier upland prairies so needs full sun and good drainage in a garden setting and is extremely drought tolerant. Like meadow blazingstar, it is suitable for prairie reconstructions but it blooms later in September and into October

Savanna blazingstar (*Liatris scariosa*) is more of a savanna species growing in a diagonal sporadic range from northwestern Arkansas to western Illinois, southern Chicago region, northern Indiana, northern Ohio, and across Lower Michigan. There are outliers in southern Indiana and Ohio. It grows in mesic sites in full sun or partial shade in well-drained soil, readily self-sowing when its cultural needs are met. It is appropriate for savanna reconstructions and blooms with meadow blazingstar around Labor Day.

Liatris punctata
Dotted blazingstar
Asteraceae (aster family)

Dotted blazingstar is one of two diminutive species, 1–2 feet tall with few or otherwise bottlebrush-looking flowers. These smaller blazingstars are native in harsher conditions (hot, dry, in poor soil) where large, competing plants are absent.

HOW TO GROW Dotted blazingstar is a Great Plains species native only in the western Midwest eastward to the western Osage Plains, Loess Hills, and into the Twin Cities region as far east as westernmost Wisconsin. It has naturalized in a few spots eastward to Michigan. It thrives in upland prairies that are rocky, gravely, steep loess or other very well drained, summer-dry substrate. It is easy to cultivate in similar situations of droughty, well-drained soil in full sun.

LANDSCAPE USE The smaller blazingstars are fine additions to rock gardens, above walls, in container gardens, or on the edges of raised perennial borders with smaller grasses.

ORNAMENTAL ATTRIBUTES The lavender-purple flowers are part of the floral color scheme indicative of the prairie, though all species occasionally produce a white-flowered form. All produce whitened tufts of fruit that also are quite attractive in the fall but they quickly disintegrate and disperse the seeds.

Liatris cylindrica (cylindric blazingstar) is usually the smallest of the Midwest's blazingstars.

This spectacular clump of *Liatris spictata* (marsh blazingstar) resided on the Gensburg-Markham Prairie, Illinois.

RELATED PLANTS Cylindric or dwarf blazingstar (*Liatris cylindrica*) is native through the center of the Midwest from central Minnesota southward through the Driftless Area, most of Illinois and into the Ozarks, and eastward across southern Wisconsin, northern Illinois, northern Indiana, Lower Michigan, and into Ontario. There are outliers in Ohio and Kentucky. It is found on dry upland hill and gravel prairies, glades, and oak openings. It grows in full sun to partial shade in well-drained soils with smaller plants that won't smother it. It often blooms best after a fire or other disturbance opens up space for it.

Liatris spicata

Marsh blazingstar

Asteraceae (aster family)

Liatris spicata is one of two wandlike blazingstars with dense flower heads. It is the most common blazingstar in cultivation with its strain 'Kobold' readily available at any midwestern nursery. It is also a popular cut flower found at florist shops as well. Personal observations have revealed that more butterflies and pollinators visit local native strains of marsh blazingstar. It's a dazzling native wild-flower blooming in midsummer on mesic and wet prairies around the Chicago region: Gensburg-Markham Prairie's marsh blazingstar stands are as beautiful as any floral display anywhere.

HOW TO GROW This species is found sporadically across eastern North America from the Midwest to the East and Gulf Coasts. It's mainly an Eastern Midwest species though it grows northwestward into southeastern Wisconsin and sporadically downstate in Illinois, but it's not native west of the Mississippi River in the Midwest. It grows in moist to wet prairies and savannas so does well in average garden soil but also thrives in wet locations in full sun or just a bit of shade. It's easy to propagate from seed and forms a corm that is easily transplanted. The corm is in widespread nursery production and widely available, but the native version is rarely sold that way.

LANDSCAPE USE The tall, wand-flowering blazingstars are lovely in traditional perennial borders, pondside or wetland gardens, and in rain gardens. The two species can be planted together as marsh blazingstar will bloom first in midsummer, then similar prairie blazingstar blooms in later summer extending the bloom season. They are easily added to prairie reconstructions in rich to wet soils and rather quickly make a spectacular display.

ORNAMENTAL ATTRIBUTES The lavender-purple flowers are part of the floral color scheme indicative of the prairie, though all species occasionally produce a white-flowered form. All produce whitened tufts of fruit that also are quite attractive in the fall but they quickly disintegrate and disperse the seeds.

RELATED PLANTS Prairie blazingstar (*Liatris pycnostachya*) is essentially a middle-of-the-country species growing from Minnesota and Wisconsin south to Texas and Louisiana. It is found in the tallgrass prairie between the Great Plains and Lake Michigan. There are rare outliers in the Eastern Midwest but it is not considered native to Michigan.

Liatris pycnostachya (prairie blazingstar) forms a sweep of purple in the prairie wave border at Powell Gardens, Missouri.

It prefers moist to seasonally wet prairies, fens, and sedge meadows and performs well in a garden in full sun, in humus-rich loamy soil.

Lupinus perennis (wild lupine) thrives at Indiana Dunes National Lakeshore.

278

Lupinus perennis
Wild lupine
Fabaceae (legume family)

Wild lupine was historically collected from a sand prairie near my hometown as the only site in Iowa. In 1982, I rediscovered the species in the state in a sand savanna near the Mississippi River.

HOW TO GROW Wild lupine is native from eastern Minnesota and the northeast corner of Iowa across Wisconsin and Michigan southward to northeastern Illinois, northern Indiana, and northern Ohio. It grows wild only in sand prairies, sand savannas, and sandy lakeshores so is limited to growing only in similar sandy, well-drained garden soils. Wild lupine requires full sun or only partial shade and can be grown from seed with proper inoculation. The full range of the species includes the Atlantic and Gulf Coastal Plains so the species is tolerant of heat and humidity.

LANDSCAPE USE Wild lupine is excellent for a traditional perennial border, prairie reconstruction, or other natural landscape but only in sandy soils. It grows just 1–2 feet tall. It is the only host plant for the rare butterfly Karner blue and the flowers are rich in nectar for many pollinators.

ORNAMENTAL ATTRIBUTES The spikes of deep blue flowers in late spring are among the most beautiful of our wildflowers. The palmate leaves also stand out with their uniqueness, and seeing them with water droplets is pure magic.

Monarda fistulosa
Wild bergamot
Lamiaceae (mint family)

Some of my fondest memories outdoors are of perusing stands of wild bergamot while looking for eastern tiger and giant swallowtail butterflies. Wild bergamot blooms in late summer in the Upper Midwest when the larger, second brood swallowtails are at their peak flight and they readily congregate on the flowers for nectar. Bergamot's aroma brings me back to those summer days, in the heat and humidity combined with bergamot's showy flowers with swarms of colorful and intriguing insects.

HOW TO GROW Wild bergamot is native through much of temperate North America and is found throughout the entire Midwest as the core of its range. It grows wild in disturbed prairies, meadows, savannas, and other open lands. It is easy to grow in any good garden soil in full sun or partial shade. It is easy to divide groupings of plants or to grow plants from seed.

LANDSCAPE USE Wild bergamot lightly spreads into a clump and reaches about 4 feet tall. It's suitable for a perennial border but the leaves' susceptibility to mildew make it unsightly some seasons to some gardeners. It is a must for a

The lavender flowers of *Monarda fistulosa* (wild bergamot) are among the best to attract bees and butterflies.

Oenothera macrocarpa (Missouri primrose) grabs attention with large sunny yellow flowers on decumbent plants.

ORNAMENTAL ATTRIBUTES The lipped, light lavender-purple flowers are elegantly exquisite arranged in a whorl around a rounded flower head. The foliage is exceedingly fragrant when rubbed or crushed. The seed heads hold well into winter, retaining their bergamot aroma.

NOTES I have witnessed gardening friends try all sorts of ridiculous concoctions to control mildew on this plant, it is purely a cosmetic disorder that I don't let bother me. If you forgo a wild bergamot solely because of mildew on its leaves, you will miss out on a wealth of experiences that define our region.

Oenothera macrocarpa
Missouri primrose

> Syn. *Oenothera missouriensis*
> Onagraceae (evening primrose family)

HOW TO GROW Missouri primrose grows wild on rocky prairies and glades in the central and southern Great Plains from southern Nebraska to Texas and eastward through the Ozark Highlands of Missouri and Arkansas. In a garden it prefers well-drained soil in full sun. It may be propagated by cuttings or from seed.

LANDSCAPE USE The plant grows under 1 foot tall but may spread as much as 3 feet. It is a good choice for the edge of a perennial border, rock garden, or where allowed to spill over a rock wall. It is in the hottest, driest, rockiest locations where other perennials languish.

ORNAMENTAL ATTRIBUTES Missouri primrose is a day-flowering sundrop type in its genus, opening in the evening but holding well through a sunny day. Its huge yellow flowers occur in early to midsummer and are as brilliant yellow as the summer solstice sun. The long floral tube is pollinated by sphinx moths and produces very unique fruits about 3 inches long with papery wings that allow them to blow in the wind.

RELATED PLANTS The silvery-leaved Great Plains variety *incana* is highly ornamental and flowers well into fall. Cultivars 'Comanche Campfire' and 'Silver Blade' are part of that variety.

Oligoneuron spp.
Goldenrods
Asteraceae (aster family)

Once these plants were listed under both *Aster* and *Solidago*, but now these distinctive flat-topped flowers species are included in their own genus.

natural landscape or prairie restoration, as it thrives in disturbed ground so functions as a nurse plant. It attracts many bees and butterflies to its nectar-rich flowers. The foliage is host to the hermit sphinx moth, a very important pollinator on prairies—including the main pollinator to the endangered western prairie fringed-orchid. Wild bergamot is a native herb with aromatic foliage that can be brewed into a tea.

Oligoneuron album
Upland white aster

Syn. *Solidago ptarmicoides, Aster ptarmicoides*
Asteraceae (aster family)

I first learned this plant as an aster but it is proven to be a goldenrod—prairie goldenrod is its current promoted common name. You may wonder why I still use its common name aster—I'll use bird names as an analogy: the popular bird scarlet tanager is now known to be in the cardinal family and not a tanager—its common name won't change to scarlet cardinal. Upland white aster is very unlike the Midwest's other goldenrods as its flowers are larger, more like an aster, and white in color atop diminutive plants, unlike most goldenrods. It even hybridizes with other goldenrod species, and the hybrids were formerly known as ×*Solidaster*—a combination of *Solidago* (goldenrod) and *Aster* (asters). Solidasters have very showy soft yellow disc and ray flowers, and I have photographed them in the wild at Chiwaukee Prairie.

HOW TO GROW Upland white aster is native across the Upper Midwest from northwestern Ohio westward to South Dakota with a disjunct population in the Ozark Highlands. It is found in dry upland prairies, glades, and savannas in the company of smaller plants able to survive in harsh conditions. It grows in well-drained soil in full sun and is easy to propagate from seed.

LANDSCAPE USE The plant is suitable for a traditional perennial border providing the soil is very well drained. This aster grows just 1 foot tall so is perfect for the edge of a border where larger perennials are not allowed to crowd it out. It also is a great choice in rock gardens, atop rock walls, or in harsh conditions of droughty or rocky soils in hot afternoon sun. As with other goldenrods, the flowers are nectar rich and great for pollinators while songbirds seek the seeds.

ORNAMENTAL ATTRIBUTES Upland white aster blooms in late summer with airy white flowers in an open dome above the basal foliage.

NOTES In my Missouri garden, an upland white aster plant I brought from the Upper Midwest blooms in June, more than a month before a Missouri-sourced plant I have. Both set no seeds so I also suspect they need cross-pollination. They're a prime example of why it's important to cultivate locally sourced plants.

Oligoneuron rigidum
Stiff goldenrod

Syn. *Solidago rigida*
Asteraceae (aster family)

On a remnant prairie, I once came across an indelible sight: a roost of monarchs nectaring on a healthy stand of this goldenrod. Stiff goldenrod is very rich in nectar for all sorts of pollinators and produces copious amounts of seed that songbirds also savor.

HOW TO GROW Stiff goldenrod is native across the entire Midwest and is found on dry to mesic upland prairies and

The starry white flowers of *Oligoneuron album* (upland white aster) illuminate Sumner Prairie, Iowa.

Migrating monarchs mob a wild stand of blooming *Oligoneuron rigidum* (stiff goldenrod) on a private prairie remnant in Iowa.

The airy panicles of switchgrass make it garden standout in fall and winter. Shown here is *Panicum virgatum* 'Cloud Nine'.

The chunky star-shaped flowers of *Parthenium integrifolium* (wild quinine) are somewhat reminiscent of a cauliflower head.

disturbed ground. It grows in full sun to partial shade in any well-drained soil and is easy to grow from seed.

LANDSCAPE USE This goldenrod makes a fine garden plant if grown in poor, dry soil with established plants but becomes too gangly and will self-sow into a nuisance in rich garden soil without competing plants. It is best in natural gardens and prairie restorations where it has very high value to all sorts of wildlife.

ORNAMENTAL ATTRIBUTES The large, strappy gray-green basal foliage is quite handsome and turns shades of orange and red in the fall. The flowers are true yellow on up-facing flat inflorescences that are quite distinctive. Stiff goldenrod also is ornamental when the seed fruits mature, each with a white crownlike tuft atop the fruit that helps carry it in the wind and gives the plant a frothy almost floral look.

Panicum virgatum
Switchgrass

Poaceae (grass family)

Switchgrass is one of the tall grasses of the prairie and has been planted extensively in conservation reserve program lands and is tested as a biofuel. It's become a rather popular garden perennial with many cultivars of the grass found in nurseries across the Midwest including those with bluer leaf blades, burgundy wine red foliage, upright growth or airy but magnificent size.

HOW TO GROW Native across much of the North American continent from the Rocky Mountains eastward, switchgrass grows throughout the Midwest on wet to mesic prairies, from floodplains to uplands. In a garden it is tolerant of all types of soils in full sun. It is easy to grow from seed and established clumps can be divided.

LANDSCAPE USE Local strains of the plant are appropriate in prairie reconstructions and only in small quantities as individual plants can become large and invasive and overtake many smaller grasses and forbs. Switchgrass is highly flammable and burns very hot, so is not recommended as a foundation plant. It stays in a clump, usually grows 4–5 feet tall (some strains reach 8 feet), and self-sows lightly but can be readily maintained in a traditional perennial border. It's also used in rain gardens and swale gardens that collect storm water runoff. Switchgrass also makes a good hedge or seasonal screen.

ORNAMENTAL ATTRIBUTES This classic warm-season grass emerges with warm weather and flowers in late summer with a very airy open panicle of florets. It looks wonderful through fall when it goes to seed and even looks fine into winter. The foliage usually turns amber yellow to orange in the fall and retains a warm golden blond look in the winter landscape.

Parthenium integrifolium
Wild quinine

Asteraceae (aster family)

HOW TO GROW Wild quinine is native almost throughout Missouri and Illinois and into adjacent states through eastern Iowa, southeastern Minnesota, southern Wisconsin, northwestern Indiana, and western Kentucky. It is found in mesic to dry upland prairies and savannas and in the garden grows well in most well-drained soils in full sun to partial shade. It is easy to propagate by seed.

LANDSCAPE USE Because it self-sows abundantly, wild

quinine can be seedy in a traditional landscape though it otherwise stays in a sturdy clump of foot-long, crimped-edged leaves. The flower stalks rise 3–4 feet above the foliage. The plant is best used in mass in natural landscapes and as a component in a prairie reconstruction.

ORNAMENTAL ATTRIBUTES The long-lasting, somewhat cauliflower-like white flower heads are sturdy and resist the extremes of weather. Individual flower heads have five cute little "petals" that are actually ray flowers. The seed heads turn almost black in late fall and are quite architectural as they sturdily stand into winter.

The white flowers of *Penstemon digitalis* (foxglove penstemon) sparkle in the Perennial Garden at Powell Gardens, Missouri.

Penstemon digitalis

Foxglove penstemon

Plantaginaceae (plantain family)

HOW TO GROW Foxglove penstemon is native from New England westward to the Midwest and southward to Louisiana. It does not occur in most of Minnesota, Iowa, or Nebraska and points northwest. It's found wild in prairies, savannas, and other open lands. It readily grows in well-drained garden soil in full sun or partial shade. It is easy to propagate from seed.

LANDSCAPE USE This penstemon is suitable for a traditional perennial border where several of its cultivars are readily grown. It grows about 30 inches tall and stays in a clump. It should be considered for an evening garden where its white flowers glow and attract pollinating sphinx moths. Foxglove penstemon also is a fitting plant for a natural landscape, in prairie reconstructions, or in the company of open trees. It attracts many species of pollinators from bees to hummingbirds including a mason bee (*Osmia distincta*) that is a penstemon specialist.

ORNAMENTAL ATTRIBUTES The showy white flowers are shaped like foxgloves on upright stems above the mainly basal foliage. They bloom in early summer, welcome flowers in the greens of June. The seedpod fruits often turn reddish in late summer, dry brown in fall, and hold well into winter.

RELATED PLANTS The cultivar 'Husker Red' (from Nebraska) has purplish leaves and the Perennial Plant Association awarded it Perennial Plant of the Year in 1996.

Calico penstemon (*Penstemon calycosus*) is very similar to foxglove penstemon but more moisture demanding and its range is centered on Illinois, Indiana, Ohio, Kentucky, and Tennessee, sparingly beyond. It grows in moist prairies, savannas, rocky woodland, and adjacent disturbed areas. It is easy to cultivate in rich, moist soils but is short-lived in dry conditions. It grows in full sun to partial shade and may be propagated from seed. Its flowers are fused with

Fine white hairs stand out on the sepals, buds, and flowers of *Penstemon calycosus* (calico penstemon).

a lovely violet pink and have longer green sepals than foxglove penstemon.

Trumpet penstemon (*Penstemon tubaeflorus*) also is very similar with gorgeous pure white flowers on short stalks in whorls along the flower spike. It is easy to grow from seed but nigh impossible to grow in a garden and demands good drainage in sandy loam to gravelly soils. I have never had the plant persist at home or at Powell Gardens and never see it in gardens, though it may be spectacular along harsh roadsides or in its dry prairie haunts. It's native from eastern Kansas and Oklahoma eastward sparingly to southern Illinois and has occasionally naturalized elsewhere as far away as Wisconsin and Ontario.

Penstemon tubaeflorus (trumpet penstemon) produces whorls of beautiful white trumpet-shaped flowers.

Flowering groups of *Penstemon grandiflorum* (large-flowered penstemon) often have white or variable lilac-colored flowers.

Penstemon grandiflorum
Large-flowered penstemon

Plantaginaceae (plantain family)

HOW TO GROW Large-flowered penstemon is native through the Great Plains eastward into the Flint Hills, Loess Hills, and across Minnesota, through Wisconsin, and into northern Illinois. It is found in dry loess, sandy, or gravelly prairies and grows in exceedingly well-drained soils in full sun. It is short-lived, very much so in typical garden soil. It is easy to propagate from seed.

LANDSCAPE USE Large-flowered penstemon is a 2- to 3-foot-tall, short-lived perennial for a rock garden, above a wall, or in a container garden. It is suited to natural gardens with appropriate sandy, loess, or gravelly settings where it can self-sow and persist. Bumblebees pollinate its large flowers.

ORNAMENTAL ATTRIBUTES The flowers bloom in early summer and are in clear shades from lilac to white. The paired leaves are smooth and blue-gray-green contrasting with many plants—looking like open clamshells along the upright stem. The seedpod fruits hold well into winter.

RELATED PLANTS Showy penstemon (*Penstemon cobaea*) is similar to large-flowered penstemon but is overall fuzzier and its flowers are light lilac to white. It's native in dry prairies and glades in the southern Great Plains and throughout the Flint Hills, western Osage Plains and northward into

Penstemon cobaea (showy penstemon) has the largest and most colorful flowers of midwestern penstemons.

southeast Nebraska and the southwestern corner of Iowa. The plants native to the Ozark Highlands in southwestern Missouri and northwestern Arkansas have deep purple flowers and are sometimes considered a separate variety *purpureus*. Showy penstemon is easier than *P. grandiflorum* to grow in average garden soil but prefers rocky, well-drained soils so is best cultivated in a rock garden.

Phlox pilosa
Prairie phlox, downy phlox
Polemoniaceae (phlox family)

The tallgrass prairie remnant, Hayden Prairie in northeast Iowa, was where I really got to know this plant a long time ago. That prairie's diversity of this flower is like no other population I have ever witnessed because the flowers there can be white, pink, lavender, or purplish, and each of these variously colored flowers may be with or without eyemarks around the flower's center. The local prairie phlox in openings (long since swallowed up by brush) were always fuchsia pink.

HOW TO GROW Downy phlox is native in upland prairies and savannas throughout the Midwest and scattered eastward and southward to the Gulf. It is somewhat of a mystery to grow in a traditional landscape or prairie restoration, often surviving and blooming for a season but not returning. I have just one strain surviving after many, many plantings and it's the cultivar 'Eco Happy Traveler' that originated outside the Midwest. Downy phlox is easy to propagate by cuttings or grown from seed. Seed is challenging to collect, unless fruiting heads are bagged as they explode when ripe.

LANDSCAPE USE This phlox can be used as a trial plant for a perennial border with shorter grasses and in natural landscapes and prairie restorations. Its early summer bloom is a welcome sight atop a 1- to 2-foot-tall plant.

ORNAMENTAL ATTRIBUTES The early summer bloom is usually a fuchsia pink but local populations may be any color from white to pink, lavender, and purplish. The foliage is evergreen and spiky but sparse so the flower celebrates the plant's presence.

Phlox pilosa (prairie phlox) grows with leadplant, prairie onion, and pussytoes on the gravelly edge of a driveway.

Physostegia virginiana
False dragonhead, obedient plant
Lamiaceae (mint family)

I know of no other plant that is so obedient—you can move the flowers side to side and they stay where you put them.

HOW TO GROW This species is native through much of temperate eastern North America including the entire Midwest. It is found in mesic to wet prairies, open floodplain woodlands, and other wetland edges. It readily grows in good garden soil and wet soils in full sun or partial shade. It is easiest to propagate by dividing a clump.

LANDSCAPE USE False dragonhead grows 3 to occasionally 4 feet tall and runs by underground rhizomes to form an extensive thicket in moist, rich soils—becoming aggressive

Physostegia sp. (false dragonhead) produces stunning towers of flowers. This plant was labeled as *P. virginiana* but is probably *P. angustifolia.*

Pulsatilla patens (Pasque flower) is a harbinger of spring at Harlem Hills Prairie, Rockford, Illinois.

in gardens without competing plants. It's a better choice for natural gardens than for traditional gardens. Insects with long proboscises, including sphinx moths, pollinate the tubular flowers.

ORNAMENTAL ATTRIBUTES The tubular flowers arranged on a spike at the top of a stem are very showy when in bloom in late summer. The flowers may be white to soft pink or pale lavender.

RELATED PLANTS *Physostegia angustifolia* is a closely related plant and sometimes impossible to separate without looking at its minute inflorescence hairs. It appears to spread less and in looser clumps in a garden setting.

Pulsatilla patens

Pasque flower

> Syn. *Anemone patens*
> Ranunculaceae (buttercup family)

This spring wildflower is Eloise Butler's "crocus in chinchilla fur" and a harbinger of spring across rocky savannas and prairies of the Upper Midwest. I recall that the first one I saw was much larger than I envisioned. Pasque flower dwarfs even the largest Dutch bulb crocus and is the state flower of South Dakota.

HOW TO GROW This species grows in dry upland prairies and savannas and requires perfectly draining, lean soil and springtime sun. It cannot compete with robust, let alone larger plants. It is challenging to grow from seed and to

transplant, which is why the European species (*Pulsatilla vulgaris*) is more readily available at nurseries.

LANDSCAPE USE Pasque flower makes a premier rock garden plant and thrive in dry, rocky garden locations where few other plants would. It grows just 12 inches tall.

ORNAMENTAL ATTRIBUTES The light mauve or lavender flowers wear a coat of silky hair that feels as soft as chinchilla fur. The inside of the flowers is smooth and surrounds a warm center of yellow stamens. The flower produces a wild hairdo (recall Einstein?) seed head fruit that adds interest after the flowers have faded in late spring. The soft, delicate leaves are deeply divided and also adorned with silky hairs and provide ornamental appeal after the seed heads shatter.

Pycnanthemum spp.
Mountain-mint
Lamiaceae (mint family)

HOW TO GROW Mountain-mints thrive in good garden soil, blooming best in full sun. They are easy to grow from seed but form clumps that also may be readily divided. When the species listed below are grown together in a garden, they may hybridize.

LANDSCAPE USE The plants reach 24 inches tall and are superior insectaries providing nectar for more pollinators and beneficial insects than virtually any other group of plants. I have found these to be perfect plants for a traditional perennial border though they are rarely utilized that way. They epitomize a healthy landscape as they provide nectar for so many beneficial insects.

ORNAMENTAL ATTRIBUTES The off-white tiny flowers appear in showy flat clusters atop the plant in late summer. The stout seed heads hold well through winter, retain a modest mint smell, and are some of the best for winter interest.

Pycnanthemum tenuifolium (slender mountain-mint) is native across the Eastern and Lower Midwest northwestward into Wisconsin, Iowa, and the southeastern corner of Nebraska. It has the narrowest leaves showing a refined

A regal fritillary clings to *Pycnanthemum tenuifolium* (slender mountain-mint) flower on windswept Friendly Prairie, Missouri.

texture but lacks a minty aroma. This species is the most drought tolerant of the mountain-mints.

Pycnanthemum verticillatum var. *pilosum* (whorled mountain-mint) is found across most of Missouri and Illinois and sparingly into all adjacent states. This species has slightly larger flowers with larger purple speckles that may produce more nectar, as pollinators seem to choose it over the other two. I had a large clump of this perennial outside my previous home's kitchen window where I would enjoy

butterflies nectaring on the flowers—the place I first spied a dazzling white M hairstreak in the garden.

Pycnanthemum virginianum (Virginia mountain-mint) is native across the entire Midwest but scattered in peripheral areas. It is found in moist to wet prairies, fens, and sedge meadows but grows well in good garden soil and tolerates being wet better than the other species. It is a marvelous native herb with a fresh minty aroma that can be made into a tea.

Pycnanthemum virginianum (Virginia mountain-mint) shows flat-topped white flower clusters indicative of all mountain-mints.

Pycnanthemum verticillatum var. *pilosum* (whorled mountain-mint) produces the largest individual flowers of the midwestern mountain-mints.

Ratibida pinnata
Gray-headed coneflower, yellow coneflower
Asteraceae (aster family)

Gray-headed coneflower is named after its distinctive oval fruiting heads ("cones") that mature gray and gradually disintegrate through the winter. These cones have a distinctive and pleasant aroma (reminding some noses of black licorice).

HOW TO GROW Gray-headed coneflower is native almost throughout the Midwest, absent only from some peripheral edges, and is found in disturbed upland prairies, meadows, and old fields. It requires well-drained soil in full sun and self-sows readily into bare soil. It is somewhat short-lived and cannot compete with dense native prairie grasses and forbs—though seed remains in the seed bank until a burrowing animal or other force exposes it and allows it to germinate. This species is easy to propagate from seed.

LANDSCAPE USE Gray-headed coneflower is perfect for a prairie reconstruction as it grows quickly, readily flowering

Ratibida pinnata (gray-headed coneflower) epitomizes the rich yellow flowers of the prairie.

Rudbeckia missouriensis (Missouri black-eyed Susan) paired with *Silene regia* (royal catchfly) was a stunning combo in my former garden in Odessa, Missouri.

by the second year. As other plants mature, its abundance wanes but it always fills in if the planting is disturbed. Gray-headed coneflower grows about 4 feet tall but is usually too seedy to be included in a traditional perennial border.

ORNAMENTAL ATTRIBUTES The showy flowers consist of pendant, pure yellow ray flowers hanging below a brown oval cone of disc flowers that have golden-orange pollen. The flowers bloom in mid to late summer and are heavily visited by pollen-collecting bees.

NOTES Powell Gardens' staff have found two double-flowering gray-headed coneflowers with a dense cluster of ray flowers but still adorned with a lovely cone of disc flowers.

Rudbeckia spp.
Black-eyed Susans
Asteraceae (aster family)

There are several yellow ray-flowered "petaled" dark center disc-flowered "coned" wildflowers called black-eyed Susan. The classic black-eyed Susan (*Rudbeckia hirta*) is an annual or biennial (see description in the annuals chapter). Below are two true perennial black-eyed Susan species. A third perennial species, often called orange coneflower (*Rudbeckia fulgida*), self-sows so prolifically that it is best planted in a mass (see description in the groundcovers chapter).

Rudbeckia missouriensis
Missouri black-eyed Susan
Asteraceae (aster family)

HOW TO GROW Missouri black-eyed Susan is endemic to the Ozark Highlands growing barely into Illinois with a few populations southward to the Gulf. It is found in mesic, rocky prairies, savannas, and open woodland and grows well in moist garden soil in full sun or partial shade. It is hardy throughout the Midwest and is easy to grow by dividing clumps or from seed.

LANDSCAPE USE Missouri black-eyed Susan is a premier plant for a traditional perennial border. It grows 24–30 inches tall and is appropriate for savanna and prairie restorations within its limited native range. It's a fine nectar and pollen plant for any insectaries garden.

ORNAMENTAL ATTRIBUTES The plant blooms in late summer with classic, glowing orange-yellow black-eyed Susan-style flowers but is a long-lived perennial. The seed heads hold well into winter on stems over the evergreen basal foliage.

Rudbeckia subtomentosa (sweet black-eyed Susan) creates a mass of gold in the Perennial Garden at Powell Gardens, Missouri.

Rudbeckia subtomentosa
Sweet black-eyed Susan
Asteraceae (aster family)

HOW TO GROW Sweet black-eyed Susan is found across most of Missouri and Illinois and into contiguous states in prairies or disturbed open lands. It grows in moist, well-drained soil in full sun or partial shade. The species is easy to propagate from cuttings or seed and readily self-sows if not deadheaded.

LANDSCAPE USE This tall, sturdy perennial grow to around 5 feet. It makes an acceptable back-of-the-border perennial but can self-sow to the point of being a nuisance in meticulous landscapes. It is best suited to a natural landscape or prairie reconstruction where other plants can compete with it.

ORNAMENTAL ATTRIBUTES The very showy flowers are in the classic black-eyed Susan format of orange-yellow ray

flowers surrounding a brown cone of disc flowers. The seed heads hold well into winter.

RELATED PLANTS This species shows variation. A plant with fluted ray flowers like a spoon chrysanthemum was found in Illinois and is sold as the cultivar 'Henry Eilers.' I found a semidouble-flowering plant with extra ray flowers surrounding its cone on prairie remnants at Powell Gardens and it's now on display in the gardens there.

Salvia azurea (pitcher's sage) is a beautiful component of this traditional perennial border at the Chicago Botanic Garden.

Salvia azurea

Pitcher's sage

Lamiaceae (mint family)

HOW TO GROW Pitcher's sage is mainly native in the central and southern Great Plains into southeastern Nebraska, southwestern Iowa, western Missouri, western Arkansas, and points south and west in dry to mesic upland prairies. It has naturalized sparingly northward and eastward across the Midwest. It is easy to grow in almost any nonwet garden soil in full sun. Gardeners may propagate it by cuttings or seed.

LANDSCAPE USE Pitcher's sage is occasionally seen in traditional perennial borders where its selection 'Grandiflora' is quite showy and grows around 3 feet tall. Most strains are ganglier looking and are at their best in a natural landscape where they are mixed with other native grasses and forbs. Within its native range, pitcher's sage has a specialist species of bee (*Tetraloniella cressoniana*). It is nectar rich and visited by many pollinators and a great choice for an insectaries garden.

ORNAMENTAL ATTRIBUTES Flower color is close to sky blue, a very rare hue in midwestern native plants, so the flowers really standout when in bloom in late summer into fall.

The fall color and illuminated backlit awns of *Schizachyrium scoparium* (little bluestem) make it one of the showiest prairie grasses.

Schizachyrium scoparium

Little bluestem

Poaceae (grass family)

HOW TO GROW Little bluestem is native across much of temperate North America east of the Rocky Mountains but includes the southern Rockies. With such a wide range it is important to grow only regionally sourced plants to ensure that they are adapted to a particular growing season. Little bluestem grows best in full sun and about any well-drained soil; some strains can be floppy in soils that are too rich. The species is easy to propagate from seed.

LANDSCAPE USE This upright clump-forming grass only lightly self-sows so is suitable for traditional landscapes. It grows 30–40 inches tall and makes a fine mass planting or a component in a perennial border. It is integral to a prairie planting, especially in drier soils.

ORNAMENTAL ATTRIBUTES Little bluestem is most beautiful in fall when the seed fruits ripen with whitened feathery hairs that are illuminated when backlit. The plant turns rich reddish orange shades in autumn and holds this warm color well through winter without bleaching out. Some plants are noticeably bluish in the growing season.

RELATED PLANTS Several selections are available, most with bluer leaves.

Senna hebecarpa / *Senna marilandica*
Wild senna / Maryland wild senna

Syn. *Cassia hebecarpa* / *Cassia marilandica*
Fabaceae (legume family)

These two wild sennas are sister species and nearly identical; you likely have to be a botanist to tell them apart. They have different native ranges yet overlap east of the Mississippi River, require the same horticultural requirements, and have similar landscape uses and ornamental attributes.

HOW TO GROW Wild senna is present east of the Mississippi River where it is found northward into Michigan and Wisconsin. Maryland wild senna is native across the Lower Midwest and northward into southeastern Nebraska, southwestern Wisconsin, and northeastern Ohio. Both species are found in open woodland, savannas, and prairies from floodplains to uplands and grow in almost any soil moist to dry in full sun or partial shade. They are easy to propagate from seed.

LANDSCAPE USE These dramatically tall perennials are 4–6 (occasionally 8) feet tall and magnificent. They can self-sow to the point of being a nuisance in rich soils so are shunned in traditional perennial borders though make a spectacular back-of-the border plant. They're perfect for a natural landscape on the edge of woodland or in a prairie reconstruction. They also make a fine perennial hedge because they're so stout and stand tall through winter. In the Lower Midwest (sometimes northward) they're a butterfly magnet, host to summer's flying sunshine, cloudless sulphur, and cantaloupe-colored sleepy orange.

ORNAMENTAL ATTRIBUTES The foliage is pinnately compound and lines the stem as if to show off the candle of bright yellow, late-summer flowers at the top. In fall, sickles of downward-curving fruits mature black and hold steadfast through the winter to continue their drama. The seed pods of wild senna split open in winter while those of Maryland senna remain closed.

Silene regia
Royal catchfly

Caryophyllaceae (pink family)

HOW TO GROW Royal catchfly is native from the Ozark Highlands eastward to central Ohio and at one time northward to northern Illinois, where it was mainly found native in moist, well-drained savannas and occasionally prairies—habitats that have largely been destroyed so it is currently rare or extirpated from much of its range. It grows very well and is long-lived in a garden setting in moist, well-drained soil in full sun to partial shade. It is easy to propagate from seed.

LANDSCAPE USE The plant is upright and sturdy, but looks rather lanky, growing 3–5 feet tall. It looks best as a companion with other plants where its late-summer flowers

289

Senna hebecarpa (wild senna) proved a dramatic perennial along the back fence of my former garden.

A stand-alone plant of *Silene regia* (royal catchfly) in bloom depicts the tall open form of the plant.

Silene stellata (starry campion) blooms among wild sunflowers at Hayden Prairie, Iowa.

can simply wow you. It is pollinated by the ruby-throated hummingbird.

ORNAMENTAL ATTRIBUTES Royal catchfly plants are few stemmed and lanky with paired foliage so it's all about the screaming, true intense red, five-petaled flowers that are borne in loose clusters atop the plant.

Silene stellata

Starry campion

Caryophyllaceae (pink family)

HOW TO GROW Starry campion ranges from the Great Plains to the Mid-Atlantic and is found across the entire Midwest in open woods, savannas, and upland prairies. It is easy to grow in upland soils in full sun to light shade and is easy to transplant and grow from seed.

LANDSCAPE USE The plant grows 3 feet tall and is rather fine-textured and delicate looking—best combined with companion plants including grasses such as bottlebrush grass, silky wild rye, and American beakgrain and forbs including bracted spiderwort and purple coneflowers.

ORNAMENTAL ATTRIBUTES The midsummer white flowers consist of delightfully fringed petals radiating out from (upon close inspection) a stunningly circular center. The leaves are attractive in whorls of four, spaced along the stem.

Silphium spp.
Silphiums
Asteraceae (aster family)

Silphiums produce deep yellow flower heads in late summer that look like a sunflower and attract wide diversity of pollinators. The flower heads differ from sunflowers in that the central disc flowers are staminate (male) while the "petaled" ray flowers around the flower head are pistillate (female). The female flowers produce seed fruits that are round and flat and look like petals. Silphium fruits are oily and relished by songbirds. I ate one once and it gave me a horrible stomach ache—they are not edible like sunflower seeds. Compass plant and prairie dock are long-lived and develop huge, almost tuberous, deep roots.

Silphium integrifolium

Rosinweed silphium

Asteraceae (aster family)

HOW TO GROW Rosinweed silphium is native from the southeastern corner of South Dakota eastward to southern Wisconsin, and southwestern Michigan, southward to western Kentucky and Oklahoma. It is found in moist to dry upland prairies and savannas and grows readily in average garden soil in full sun or partial shade. It is easy to propagate from seed but does not self-sow as aggressively as cup plant or prairie dock.

LANDSCAPE USE Rosinweed silphium makes a stout, 4- to 5-foot tall perennial for the back of a perennial border and

The bright yellow flowers of Silphium integrifolium (rosinweed silphium) show their similarity to sunflowers.

is an important component of prairie reconstructions and other natural landscape plantings.

ORNAMENTAL ATTRIBUTES The foliage is beautiful and paired along the stem like mini-cup plant leaves. The flowers are true yellow and crown the plant in late summer, producing typical silphium seed fruits that often hold like flower petals.

RELATED PLANTS Whorled rosinweed silphium (*Silphium trifoliata*) is virtually identical to rosinweed except that the leaves are whorled along the stem in groups of three and the species is found only in the Eastern Midwest in Ohio, eastern and southern Indiana, Kentucky, and barely into southern Illinois.

Silphium laciniatum
Compass plant silphium

Asteraceae (aster family)

Can you imagine crossing a flat prairie that stretches as far as the eye can see in all directions and it's a cloudy day and you need to know your direction? Compass plant, with its leaves that are generally aligned north to south, will guide your way.

HOW TO GROW Native across most of the Midwest, though rare and local in the Eastern Midwest, compass plant is found on moist to dry prairies. It grows well in almost any well-drained soil in full sun. If planted in too much shade, its flower stalk will not be sturdy and upright. It is easy to propagate from seed but can take many years to reach flowering size, especially in poor soils.

LANDSCAPE USE Compass plant makes a dramatic foliage plant as its naturally lacerated leaves reach upward around

3 feet tall. Its vertical flower stalks can be as short as 5 feet and occasionally well over 8 feet tall so make a fine sentinel best reserved for the back of a perennial border. Its foliage and form make it a good gateway plant—paired on either side of a walk or driveway between garden spaces. As a classic and widespread prairie plant, it's obviously an important component of prairie reconstructions.

ORNAMENTAL ATTRIBUTES The divided leaves held vertically in a north–south configuration remind me of distinctive sails and they capture light as if to be illuminated, showing shadows equally well, though in piecemeal fashion. Leaves often turn yellow in the fall, becoming flaming yellow when backlit by a sunset.

Silphium perfoliatum
Cup plant silphium

Asteraceae (aster family)

HOW TO GROW Cup plant's native range almost perfectly defines the Midwest. It is found in moist meadows, woodland edges, and open riparian lands. It grows gargantuan in rich garden soils without competing vegetation and self-sows with abandon. It prefers full sun but will flower well in partial shade and is easy to grow from seed.

LANDSCAPE USE This striking and sturdy large plant grows 5–8 feet tall. It can be carefully maintained as a mass in a bed but is best in a natural landscape where other plants will compete with it and slow its aggressive, self-sowing nature. Its perfoliate leaves form a V-shaped cup at the stem that collects rainwater—and dew during a drought when songbirds heavily utilize it as a water source. The flowers are rich in nectar and pollen and attract a diverse group

The intricately divided, large basal leaves of *Silphium laciniatum* (compass plant silphium) align themselves north to south.

A massive clump of blooming *Silphium perfoliatum* (cup plant silphium) commands attention in a traditional flower border at Lincoln Park in Chicago.

of pollinators. Many songbirds also seek the seed fruits in autumn, more so than sunflowers.

ORNAMENTAL ATTRIBUTES The form of the plant is very architectural with its synchronized, paired leaves alternating 90 degrees from a sturdy squarish stem that is sometimes red. The deep yellow flowers look like 3-inch-wide sunflowers and are produced in clusters above the foliage. The plant blackens after a freeze and holds its interesting architecture into winter.

Silphium terebinthinaceum
Prairie dock silphium

Asteraceae (aster family)

HOW TO GROW Prairie dock is native in the Ozark Highlands and northward across almost all of Illinois into southern Wisconsin and eastward throughout most of the Eastern Midwest. It is found in moist to dry prairies, glades, savannas, and fens, and has been planted in most prairie reconstructions so has naturalized beyond its native range. It grows in almost any soil type in full sun to partial shade and is easy to propagate from seed.

LANDSCAPE USE The spectacular large leaves make prairie dock a great foliage plant 24–30 inches tall but it readily self-sows so is challenging to maintain in traditional landscapes without soon becoming a mass planting. Seedlings

The large leaves of *Silphium terebinthinaceum* (prairie dock silphium) provide a shadow box canvas to its lengthening flower stalk and developing flower buds.

quickly form deep, fleshy carrotlike roots. The naked flower stalks tower above the leaves growing 5–8 feet tall; the flower stalks can be floppy in shady conditions.

ORNAMENTAL ATTRIBUTES The leaves break the mold of prairie plant leaves which are usually cut or divided to reduce heat gain in full sun. Each prairie dock leaf has short, raspy hair tufts that hold a layer of air against its surface so that the leaves feel cool even on a hot sunny day. The leaves also capture light and shadows like no other, setting a beautiful prairie stage. They blacken in a freeze and curl up, but remain interesting in the winter landscape when hairs on the leaves are noticeably whitened, creating a polka-dotted "guinea hen" look. The flower heads are set atop sleek, tall, nearly leafless stems and open from elegant round buds that produce bright yellow ray flowers.

Sisyrinchium spp.
Blue-eyed grass

Iridaceae (iris family)

Every plant enthusiast whether botanist, gardener, or novice has a plant that tripped their trigger of interest in pursuing the company of plants. The common name blue-eyed grass describes the plant well. This was the plant that inspired me to delve into the marvelous realm of native plants.

HOW TO GROW Blue-eyed grass species are native across the entire Midwest in prairie and savanna remnants as well as in disturbed old-field meadows, roadsides, and other open land. Oddly, they often are challenging to cultivate and have their own idea where they want to grow. They are easy to grow from seed but seedlings do not always transplant or establish well.

LANDSCAPE USE Plants grow no more than 12 inches tall and remain grassy tufts, perfect edging to a flower border were they not so fickle to grow. They are overlooked in prairie reconstructions and usually absent from such landscapes though present in nearly all native prairie remnants.

ORNAMENTAL ATTRIBUTES Blue-eyed grass is obviously not a grass but an iris relative and the mid to late spring flowers adorn the upper parts of the leaf bladelike stems. The flowers open only in sunshine, and are often white and various shades of blue and always with a touch of yellow in the flower center. The flowers produce tiny round seed capsule fruits. There are four widespread midwestern species: *Sisyrinchium albidum* (common blue-eyed grass), *Sisyrinchium angustifolium* (stout blue-eyed grass), *Sisyrinchium campestre* (prairie blue-eyed grass), and *Sisyrinchium montanum* (prairie blue-eyed grass).

A floriferous, cultivated mass of *Sisyrinchium angustifolium* (stout blue-eyed grass) edges a walk at the Missouri Botanical Garden in St. Louis.

Solidago nemoralis (gray goldenrod) flowers peek through shorter prairie grasses.

Solidago spp.
Goldenrods
Asteraceae (aster family)

Along with asters (*Symphyotrichum*), blazingstars (*Liatris*), bonesets (*Eupatorium*), and sunflowers (*Helianthus*), goldenrods are truly icons of the prairie, meadow, old-field, or roadside in late summer and early fall. As with sunflowers, their golden-yellow flowers epitomize the glorious brilliant midwestern sun and anchor the color scheme. They still have a bad reputation as a source of allergens which is not true as they are pollinated by insects and don't throw their pollen to the wind. Goldenrods also are reputed garden thugs, which is true of some species, but several are distinctive additions to a traditional landscape. They all are easy to propagate from seed. Note that several species of goldenrods have been moved into the genus *Oligoneuron*.

The most common species, Canada goldenrod (*Solidago canadensis*), and closely related tall goldenrod (*Solidago altissima*), are too invasive for a garden but suited for a ruderal area where their nectar-rich flowers attract an abundance of pollinators. They were a godsend for many pollinators during the historic drought of 2012 when few other plants flowered. Goldenrods are rare or absent from high-quality native prairie remnants but thrive in disturbed ground. This adaptation means they must be controlled in prairie reconstructions or they will take over in the short run—which is what our short lives witness. *Solidago* species are the state flower of both Kentucky and Nebraska.

Solidago missouriense (Missouri goldenrod) creates a rather full floral stand on Roscoe Prairie, Minnesota.

Solidago nemoralis
Gray goldenrod, dyersweed goldenrod, old field goldenrod
Asteraceae (aster family)

HOW TO GROW This species is native across most of North America from the Rocky Mountains eastward through the entire Midwest. It is found in very poor, open soils that are often sandy, gravelly, or rocky. It grows in nearly any well-drained soil and becomes almost unrecognizably large in rich soil. Gray goldenrod requires full sun or partial shade and is easy to grow from seed.

LANDSCAPE USE Gray goldenrod stays in a clump and doesn't overly self-sow. It is a solution to hot, dry "hell-strips," rock gardens, and other challenging sites. It is usually less than 2 feet tall in such sites but can be 3 feet tall in good soil.

ORNAMENTAL ATTRIBUTES Gray goldenrod has the classic goldenrod flower shape of a plume whose leader arches sideways. The flower is typical golden yellow and the leaves are contrasting gray-green turning orange to red in the fall.

RELATED PLANTS Early goldenrods (*Solidago juncea* and *S. missouriensis*) are two prairie species with the same common name. They produce classic nodding goldenrod flower heads but bloom in mid to late summer ahead of all other species. Early goldenrods form patches comprised mainly of basal rosettes of leaves—three parallel leaf veins in *S. missouriense* and a single vein in *S. juncea*. Few flowering stems are usually produced in each patch which may be why these two lovely species are rarely cultivated.

Solidago speciosa (showy goldenrod) produces distinctive pyramidal flower clusters.

Solidago speciosa
Showy goldenrod

Asteraceae (aster family)

HOW TO GROW This species is most common across the Upper Midwest, but is native throughout the entire Midwest. It is found on mesic to dry upland prairies and grows well in almost any well-drained soil in full sun to partial shade.

LANDSCAPE USE Showy goldenrod stays in a clump and rarely self-sows so is suitable for a traditional perennial border where it will grow around 3 feet tall. It is an important component of natural landscapes and prairie restorations and a key late-season nectar source.

ORNAMENTAL ATTRIBUTES The very showy pyramidal domes of golden-yellow flowers in mid-fall are among the last to bloom on a prairie. Plants often have phenomenal fall color as the whole plant may turn red or burgundy wine colored after blooming.

A stand of *Sorghastrum nutans* (Indian grass) looks coppery in early autumn.

Sorghastrum nutans
Indian grass

Poaceae (grass family)

HOW TO GROW Indian grass is a warm-season grass native across the entire Midwest and an important component of the tallgrass prairie. It's found in moist to dry upland prairies and grows exceedingly well in average garden soil in full sun. It's easy to propagate from seed.

LANDSCAPE USE This grass reaches 4–6 feet tall and occasionally taller. It self-sows abundantly so is challenging to use in a traditional landscape. It makes a marvelous mass planting with big bluestem and switchgrass. Although it is an integral grass for a tallgrass prairie reconstruction, it should be sown in low quantity so it doesn't overtake a planting.

ORNAMENTAL ATTRIBUTES In late summer into early fall a golden-yellowish "plume" of flowers adorned with showy yellow stamens appears atop the stem. The fruiting heads age from coppery to tawny and these hairy awn-tipped seed fruits are stunningly illuminated when backlit.

The clump of fine leaf blades makes *Sporobolus heterolepis* (prairie dropseed) an attractive ornamental through all seasons at Missouri Botanical Garden in St. Louis.

Symphyotrichum laeve (smooth aster) blooms at Chiwaukee Prairie in the southeastern corner of Wisconsin.

Sporobolus heterolepis
Prairie dropseed
Poaceae (grass family)

HOW TO GROW Prairie dropseed is a heart-of-the-continent species native in every midwestern state, but is more localized in the Eastern Midwest. It's a warm-season grass that grows in full sun and about any well-drained soil. It can be propagated from seed or by dividing clumps.

LANDSCAPE USE This is the most widely cultivated of our native grasses as it is suitable for a traditional landscape and almost never self-sows. It grows in tidy clumps 18–24 inches tall consisting of outward arching leaf blades. Flowering occurs in late summer into early fall with airy inflorescences atop the foliage reaching 30–40 inches tall. It makes memorable mass plantings or additions to perennial borders. It is difficult to establish by seed into restorations so planting plugs of this grass ensures its rightful place in a high-quality restoration.

ORNAMENTAL ATTRIBUTES The finely textured tuft of leaf blades reminds me of Tina Turner's classic hairdo. This grass looks fine alone or in mass plantings. The airy inflorescences are especially sparkly when backlit and emit an aroma reminiscent of popcorn that is liked or hated depending on the nose. Plantings of this grass on Drake University campus in Des Moines, Iowa were removed because people were complaining about the smell. The plant turns golden in autumn and holds a yellowish tan color all through winter.

Symphyotrichum laeve
Smooth aster
Syn. *Aster laevis*
Asteraceae (aster family)

HOW TO GROW Smooth aster is native from New England to the Great Plains In the Midwest it is most common in areas dominated by tallgrass prairie. It grows on dry to mesic prairies and savanna remnants and less so in disturbed meadows and roadsides.

LANDSCAPE USE This aster usually grows 2–3 feet tall and makes a fitting addition to a perennial border. It also is handsome in a natural landscape and prairie restoration mixed with the shorter grasses.

ORNAMENTAL ATTRIBUTES The waxy smooth, bluish green foliage is handsome all season but the light lavender-blue flowers with yellow centers that turn reddish at maturity are its peak of beauty in late summer into fall.

Symphyotrichum novaeangliae
New England aster
Syn. *Aster novaeangliae*
Asteraceae (aster family)

HOW TO GROW Native across all of northeastern North America including the entire Midwest, New England aster is found in sedge meadows, mesic to wet prairies, moist meadows, and roadsides and can be overly exuberant in rich

The royal flowers of *Symphyotrichum novaeangliae* (New England aster) contrast well with autumn grasses.

Symphyotrichum oblongifolium (aromatic aster) creates a living wall of purple on the Island Garden at Powell Gardens, Missouri.

garden soils without competition. If plants grow too tall and gangly, they can be cut back before late summer and still bloom. New England aster prefers moisture and can lose its lower leaves and look scraggly if in too dry of site. It's also really easy to grow from cuttings.

LANDSCAPE USE New England aster and its cultivars are well-known perennials and may be utilized at the back of a traditional perennial border or as an informal hedge. They grow 4–6 feet tall and are at their best in wetland or rain gardens where they have extra moisture. They produce seedlings abundantly in some settings that make them a better choice for natural gardens. They are, along with aromatic aster, premier insectaries and butterfly garden asters.

ORNAMENTAL ATTRIBUTES New England aster is a regal sight from early fall to around the first frosts of autumn: its purple flowers with yellow centers are beautifully covered with migrant monarchs and other late-season butterflies in contrasting yellows and oranges.

RELATED PLANTS The cultivar 'Purple Dome' stays smaller but does best in USDA zone 5 and southward. Other cultivars may have hot pink to purple flowers but from my experience appear to be less visited by pollinating insects than the wild forms, which are on rare occasion, pink or white-flowering.

Symphyotrichum oblongifolium
Aromatic aster

Syn. *Aster oblongifolius*
Asteraceae (aster family)

HOW TO GROW Aromatic aster is native across most of the Midwest west of Lake Michigan; it's found eastward into the Bluegrass Region of Kentucky but only locally in Indiana and Ohio, and is absent from Michigan. It grows in dry, gravelly or rocky prairies, savannas and glades but is easy to cultivate in any well-drained garden soil in full sun or partial shade. In some settings, it can self-sow to the point of overtaking less-vigorous perennials. It can be trimmed for tidiness through midsummer and still bloom by fall.

LANDSCAPE USE The plant reaches just 24–30 inches tall in a full, mounded form so is accepted as a perennial for a traditional flower border or informal hedge where its late fall bloom is treasured. The flowers are exceptionally rich in nectar and pollen and visited in earnest by hordes of pollinators including the last-of-the-season butterflies.

ORNAMENTAL ATTRIBUTES The flowers are silvery lavender-blue and produced in mass over the entire plant creating a carpet of color. The foliage is pretty handsome and aromatic too, but can be mired by lacebugs.

The needlelike leaves identify this as *Ionactis linariifolius* (savory-leaved aster).

Symphyotrichum sericeum (silky aster) shines among prairie grasses at Harlem Hills Prairie in Rockford, Illinois.

Flower color gives *Symphyotrichum oolentangiense* (sky blue aster) its common name.

Symphyotrichum oolentangiense
Sky blue aster

> Syn. *Aster azureus*
> Asteraceae (aster family)

This aster's scientific name change from *azureus* (as in azure blue) to *oolentangiense* was a shock, but being told that the latter means "of the wind" helped me accept it. The plant is really named after the Olentangy River in Ohio.

HOW TO GROW Sky blue aster is native throughout the core of the Midwest, absent only from Kentucky and North Dakota. It is found on mesic to dry prairies and savannas and rarely in disturbed sites. It is easy to cultivate in well-drained soil in full sun or partial shade.

LANDSCAPE USE The plant reaches 2–3 feet tall and tends to be somewhat asymmetrical and open. It looks best in a natural landscape where it is mixed with little bluestem, sideoats grama, and prairie dropseed and is an important component in a prairie restoration as its airy flowers are welcome in late summer into early fall.

ORNAMENTAL ATTRIBUTES The flowers are light lavender-blue, pinker than sky or azure blue but lovely nonetheless.

RELATED PLANTS Silky aster and flax-leaved aster are two of my favorite asters that usually prefer to stay free in their native haunts. Silky aster (*Symphyotrichum sericeum*, syn. *Aster sericeus*) has the cutest leaves covered in silky silvery hairs and produces blooms of deep silky violet. It's found native only on sand, gravel, or hill prairies throughout all the Midwest west of Lake Michigan, but only in northern Indiana and southwestern Michigan in the Eastern Midwest. Try silky aster in a rock garden or other perfect drainage.

Flax-leaved aster (*Ionactis linariifolius*, syn. *Aster linariifolius*), with its needlelike leaves and short stature under 1 foot tall, is adorned with light violet flowers that make it a perfect plant for a rock garden. It is native mainly on the Atlantic and Gulf Coast states but has inland populations in the Ozark Highlands, Driftless Area, much of the Chicago region, Kentucky, southern Indiana, and south central Ohio. I know it from gravel and hill prairies as well as in sand savannas.

A female plant of *Thalictrum revolutum* (waxy-leaved meadow-rue) in flower makes an elegant silhouette against a fence.

Tradescantia ohiensis (Ohio spiderwort) has self-sown in abundance in the prairie border at Powell Gardens, Missouri.

Thalictrum spp.
Tall meadow-rues
Ranunculaceae (buttercup family)

Thalictrum dasycarpum (purplish meadow-rue) and *T. revolutum* (waxy-leaved meadow-rue), two tall species of meadow-rue, are nearly identical: purplish meadow-rue has leaf undersides that are smooth or with nonglandular tipped hairs, while waxy-leaved meadow-rue has glistening glandular-tipped hairs under its leaves and its leaves are pungent smelling if crushed.

HOW TO GROW Purplish meadow-rue is native through the center of North America from Ontario through the Midwest though is localized eastward in Kentucky and Ohio. Waxy-leaved meadow-rue is less widespread growing wild in the Ozark Highlands northward to southeast Iowa, southeast Wisconsin and southeast Michigan and eastward to southern Ohio. Both species are found in moist to wet prairies, savannas, and open woodlands in full sun to partial shade. Clumps may be divided or plants grown from seed.

LANDSCAPE USE The nonnative meadow-rues like *Thalictrum flavum* (yellow meadow-rue) and *T. rochebrunnianum* (lavender mist meadow-rue) are embraced in shady perennial gardens yet the native species are seldom seen, though equally garden worthy. Tall meadow-rues make an interesting subject in a moist traditional perennial border and grow 3–5 feet tall (rarely to over 6 feet). They're a good choice for a partially shaded rain garden, the edge of a water garden, or planted in a wet prairie reconstruction.

ORNAMENTAL ATTRIBUTES Tall meadow-rues are dioecious and male flowers, with their pendant creamy yellow stamens, are showier than the female flowers that look more like creamy micro starbursts. Male and female flowers are produced in a billowing, delicate, and frothy mass atop the plant in early summer. The stamens of male flowers jiggle in the slightest breeze and are wind pollinated. The plant's leaves are gracefully compound and turn yellow in the fall.

Tradescantia ohiensis
Ohio spiderwort
Commelinaceae (spiderwort family)

As horticulture director of a public botanical garden, I can relay this message from gardeners about spiderworts (*Tradescantia* spp.): they bloom spectacularly on sunny mornings then wither by afternoon at which time you better not brush against the spent flowers or any clothing becomes permanently stained purple. I doubt we'll ever see a commercial on how a stain remover cleans an encounter with spiderwort blossom. Spiderworts usually have violet-blue flowers, occasionally more fuchsia or blue, and sometimes white. The native species have been hybridized into a menagerie of cultivars with various flower colors, some with purplish or yellow foliage.

HOW TO GROW Ohio spiderwort is native across Ohio and much of the temperate eastern North America but it's most widespread across all but the northwestern Midwest (absent only from the Dakotas). It's found in prairies, savannas, and other open lands and grows well in moist, well-drained soils in full sun to partial shade. It is easily propagated from seed.

LANDSCAPE USE This species is best in a natural landscape or prairie reconstruction where it may be allowed to self-sow with companion plantings into an open mass. It's an aggressive self-sower without competition. The plant reaches around 3 feet tall with rather open stems. The flowers contain no nectar, but many insects gather yellow pollen from the flowers.

ORNAMENTAL ATTRIBUTES The three-petaled flowers are gorgeous violet-blue and bloom in the morning, remaining open in the afternoon only on cloudy days. The flowers bloom over a couple of weeks from a cluster atop the bare but elegant and bluish gray-green stems. A self-sown mass of the stems protruding from other plants and adorned with bloom is a memorable sight.

RELATED PLANTS Prairie spiderwort (*Tradescantia bracteata*) is native in the central and northern Great Plains eastward into Wisconsin, Illinois, and Missouri. It grows just 12–18 inches tall and is found in dry prairies and savannas, usually on hill or sand prairies eastward as larger plants in rich soils smother it. It thrives in well-drained poorer soils in full sun or partial shade. It has glandular hairs (hairs with a swollen tip) on its flowering stems and flower sepals.

Western spiderwort (*Tradescantia occidentalis*) is mainly a Great Plains species but found eastward into Arkansas, Kansas, Iowa, and Wisconsin. It is very similar to prairie spiderwort but usually has violet-pink flowers.

Virginia spiderwort (*Tradescantia virginiana*) is native from the Mid-Atlantic westward to the Midwest and southward to the Gulf, mainly across the Lower Midwest from Missouri to Ohio and scattered northward. Some population are possibly naturalized from cultivation. The species is found in dry to mesic open woodlands, forest edges, savannas, cliffs, and in prairies in the western part of its range. It is easy to cultivate in moist soil in full sun or partial shade. The plant is 24–30 inches tall, has hairy flower stems are hairy, and is a darker or olive green unlike the blue-green of Ohio spiderwort.

Vernonia spp.
Ironweed
Asteraceae (aster family)

There are several very similar species of ironweeds found across the Midwest. Botanists identify ironweeds by the various shaped bracts around their flower heads.

HOW TO GROW Ironweeds are found in open floodplain woods to wet prairies, sedge meadows, mesic upland savannas, and mesic prairies. Cattle do not eat these plants so they persist in pastures. Ironweeds grow just fine in rich garden soil in full sun or partial shade. They are easy to start from seed.

LANDSCAPE USE These rugged perennials are quite drought tolerant once established but most species grow very tall, even more so in cultivation. They are best in natural landscapes and prairie restorations, or planted with other larger and aggressive species to keep them from flopping. They have very nectar-rich flowers that attract a plethora of pollinators including many butterflies. A long-horned bee (*Melissodes denticulata*) is a specialist on ironweed flowers.

ORNAMENTAL ATTRIBUTES The fuchsia-purple flowers are vivacious and part of the prairie color scheme in late summer. The foliage is deep green and impervious to heat and drought.

299

Here's an up-close look at the flowers of *Tradescantia bracteata* (prairie spiderwort).

Vernonia spp. (prairie ironweed) bloom vibrant purple between big bluestem and stiff sunflowers.

Vernonia arkansana (curly-top ironweed, Arkansas ironweed) is native only from the Flint Hills, Osage Plains, and Ozark Highlands. It is possibly native or naturalized locally eastward. It has the most distinctive, large flower heads somewhat reminiscent of a small thistle.

Vernonia baldwinii (Baldwin's ironweed) is a Southern Plains species native northward into southeastern Nebraska, southern Iowa, and southwestern Illinois. It is found in upland prairies.

Vernonia fasciculata (prairie ironweed) is native across the entire Midwest but absent from Michigan and is very scattered in Ohio and Kentucky.

Vernonia gigantea (tall ironweed, syn. *V. altissima*) is native mainly across the southern and Eastern Midwest from easternmost Kansas and Oklahoma eastward to southern Michigan and across all of Indiana, Ohio, and Kentucky. It is found mainly in wetland habitats but tolerates cultivation in dry soil well.

Vernonia missurica (Missouri ironweed) is native across the middle of the Midwest from easternmost Oklahoma, Kansas, and the southeastern corner of Nebraska eastward to northeastern Illinois, northwestern Indiana, and southern Michigan. It is found in both wetland and mesic situations, more often in upland prairies.

A honeybee visits the central spire of a *Veronicastrum virginicum* (Culver's root) inflorescence.

NOTES Culver's root is related to nonnative tall speedwells or veronicas, which are perennials that are commonly sold in hues of white, pink, or blue; but most veronicas are not long-lived in the Lower or western Midwest.

Veronicastrum virginicum

Culver's root

Plantaginaceae (plantain family)

HOW TO GROW Culver's root is native across the entire Midwest and sparingly southward and eastward to southern New England. It is found in open woods, savannas, and moist to wet prairies. It thrives in average to rich garden soils in full sun to light shade. It is easy to grow from seed but only rarely self-sows in a garden.

LANDSCAPE USE This species makes a striking back-of-the-border, tall perennial growing 4–6 feet tall. It's a long-lived perennial that stays put so is suitable for a traditional landscape perennial border. It is a great addition to a natural landscape, especially in open woodland gardens where its midsummer bloom is welcome.

ORNAMENTAL ATTRIBUTES The narrowly pyramidal, striking spires of white flowers complement the protruding salmon-colored stamens and a corresponding cloud of pollinators busy climbing the flower towers. The foliage is whorled and evenly spaced along the stem creating a very architectural look. The seed heads also are handsome and hold well into the winter landscape.

Viola pedatifida

Prairie violet

Violaceae (violet family)

HOW TO GROW Prairie violet is native across most of the Midwest though it's localized in the Eastern Midwest and absent in parts of Missouri and all of Kentucky. It is found in undisturbed prairies that are moist to dry. It is challenging to cultivate but easier than bird'sfoot violet (*Viola pedata*) and will take hold, sometimes hybridizing with other violets. It grows in almost any well-drained soil in full sun to light shade. It is easy to propagate by dividing clumps into multiple plants or by sowing seed.

LANDSCAPE USE A forgotten spring wildflower for prairie reconstructions, prairie violet also makes a cute addition to the edge of a natural garden where its diminutive, 6-inch size can be easily observed. It's the main host plant of the regal fritillary, a keystone butterfly of the midwestern tallgrass prairie, now very rare or absent from most areas east of the Mississippi River.

ORNAMENTAL ATTRIBUTES The divided leaves are aptly named as they are deeply cut and look like delphinium leaves. The flowers are rich to medium blue-violet.

Dark purple flowers and deeply lobed leaves characterize *Viola pedatifida* (prairie violet).

Zizia aptera (heart-leaved alexanders) has distinctive basal leaves that give the plant its common name.

Zizia aurea
Golden alexanders

Apiaceae (carrot family)

HOW TO GROW Golden alexanders is found in upland open woods, savannas, and prairies and is easy to grow in about any well-drained soil in full sun to light shade. It self-sows to the point of being a nuisance in rich soils without competing plants. It is easy to propagate from seed and hardy throughout the Midwest.

LANDSCAPE USE This premier late-spring pollinator plant reaches 2–3 feet tall and should be used carefully in a garden setting so that it doesn't take over. Plant it in poorer soils among established plants; it makes a beautiful companion with wild columbine in rocky sites. Golden alexanders is valued for its early bloom in prairie restorations where intense competition from other plants helps control it. It is a native caterpillar host plant for the black swallowtail and for the rare endemic Ozark swallowtail.

ORNAMENTAL ATTRIBUTES The tiny flowers are in dense umbels that create flat lenses of intense yellow that are difficult to photograph. The umbels become full of fruits by late summer and the lush deep green foliage also turns orange and red shades in the fall.

RELATED PLANTS Heart-leaved alexanders (*Zizia aptera*) is a close relative with beautiful heart-shaped basal leaves and a paler, sulfur yellow flower umbel. It is also a great pollinator plant but less common as it's found on dry to mesic prairies. It is less rambunctious than golden alexanders as it does not self-sow so abundantly.

Zizia aurea (golden alexanders) flowers are reminiscent of the familiar, though nonnative, Queen Anne's lace.

Woodland Perennials

Perennials native to woodland and forest have one main adaptation: thriving in shade or at least partial shade. Chief among such plants are ferns—primitive, nonflowering vascular plants in multiple families that reproduce by dustlike spores. They have few insect hosts but most are found in rich, moist woodlands in deep leaf litter where they attract many invertebrates and provide habitat for birds including the wood thrush and ovenbird.

Perennials listed for partial shade tolerate quite a bit of sun but those listed as requiring shade may take morning or late evening sun but midday sun would scorch them. They all will bloom well in light shade, though many of the spring-flowering species require full sun before the overhead trees leaf out. Dense shade of evergreens, dense deciduous shrubs, and some trees create too shady of an environment for many of these forest and woodland perennial species.

Woodland perennials are the structural elements of a shade or woodland garden and most hold their foliage through the entire growing season. They are usually thought of as blooming mainly in spring but many of our beloved spring wildflowers are known as spring ephemerals that grow, bloom, set seed, and die by midsummer. Note that most spring ephemeral shade-loving plants will be found in the bulbs chapter. Select the shade perennials in this chapter first to create the "bones" or structure of a beautiful woodland garden. Groundcover and bulb plants may then be integrated through these perennials to complete the composition. The results are that you will never have a bare garden after spring, nor a garden overwhelmed by aggressive groundcovers.

PERENNIALS FOR A TRADITIONAL PERENNIAL BORDER IN SHADE

Adiantum pedatum (maidenhair fern)
Amsonia illustris (shining bluestar)
Amsonia tabernaemontana (bluestar)
Aruncus dioicus (goatsbeard)
Asclepias purpurascens (purple milkweed)
Athyrium filixfemina (lady fern)
Cimicifuga racemosa (bugbane)
Diplazium pycnocarpon (glade fern)
Dryopteris spp. (wood ferns)
Gillenia stipulata (American ipecac)
Gillenia trifoliata (Bowman's root)
Monarda bradburiana (Bradbury bergamot)
Polystichum acrostichoides (Christmas fern)
Symphyotrichum shortii (Short's aster)

Actaea rubra (red baneberry) fruits glisten in the dappled sunlight of an open woodland garden.

The fruits of *Actaea pachypoda* (white baneberry) glow in the dark shade of a woodland.

White-fruited forms of red baneberry are known as *Actaea rubra* f. *neglecta* but otherwise resemble the species.

Actaea spp.
Baneberries
Ranunculaceae (buttercup family)

Baneberries brighten rich moist woodlands and forests with their white flowers in spring and showy berries in late summer into fall. There are two similar species, one with white berries and the other with red or white berries.

HOW TO GROW Baneberries grow best in moist, organic, well-drained soils in shade and languish in heavy clay or droughty sites. They are challenging to grow from seed. A mature clump could be divided but this plant is usually too treasured in a garden to disturb.

LANDSCAPE USE Baneberries are suited to traditional perennial borders in shade or in woodland gardens. They mature around 30 inches tall and are long-lived once established. The berries are poisonous if ingested.

ORNAMENTAL ATTRIBUTES The divided foliage is handsome all year from spring through fall, turning a nice gold in autumn. The midspring flowers are not exceedingly showy but are a charming mini-sparkler of creamy white held above the emerging foliage that brightens woodlands along with spring ephemerals. The foliage holds well through summer and the berries are showy in late summer into fall, making a beautiful companion to blooming woodland bonesets, goldenrods, and asters. Red and white baneberries planted with blue cohosh create a patriotic combination.

Actaea pachypoda (white baneberry) is native throughout the Eastern Midwest and more localized westward in its appropriate habitat to eastern Minnesota, southeastern Nebraska, and the Ozark Highlands. The berries are white

with a black dot on the end giving them the colloquial name "doll's eyes." The berries are set on a thickened red stem and stalk (pedicels) that enhances their showiness and causes the berries to glow in the shade of the forest.

Actaea rubra (red baneberry) is a widespread native across the western mountains of North America eastward across the Upper Midwest to New England. It has a few outlying populations in the Lower Midwest including the Shawnee Hills of Illinois. Red baneberry is more challenging to cultivate and produces lustrous red berries on thin green stems. Some forms of the species have white berries: *A. rubra* forma *neglecta*, the white red baneberry, which is unlike white baneberry, bears its berries on thin green stems. Be sure to grow midwestern sources of red baneberry to be successful.

Adiantum pedatum
Maidenhair fern
Pteridaceae (maidenhair fern family)

HOW TO GROW Maidenhair fern is native east of the Great Plains in temperate Eastern North America and found throughout the Midwest. It's a denizen of mesic upland forests and grows in moist, well-drained soil in shade or at least afternoon shade. Gardeners can divide plants to create still more plants.

LANDSCAPE USE This fern is a popular choice for traditional shade and woodland gardens and grows very well in containers too. It is a prime choice to plant with spring ephemeral woodland wildflowers to hold their place after they go dormant.

Adiantum pedatum (maidenhair fern) is like no other and is possibly the most beloved of all native ferns.

Agastache foeniculum (anise hyssop) blooms with butterflyweed at White River Gardens in Indianapolis.

ORNAMENTAL ATTRIBUTES The fan-shaped leaf is doubly compound and finely sublime. The dark purplish-black stalk splits into two, each half circling around with radiating fine stems of fin-shaped leaflets. The plant is always a standout in any setting.

RELATED PLANTS Venus maidenhair fern (*Adiantum capillusveneris*) is widely native in much of the Ozark Highlands and sparingly into Kentucky and scattered in points southward into the tropics. The midwestern Ozark strain of the plant is hardy but very difficult to procure even though tropical and nonhardy (tender) strains of this fern are popular container and houseplants. It's an untapped member of our floral resources.

Agastache foeniculum
Anise hyssop, blue giant hyssop
Lamiaceae (mint family)

HOW TO GROW Anise hyssop is barely native into the Midwest as it's a more northern species reaching southward into the western Upper Midwest in the Dakotas, extreme northern Iowa, and northwestern Wisconsin. It naturalizes almost anywhere, as it is currently a very popular herb and edible flower that tastes just like licorice. It is hardy and grows well throughout the Midwest in moist, well-drained soils in full sun to partial shade. It's easy to propagate from seed and can self-sow abundantly when given ideal conditions.

LANDSCAPE USE The plant is in flower over a long period from early or midsummer into autumn and grows 3–4 feet tall. It blooms itself to death like an annual so is a rather short-lived perennial that must be allowed to self-sow to carry on. It attracts pollinating insects exceedingly well so makes a premier insectaries garden plant that attracts beneficial insects. It's used as a companion plant in orchards for this reason. It is best in more informal landscapes, utilized in an herb border, edible landscape, or included in a food forest. Its fruiting heads attract songbirds too.

ORNAMENTAL ATTRIBUTES The violet blue florets stand out from more purple calyxes for a strikingly showy flower that produces a seed head that holds well into winter. The aromatic nature of the plant makes it always fun to rub the foliage if you like the scent of anise.

RELATED PLANTS A close relative from Asia, the Korean anise hyssop (*Agastache rugosa*), is sometimes sold as native, and hybrids between the two species also are sold.

Agastache nepetoides
Yellow giant hyssop
Lamiaceae (mint family)

Here's a flower where a photograph captures it better than the naked eye, imaging the contrast between floret and calyx more strikingly than what the eye alone can see. The flower is not overly showy in bloom but very nectar rich with hummingbirds and other pollinators vying for

307

Agastache nepetoides (yellow giant hyssop) flowers show best where they will be backlit in the garden.

Agastache scrophulariifolia (purple giant hyssop) flowers provide great contrast to late summer's yellow-flowering composites.

position. I never would have planted this sleeper, but inherited it as a native plant on my property and at Powell Gardens and now wouldn't be without it.

HOW TO GROW Yellow giant hyssop is found in moist to summer-dry upland savannas, open woods, and woodland edges. It grows exuberantly 5–7 feet tall in well-drained moist soils in full sun to light shade. Deer love this plant (unusual for a mint) so it needs protection from them. Individual plants are rather short-lived but readily self-sow and are easy to propagate from seed.

LANDSCAPE USE We rarely design a landscape with dormant perennials in mind and that's when yellow giant-hyssop is at its showiest. Plant it at the back of a perennial border with woodland backdrop. It's best in a natural landscape, insectaries or bird garden as it brings in creatures exceedingly well. When in bloom in late summer, it is dancing with swallowtails and hummingbirds while later goldfinches vie for position on its seed heads.

ORNAMENTAL ATTRIBUTES The fruiting heads are marvelous atop tall sturdy stems like torches in winter. The late-summer flowers are interesting on close inspection as the florets are more cream than yellow and set against a light yellowish green calyx.

RELATED PLANTS Purple giant hyssop (*Agastache scrophulariifolia*) is quite similar to yellow giant hyssop only with pale bluish or purplish florets. Its common name overstates the reality of its flower color but the flowers are as nectar rich as anise hyssop and yellow giant-hyssop with an equal cloud of pollinators. Purple giant hyssop grows as large as and in similar habitats as yellow giant-hyssop. It looks great in the winter landscape with showy fruiting heads as well.

Ageratina altissima
White snakeroot

Syn. *Eupatorium rugosum*
Asteraceae (aster family)

HOW TO GROW White snakeroot is native across temperate Eastern North America and found across the entire Midwest. It grows in disturbed woodlands and forests and can become quite aggressive under continual disturbance or through a phase of establishment in a newly planted woodland garden. Established plants can be divided and seeds are easy to germinate.

LANDSCAPE USE This plant came in on its own when I removed turf from under the trees around my home. It seeded in aggressively and formed spectacular masses 24–30 inches tall for several years. Now, as longer-lived plants have established and the soil is not disturbed, white snakeroot plants have dwindled. It is a garden thug in a traditional landscape but a useful component in a natural woodland garden. It is extremely nectar rich and attracts hordes of pollinators while in bloom in early fall. White snakeroot makes a memorable fall composition with woodland goldenrods and asters. It is a toxic plant and the poison can be transferred into cow's milk. There is strong evidence that this is what killed President Lincoln's mother.

ORNAMENTAL ATTRIBUTES The flowers are bright milky white in flat clusters atop clean, dark green, paired leaves. The plant is also beautiful in fruit, covered with white tufted seeds in late fall.

RELATED PLANTS The cultivar 'Chocolate' has dark purplish colored leaves in spring that age to dark bronzy green by bloom time.

The icy light blue flowers of *Amsonia tabernaemontana* (bluestar) shine in a planting at the Chicago Botanic Garden.

This close-up shows the fine details of the milky white flower heads of *Ageratina altissima* (white snakeroot).

Amsonia tabernaemontana
Bluestar
Apocynaceae (dogbane family)

HOW TO GROW Bluestar is native across central and southern Illinois, southwestern Indiana, western Kentucky, eastern and southern Missouri, and Arkansas. It is found along fast-moving streams, flood terraces, mesic to dry open woodlands, rocky bluffs, and glades. It grows in most garden soils that are not too dry in full sun to light shade. Propagate bluestar by seed.

LANDSCAPE USE The plant reaches about 3 feet tall and is a popular traditional perennial, losing favor as of late to related species with finer leaves and the dwarf cultivar 'Blue Ice'. It has flowers rich in nectar for bees and butterflies.

ORNAMENTAL ATTRIBUTES The late spring flowers are soft blue or bluish white. The plant holds well, almost like a shrub, through the growing season and turns shades of yellow in the fall.

RELATED PLANTS Shining or Ozark bluestar (*Amsonia illustris*) is found almost exclusively in the Ozark Highlands where growing in similar habitat to the bluestar. It is a more robust plant growing to 4 feet tall with larger leaves and similar flowers and fall color to bluestar.

309

Amsonia illustris (shining bluestar) is a more open plant growing wild in a woodland in northern Arkansas.

Anemone acutiloba
Sharp-lobed hepatica

Syn. *Hepatica nobilis* var. *acuta*
Ranunculaceae (buttercup family)

Hepaticas are botanical gems native across the Northern Hemisphere, their popularity on the verge of becoming mainstream in the United States. Some botanists lump them all into one species (*Anemone nobilis*) but we'll keep our two native types separated as they occur in different habitats. The variation in flower colors and leaf patterns in hepaticas makes them one of my most beloved of all perennials—and they're one of the first wildflowers of spring.

HOW TO GROW Sharp-lobed hepatica grows across the Eastern Midwest westward to central Minnesota, central Iowa, eastern Missouri, and a couple of sites in northern Arkansas. It's found in mature mesic forests over limestone, which translates into cultivating the plant in moist, shaded sites in calcareous soils. Plants are best grown from seed, though a larger plant can be divided. The species is hardy throughout the Midwest and needs a more sheltered site in the western Midwest.

LANDSCAPE USE This is a premier evergreen perennial for an open shady woodland garden or edge of a woodland perennial border. As it grows just 6 inches tall, it's best displayed on rock walls or outcrops and planted with diminutive companions like moss, sedges and smaller ferns. It makes a prime container plant for shade.

ORNAMENTAL ATTRIBUTES The early spring flowers are something to look forward to and are usually white but can be any shade deepening to true blue or pink. Young plants will produce a few flowers but a mature clump can be crowned with many flowers. The liver-shaped three-parted leaves with pointed tips are how the plant got its colloquial name "liverleaf." The leaves are evergreen but lie on the ground usually turning burgundy in winter; some plants have variably patterned, though exquisite markings of burgundy.

RELATED PLANTS Round-lobed hepatica (*Anemone americana*, syn. *Hepatica nobilis* var. *obtusa*) has a range that is more localized in the Midwest across Michigan, eastern

The leaves of *Anemone americana* (round-lobed hepatica) may have lovely patterning, silvered on this plant at the Indiana Dunes National Lakeshore.

The very showy flowers of *Aquilegia canadensis* (Eastern red columbine) look like a golden honeycomb from below.

Anemone acutiloba (sharp-lobed hepatica) flowers may be white, pink, or most-sought-after blue.

Ohio, and the core of the Ozark Highlands with scattered locales in southeastern Minnesota, northeastern Iowa, and regions close to Lake Michigan. The species is usually found in more acidic soils that are sandy, or sandstone based, and cherty as in the Ozarks. This hepatica produces flowers virtually identical to its sharp-lobed relative. The leaves are similarly three-parted but with rounded lobes and they retain a fuzzier look through the season, sometimes showing an imaginative array of burgundy and silver patterning that is unmatched by any woodland evergreen perennial.

Aquilegia canadensis
Eastern red columbine
Ranunculaceae (buttercup family)

HOW TO GROW Columbine is native across temperate eastern North America including the entire Midwest. It grows on rock outcrops, rocky or sandy woodlands, and savannas in almost any well-drained garden soil. Columbine persists in a garden only where the soil has the extra drainage of rock, gravel, or sand. There it self-sows, which is important, as it is not a long-lived perennial (rarely living more than three years). It thrives in full sun to light shade.

LANDSCAPE USE Columbine is best used in a rock garden, raised bed, rock wall, in between stepping-stones, or other stone or concrete work. Here it readily self-sows and persists, forming basal foliage less than 1 foot tall and a candelabra of flowers that is usually about 2 feet tall but a robust plant can produce a 4-foot-tall magnificent display of flowers. The plant also is a great choice for a container garden. It is pollinated by ruby-throated hummingbirds and occasionally some butterflies like the spicebush swallowtail with its long proboscis. The floral tube fits the head and bill length of the hummingbird to a tee though children and stealing bees can cut the nectary for its sweet treat too.

ORNAMENTAL ATTRIBUTES The scarlet red pendant flowers bloom in late spring over several weeks then produce up-facing fruiting capsules by midsummer that disperse the black, pepperlike seeds and quickly disintegrate. The basal tuft of compound leaves remains evergreen and are food for a leaf mining micro moth that creates its own cursive style markings, unsightly to some gardeners but an awe-inspiring piece of artwork by a moth whose caterpillar is actually small enough to feed inside a thin leaf.

The fruits of *Aralia racemosa* (American spikenard) start out green, then mature to red and ripen purplish. All colors may be present on a ripening infructescence.

Aralia racemosa
American spikenard
Araliaceae (aralia family)

HOW TO GROW American spikenard is native across much of central and northeastern North America as well as in the southern Rocky Mountains. It is found in sheltered ravines and shaded cliffs, or in rich, moist forests across the Midwest, absent only from Oklahoma and North Dakota. It grows in moist, well-drained, humus-rich soils (often from crevices in limestone) in partial to full shade. It is best propagated from seed.

LANDSCAPE USE American spikenard makes a large, shrublike perennial (3 feet tall and wider than tall) for the shaded north or east side of a home, or a dramatic large perennial for a woodland garden. It is great for wildlife friendly gardens as its flowers are rich in nectar and attract many small insect pollinators, while birds seek the fruits.

ORNAMENTAL ATTRIBUTES The round umbels of creamy greenish flowers are produced in large panicles in late spring. The foliage is bipinnately compound, consisting of heart-shaped leaflets and very attractive through the growing season. The fruits ripen in early fall, turning purplish-black when ripe and are the showiest aspect of the plant until eaten by songbirds.

Aruncus dioicus
Goatsbeard
Rosaceae (rose family)

HOW TO GROW Goatsbeard is native around the Northern Hemisphere, though most goatsbeard plants in cultivation are the nonnative variety *acuminatus* from Eurasia and the Pacific Northwest (naturalized in Wisconsin and Michigan). There are two midwestern native varieties: variety *pubescens* from the Ozark Highlands northward to east central Iowa, and variety *dioicus* from the Shawnee Hills and Appalachian Highlands. Goatsbeard is found in rich mesic, forested ravines and grows well in moist, well-drained soils in partial to light shade. It is best propagated from seed as it has a stout, woody rhizome that is difficult to divide. Note that goatsbeard is dioecious so plants are either male or female.

LANDSCAPE USE Goatsbeard grows into a large shrublike perennial 3–4 feet tall and wider than tall. It is suitable for a traditional perennial border, edge of a woodland, or replacement for foundation shrub. It attracts a wide array of pollinators and is the sole host plant for the rare dusky azure.

ORNAMENTAL ATTRIBUTES The tiny creamy white flowers are produced in frothy abundance on groovy plumelike panicles that tower over the attractive compound leaves. The flowers on male plants are a tad showier with their stamens but disintegrate after bloom. The flowers on female plants produce seed capsule fruits (follicles) that ripen brown and hold into winter adding interest to the plant. The foliage reliably turns yellow in the fall.

Shrublike *Aruncus dioicus* (goatsbeard) in flower illuminates this woodland edge in the deep greens of June.

The whitish flowers of *Asclepias exaltata* (poke milkweed) burst pendantly like old-fashioned fireworks.

Asclepias exaltata
Poke milkweed
Apocynaceae (dogbane family)

Milkweeds (*Asclepias* spp.) are usually thought of solely as plants of the sun but several species from woodlands and forests are adapted to shady gardens. Plants are best propagated by seed and all are host to the monarch.

HOW TO GROW Poke milkweed is native across northeastern North America and found throughout the Eastern Midwest westward across most of Illinois and Wisconsin into northeastern Iowa and central Minnesota. In the wild it grows in rich, rocky or sandy upland woods and savannas. In cultivation, it prefers moist, well-drained woodland soils.

LANDSCAPE USE This species is a wonderful subject for a woodland garden where its midsummer flowers add some interest in a season of green and provide nectar for bumblebees and butterflies like the great spangled fritillary and its host the monarch.

ORNAMENTAL ATTRIBUTES The whitish flowers burst pendantly like old-fashioned fireworks and adorn the top of the plant. On close inspection the flowers are bicolored with petals that are often greenish to purplish while the crown is white or pink tinted. The flowers do not necessarily stand out in the dappled sunlight of woodlands but are always a delight to find. Upright, lance-shaped, 6-inch fruiting pods adorn the plant into winter after the leaves have fallen.

Asclepias purpurascens (purple milkweed) blooms are not true purple, but rather a royal shade of rose.

Flowers of *Asclepias quadrifolia* (four-leaved milkweed) top stems piercing distinctive whorls of four symmetrical leaves.

Asclepias purpurascens
Purple milkweed

Apocynaceae (dogbane family)

HOW TO GROW Purple milkweed is native from the Mid-Atlantic states westward throughout much of the Midwest, though absent from most of Minnesota and the Dakotas. Its favored habitat is upland savanna and prairie openings but it has become rare as woodlands have become too dense and shaded. Purple milkweed prefers light to partial shade but will grow just fine in full sun too, growing best in moist soils including heavy clay.

LANDSCAPE USE This species is probably the milkweed most suitable to a traditional perennial border as it has no underground rhizomes and in full sun or partial shade it produces an abundance of flowers that attract hordes of pollinators.

ORNAMENTAL ATTRIBUTES The flowers are showy in full-rounded umbels of rose-purple and followed by upright, cylindrical pods.

Asclepias quadrifolia
Four-leaved milkweed

Apocynaceae (dogbane family)

HOW TO GROW Four-leaved milkweed is native in two parts of the Midwest: from southern Ontario, Ohio, and Kentucky into southwestern Indiana, and from the Ozark Highlands northward to southeastern Iowa and central Illinois. It grows in moist, rocky woodlands in partial to light shade.

LANDSCAPE USE The plant reaches 2 feet tall and adds interest in a woodland garden.

ORNAMENTAL ATTRIBUTES The flowers, which appear in spring to early summer, are comprised of up-facing pendant

Newly emerged *Athyrium filix-femina* (lady fern) commemorates the acid greens of spring.

313

white to pinkish umbels that open from rosy flower buds. The flowers crown the plant's stem, which is pierced by paired leaves and by distinctive whorls of four symmetrical leaves. The foliage is very unique through the growing season and turns yellow in the fall.

Athyrium filix-femina
Lady fern

Dryopteridaceae (wood-fern family)

HOW TO GROW Lady fern is native to moist forests across temperate North America and is found throughout the Midwest in that habitat (though very localized in the western Midwest). It is our easiest-to-grow fern in moist, well-drained soil in shade or at least afternoon shade. You can propagate it by division.

LANDSCAPE USE This is a divine fern for any traditional shade garden or natural woodland planting and grows 24–30 inches tall. It spreads slowly by rhizomes and makes a fine mass planting or integrated with spring ephemerals and other woodland plants.

ORNAMENTAL ATTRIBUTES Thin wiry stalks and lobed sub-leaflets create the fine, lacey textures of the fronds—probably the finest texture of any substantial woodland garden plant.

RELATED PLANTS There are many cultivars and hybrids of this plant, some with extra frilly or oddly shaped fronds, or showing contrasting red stalks.

Blephilia ciliata
Downy wood-mint

Lamiaceae (mint family)

HOW TO GROW Downy wood-mint is found wild in upland savannas and open woodlands, usually in shallow soils over bedrock where it can compete better with smaller plants. It grows in similar well-drained soils in a garden setting in full sun or light shade and is easily smothered by larger plants. It is not necessarily long-lived (gardeners shouldn't be scared by its mint name, as so many other mints are invasive garden thugs), so it needs a place to self-sow to carry on. It is easy to propagated from seed and is hardy throughout the Midwest.

LANDSCAPE USE This wood-mint is best in a rock garden, above a stonewall, or in a rocky woodland garden where

its flower stalks may reach 2 feet tall but its foliage is more ground-hugging. The flowers are very nectar rich and bloom in late spring with the emergence of great spangled fritillaries that seek its nectar in groups.

ORNAMENTAL ATTRIBUTES The flowers are light lilac in tiered clusters like a shish kebab. These form ornamental fruiting heads that usually disintegrate by fall. The foliage along the ground remains evergreen, turning purplish shades in winter.

Blephilia hirsuta
Hairy wood-mint

Lamiaceae (mint family)

HOW TO GROW Hairy wood-mint is found wild in moist open woodlands and forests from floodplain terraces to rich, mesic slopes. It performs well in moist, well-drained soil in full sun to light shade and is hardy throughout the Midwest. It will not grow in dense shade under a stand of maples. It is easy to propagate from seed or by dividing the plant.

LANDSCAPE USE This is a marvelous plant for a natural woodland garden where its midsummer bloom is very welcome. The flower spikes grow as tall as 30 inches and are very nectar rich and attract many pollinators.

ORNAMENTAL ATTRIBUTES The foliage smells strongly of peppermint, so it's always pleasant to brush against this plant. The flowers are white in tiered clusters like the downy wood-mint. The low foliage is evergreen, and the brown fruiting heads hold beautifully into winter.

314

A quartet of great spangled fritillaries imbibe the nectar of *Blephilia ciliata* (downy wood-mint).

This stand of *Blephilia hirsuta* (hairy wood-mint) sparkles in a shady woodland at Blackhawk Springs Forest Preserve near Rockford, Illinois.

Cardamine concatenata (cut-leaved toothwort) occasionally produces pinkish flowers.

Carex species (sedges) bloom in May in the native woodland garden at Chicago Botanic Garden.

Cardamine concatenata

Cut-leaved toothwort

Syn. *Dentaria laciniata*
Brassicaceae (mustard family)

HOW TO GROW Cut-leaved toothwort is native in forests and woodlands across the Midwest and into the Northeast and Upper South. It grows in mesic woods over limestone and is a true spring ephemeral. Plant it in fall or early spring as soon as the ground thaws. Were it easy to grow from seed and division it would be more popular.

LANDSCAPE USE This is a sweet addition to a woodland or natural shade garden where the white, nectar-rich blooms enliven an early to midspring landscape.

ORNAMENTAL ATTRIBUTES The flowers are classic mustard family consisting of four white petals, which may be pinkish or tinged with pink in cool springs. The flowers sit above a whorl of finely divided, toothed leaves that look like marijuana. The leaves turn yellow before disappearing in midsummer.

Carex spp.
Woodland sedges
Cyperaceae (sedge family)

I have to admit, I garden with many species of woodland sedges and I don't have them all identified yet. They look like tufts or masses of grass but they have V-shaped (in cross section) leaves from triangular stems. The botanical adage to remember their identification is "sedges have edges."

Many sedge species can be quite similar and challenging to identify while their cultivation, garden use, and attributes are little studied.

HOW TO GROW Woodland sedges are common in any midwestern savanna, woodland, or forest and I haven't met one I would call aggressive or otherwise unwelcome in the garden. Most transplant with ease, and when I created paths in my woods, I moved and saved every one. I also leave or move sedges that self-sow in woodland garden beds. Woodland sedges should not be confused with aggressive and weedy nutsedge or what is sometimes called yellow nutgrass (*Cyperus esculentus*).

LANDSCAPE USE Woodland sedges, whether purchased from a native plant nursery, inherited as wild plants, or appearing as self-sown plants can be planted or moved into bed edges, sweeping massing, or between stepping-stones. Just be sure to notice the ultimate size of wild unknown species before moving them. It is safe to say woodland sedges are among our most underutilized of the midwestern flora. I was pleased that Roy Diblik wrote about them in his book *The Know Maintenance Perennial Garden*. Referring to a planting of sedges, he states that "a 'carex community' is the best beginning for your shade garden" as the plants inhibit weeds. Sedges produce seeds that are consumed by many songbirds and the foliage is host to many moths and a few butterflies, including the dun skipper. Consider sedges as an environmentally sound replacement for lilyturfs (*Liriope* spp.) and mondo or monkey grasses (*Ophiopogon* spp.), which offer little to native insects and wildlife.

ORNAMENTAL ATTRIBUTES Woodland sedges are cool-season plants, growing early like cool-season grasses. They produce little-noticed flowers (a few are rather showy and

315

Carex albicans (oak sedge), photographed here on the winter solstice, produces a mound of very fine green leaves that hold well through winter.

Carex glaucodea (blue wood sedge) with its evergreen blue-green leaves grows amid the fallen leaves at Big Buffalo Creek Conservation Area, Missouri.

distinctive) in spring and are fruiting by late spring with their fruits usually gone by midsummer. Traditional gardeners find this blooming time a bit untidy as the plants may splay out until the fruits are dispersed. Most woodland sedges comprise a tuft of fine grasslike foliage that remains green through the growing season and into early winter; some are evergreen. Many of the sedges depicted in the species accounts below were photographed in December after temperatures in the single digits Fahrenheit without snow cover—a good depiction of what they look like in a winter garden.

Carex albicans (oak sedge, white-tinge sedge) is native to temperate eastern North America and found across most of the Midwest except for the Dakotas and much of Minnesota. It grows in moist second-growth woods with oaks and red cedars. In cultivation it prefers good garden soil in partial to light shade. Oak sedge grows about 1 foot tall and wide.

Carex glaucodea (blue wood sedge) is native from Ohio westward to Oklahoma but is most common in the Ozark Highlands. It is found in rocky upland savannas and forests underneath oaks and hickories. It grows only 6 inches tall with a spread of less than 1 foot.

Carex grayi (Gray's sedge) is native across the Eastern and Lower Midwest, west to southeastern Minnesota and southwestern Iowa. It is usually found in mesic floodplain forests. It grows about 2 feet tall with an equal spread and produces unique spiky rounded fruits.

Carex grisea (wood sedge) is found throughout the Midwest and eastward into New England and southward into

The spiky rounded fruits of *Carex grayi* (Gray's sedge) make it easy to identify.

A spreading tuft of dark evergreen leaves identifies *Carex grisea* (wood sedge).

Satiny blue-green, strappy leaves holding through winter define *Carex platyphylla* (blue satin sedge).

The comparatively wide, seersuckered leaves of *Carex plantaginea* (plantain-leaved sedge) are quite distinctive and evergreen.

the Southern Great Plains of Texas. It grows about 9 inches tall. A mature plant may spread as much as 18 inches.

Carex plantaginea (plantain-leaved sedge) is native to the Northeast west to central Minnesota, northeastern Iowa, northeastern Illinois, Indiana, and Kentucky. It grows in mesic forests and reaches 6 inches tall with a spread of 15 inches.

Carex platyphylla (blue satin sedge) is native from New England to the southern Appalachians and westward into the Midwest on the Appalachian Plateau, Shawnee Hills, and localized in Michigan and Wisconsin. It grows less than 1 foot tall with a spread of 1 foot.

Carex sprengelii (Sprengel's sedge) is native across the entire Upper Midwest where it's found in mesic floodplain and mesic upland forests. It grows about 18 inches tall with a spread of around 1 foot, splaying out after bloom a bit wider.

Caulophyllum thalictroides
Blue cohosh

Berberidaceae (barberry family)

HOW TO GROW Blue cohosh is native to temperate eastern North America and is found throughout the Midwest though barely into the Great Plains states. In the wild it grows in mesic to dry-mesic upland forests. In cultivation it prefers moist woodland soils in shade and is challenging to grow from seed.

LANDSCAPE USE This sturdy perennial grows into a long-lived clump that reaches 2 feet tall. It's an elegant plant for a traditional shade or woodland garden, holding well when many plants have gone dormant and really shining in the

The spherical fruits of *Caulophyllum thalictroides* (blue cohosh) remain on the stems after the leaves have dropped.

The tall spires of *Cimicifuga racemosa* (bugbane) white flowers make it one of the most spectacular of all native wildflowers.

fall. It is good for a bird garden as birds devour the fall fruit and spread the seeds; it is a food source for the declining ruffed grouse, now extirpated from most of the Midwest. **ORNAMENTAL ATTRIBUTES** The emerging stems and leaves have purplish highlights and flower in midspring with tiny up-facing greenish gold flowers that may be wonderfully tarnished with a red-brown. The compound foliage is somewhat blue-green and similar to unrelated meadow rues and baneberries. The flowers produce spherical berries that turn an amazing sage green in summer, then briefly mauve, and ripen in fall to a midnight blue with a lighter blue coating—a memorable combination with the plant's yellow fall leaves. The berries may remain on the stems after the leaves have dropped.

Cimicifuga racemosa
Bugbane, black cohosh
Ranunculaceae (buttercup family)

HOW TO GROW Bugbane is native mainly in the Appalachian Mountains but is found in much of eastern Ohio, the Shawnee Hills, and westward through the Ozark Highlands. It grows in moist, well-drained woodlands. In the garden it will grow in full sun under continually moist conditions but is best in partial to light shade. Gardeners can propagate bugbane by division or seed.

LANDSCAPE USE No plant provides more drama in the midsummer perennial border, edge of a woodland, or woodland garden than this plant. Bugbane can tower 4–7 feet tall under ideal growing conditions. It is a pollen source for many insects and is the sole host plant for the Appalachian blue whose caterpillars feed on its flowers and developing fruits. **ORNAMENTAL ATTRIBUTES** The magnificent spires of perfectly rounded buds open from the bottom up with fringed, milky white flowers—always a standout in the mid- to late-summer woodland garden when few other plants are blooming. The foliage is attractive all season and turns yellow shades in the fall.

Cypripedium parviflorum var. pubescens
Large yellow lady's slipper
Orchidaceae (orchid family)

In mid to late spring, no wildflower captured my attention more than this orchid; I was lucky it grew in rich forests protected by several neighborhood parks. Large yellow lady's slipper is highly variable, various plants blooming anywhere from mid-May to mid-June with flowers rounded to elongated. Botanical mentors Mark Leoschke and Kay (Klier) Lancaster actually came to sample all the various plants to determine and document the species's variation. **HOW TO GROW** This lady's slipper was once widespread across much of the Midwest, absent only from Oklahoma. It is in decline in most of the Midwest, no longer growing

Vigorous large plants of *Cypripedium parviflorum* var. *pubescens* (yellow lady's slipper) often produce two flowers in piggyback.

Cypripedium acaule (pink lady's slipper) often requires caging against deer browse to exist.

The long delicate fronds of a mass of *Cystopteris bulbifera* (bulblet fern) create a memorable fine texture.

in many places it once did. Whether by habitat destruction or degradation, smothering by invasive exotic shrubs, digging by unscrupulous gardeners, or repeatedly cropping by deer until death, the plant is absent in Rockford and the Kansas City region even though it was once more prevalent in those places. It is native to mesic forests, woodlands, and savannas, still rather widespread in the Driftless Area. It is relatively easy to grow in appropriate calcareous, moist, well-drained, humus-rich, loamy woodland soils. All lady's slippers take a long time to grow from seed so be sure to purchase plants from reputable growers who have not pilfered them from the wild. Carefully dividing clumps of established garden plants is the best way to propagate lady's slippers. They should never be stolen from the wild, unless from in front of a bulldozer with the landowner's permission.

LANDSCAPE USE Large yellow lady's slipper grows 18–24 inches tall and may be the shining star of a woodland garden with appropriate soils.

ORNAMENTAL ATTRIBUTES When in bloom, the 2- to 3-inch glowing yellow, slipper-shaped pouch is surrounded by two sepals and two petals that are striated brown and

exquisitely twisted. The flower emits a fragrance alluring and sweet, somewhat like candy corn.

RELATED PLANTS Pink lady's slipper (*Cypripedium acaule*) is much more challenging (nigh impossible?) to grow than large yellow lady's slipper and native only in eastern and northern portions of the Midwest from the Appalachian Plateau in southern Ohio westward to northern Indiana, northern Illinois, and central Minnesota and points north. It requires humus-rich, acidic soils (sandy or igneous based) or organic aerated sphagnum moss in bogs. Consult *Growing Hardy Orchids* by John Tullock for how to grow this plant. The overpopulation of deer has become a (the?) major threat to all lady's slippers.

Cystopteris bulbifera
Bulblet fern

Dryopteridaceae (wood-fern family)

HOW TO GROW Bulblet fern is native from the Driftless Area southward to the Ozark Highlands and all points eastward—very localized westward. It's found in moist, mossy limestone slopes, cliffs, seeps, and algific talus slopes. In cultivation it grows only in continually moist, shaded, and wind-sheltered gardens. You can propagate new plants from the little "bulblets" that form along the edge of the fronds.

LANDSCAPE USE This fern fills the niche of a very moist and shady site and also grows well in well-watered containers. The pendant fronds look particularly nice draping down a wall or when the plant is used as a spiller in a container.

ORNAMENTAL ATTRIBUTES The long-tapered fronds of very fine-textured stalks and leaflets make it a striking plant.

Diminutive and delicate *Cystopteris fragilis* (fragile fern) fronds belie this fern's rugged hardiness.

This exuberant mass of *Diplazium pycnocarpon* (glade fern) is at Fernwood, Michigan.

Cystopteris fragilis complex
Fragile fern

Dryopteridaceae (wood-fern family)

HOW TO GROW Fragile fern is native across most of temperate North America and through much of the Midwest. It is actually incredibly hardy, surviving fierce heat and drought by going dormant. In the wild it grows in moist sheltered, forested ravines that may be summer dry, and in cultivation is prefers humus-rich, loamy woodland soils. Propagate fragile ferns by dividing the slender, creeping rootstalk.

LANDSCAPE USE This small 6- to 8-inch-tall fern is mainly a spring ephemeral, emerging in a luxuriant mass in spring but dying with the dryness of summer, often reappearing with moisture in the fall. With that in mind, it is a marvelous plant for a woodland garden and mixes well with other spring ephemerals and woodland plants.

ORNAMENTAL ATTRIBUTES This fine-textured fern is a wee example of the ferns and delightful because of its small size.

Diplazium pycnocarpon
Glade fern, narrow-leaved spleenwort

Syn. *Athyrium pycnocarpon*
Dryopteridaceae (wood-fern family)

HOW TO GROW Glade fern is found throughout the Eastern Midwest westward to southeastern Minnesota, northeastern Kansas, and northwestern Arkansas. It is found in rich, mesic forests and is easily cultivated in moist, well-drained shaded soils sheltered from hot drying winds. Propagate the plant by division. This fern is not as drought tolerant and I lost it in the drought of 2012 whereas most other ferns may have gone dormant but returned.

LANDSCAPE USE This is a prime shade or woodland garden perennial growing 3 feet tall.

ORNAMENTAL ATTRIBUTES The untoothed, entire leaflets make glade fern look somewhat like the common, though unhardy Boston fern (*Nephrolepis exaltata*).

Dryopteris spp.
Wood ferns

Dryopteridaceae (wood-fern family)

Several species of wood ferns are native across forested portions of the Midwest and three of them are readily available for gardeners to plant.

HOW TO GROW Wood ferns grow in moist, well-drained soil in shade. They spread slowly by rhizomes into a multi-crowned plant that can be divided.

LANDSCAPE USE Wood ferns make beautiful additions to traditional shade and woodland gardens where they can be planted in masses or integrated with other spring ephemerals or planted among woodland plants.

ORNAMENTAL ATTRIBUTES The doubly pinnately compound fronds create a very fine texture in the garden. They are more leathery than the fronds of other species, remain evergreen, but lie flat in the winter.

Dryopteris carthusiana (spinulose woodfern, syn. *D. spinulosa*) is commonly known as the "fancy fern" and actually collected for use by florists. It's native mainly in the northeastern half of the Midwest with outliers west to north central Nebraska and southward to northern Arkansas. The fern grows about 1 foot tall with dense scaly stalks that arise in a circle from its rootstalk. In the Midwest, it is the most easily grown woodfern.

Dryopteris goldiana (Goldie's fern) is our most spectacular woodfern growing to 4 feet tall in the right

This December image of *Dryopteris carthusiana* (spinulose woodfern) explains why this evergreen fern is utilized in cut arrangements.

The fresh new fronds of *Dryopteris marginalis* (marginal woodfern) stand out between a leatherwood and Virginia creeper.

Magnificent *Dryopteris goldiana* (Goldie's fern) stands tall in front of a mass of maidenhair fern.

conditions. It's native to all midwestern states except the Dakotas, Nebraska, and Oklahoma and is most widespread in the Appalachian Plateau of Ohio and the Driftless Area with outliers to the Kansas City region where it grows in mesic sheltered ravines of loess deposits. The large, somewhat triangular fronds adorn 12- to 16-inch straw-colored, scaly stalks.

Dryopteris marginalis (marginal woodfern) is easy to identify because its spore sori line the margins of its leaflets (as described by its botanical name) and its fronds are more bluish green. It is found sparingly across the Midwest but is absent from the Dakotas and Nebraska. It reaches just 18 inches tall and is found in more acidic forests of sandstone or chert but grows well in most well-drained woodland garden soils.

Elymus spp.
Woodland wild rye grasses
Poaceae (grass family)

HOW TO GROW Several species of wild ryes are native across the Midwest and prefer open woodlands or savannas. These are cool-season grasses, easy to grow in most rich soils in light to partial shade. The plants have fibrous roots and form clumps that may be divided, or plants may easily be grown from seed.

LANDSCAPE USE Wild ryes grow 30–40 inches tall and can form attractive mass plantings in natural landscapes. They provide a good structure for companion wildflower plantings in shade and woodland gardens as their foliage and seed heads hold well into winter.

ORNAMENTAL ATTRIBUTES The basal leaf blades green up early and produce inflorescences by midsummer. Flowering can be interesting on close inspection when the anthers are present but the resulting fruiting heads dry blond/tan and look extra beautiful with a dark backdrop or when backlit by sunshine.

Elymus hystrix (bottlebrush grass, syn. *Hystrix patula*) is native from the Great Plains eastward across the entire Midwest to the Northeast and Upper South. It is well named with its inflorescences that look like stiff, open bottlebrushes that dry and hold into winter but often disintegrate by midwinter.

321

The flower and seed heads of *Elymus hystrix* (bottlebrush grass) clearly depict how the plant got its common name.

The dome-shaped flower heads of *Eutrochium purpureum* (sweet Joe-Pye weed) are comprised of softest pink flowers.

Seed heads of *Elymus villosus* (silky wild rye) glisten in rays of light from the sunset.

Elymus villosus (silky wild rye or hairy wild rye) is native across the entire Midwest from the central Great Plains eastward to the Mid-Atlantic and Lower New England. It, along with bottlebrush grass, is the showiest of several species though silky wild rye holds its fruiting heads through the entire winter. These infructescences are comprised of arching, bushy awns that are not as robust as the Canada wild rye found on prairies.

Eutrochium purpureum
Sweet Joe-Pye weed

Syn. *Eupatorium purpureum*
Asteraceae (aster family)

HOW TO GROW Sweet Joe-Pye weed is native across most of the Eastern Midwest westward to central Minnesota, eastern Nebraska, eastern Kansas, and eastern Oklahoma. It is found in rich, upland open woods and savannas. It is not a prairie plant but is weak in dense shade so thrives in moist, loamy well-drained soils in partial shade. It won't grow in my droughty clay soil. It is easy to propagate from seed.

LANDSCAPE USE This tall perennial grows at least 5 feet tall or taller and works at the back of a semishaded perennial border or the edge of woods in a natural landscape. It is a butterfly magnet and a premier plant for pollinators.

ORNAMENTAL ATTRIBUTES The cloudlike domes of dense pale pink flowers serve as butterfly landing pads atop the tall stems. The leaves also are architectural as they are arranged in dramatic whorls equally spaced along the stem.

322

Geranium maculatum
Wild geranium
Geraniaceae (geranium family)

HOW TO GROW Wild geranium is native across temperate eastern North America westward into the easternmost Dakotas and Kansas, absent only from Nebraska in the Midwest. It grows in moist to dry upland woodlands and savannas—in more mesic sites westward. In cultivation, it grows in almost any moist, well-drained garden soil where it will survive in full sun to light shade. It is easy to propagate from seed or by dividing mature clumps.

LANDSCAPE USE The plant grows 1–3 feet tall and is a classic along with columbine and woodland phlox. It makes a colorful flowering component for a dazzling mid-to-late springtime woodland garden. It self-sows in good growing conditions, so is best utilized in a mass on the edge of woodland. It can go dormant in dry years but otherwise some basal foliage will hold until fall. Wild geranium attracts many pollinators, especially bees: the mining bee (*Andrena distans*) is a specialist on this wildflower.

ORNAMENTAL ATTRIBUTES A wild geranium in full bloom and its vivacious or violet pink flower captures the color scheme spirit of our place. A mature clump will produce an abundance of flowers, as showy as a prize Asian azalea and lasting just as long. The flowers produce beaklike fruiting pods "cranesbills" that curl open from the base to disperse the seeds. The foliage is palmately divided and handsome in spring into early summer.

The vibrant pink flowers of *Geranium maculatum* (wild geranium) are striated with nectar guides for pollinating insects.

Gillenia stipulata (American ipecac) is a sturdy perennial briefly adorned with white flowers.

Gillenia stipulata
American ipecac, Indian physic
Syn. *Porteranthus stipulatus*
Rosaceae (rose family)

HOW TO GROW American ipecac is a heart-of-the-continent native found across the Lower Midwest from Kansas to Ohio and southward into the Upper South. In the wild it grows in mesic to dry upland woodlands and savannas. In cultivation, it prefers well-drained humus-rich, loamy woodland soils in partial to light shade. It is easy to propagate from seed.

LANDSCAPE USE The plant reaches around 30 inches tall and makes a good structural perennial for a woodland garden as it holds well through the whole growing season.

ORNAMENTAL ATTRIBUTES The foliage is uniquely three-parted and quite lovely all growing season then turns dull but colorful shades of orange to burnt red in the fall. The delicate flowers are comprised of five white petals that are welcome in late spring but short-lived. Flowers produce little fruit capsules that look like the dry core of an apple, and this plant is considered the evolutionary precursor to the apple (*Malus* spp.).

RELATED PLANTS Bowman's root (*Gillenia trifoliata*) is mainly a native of the East with populations reaching easternmost Ohio and a few outliers to Michigan, southern Illinois, southern Missouri, and northwestern Arkansas. It is a popular traditional perennial available at most nurseries being very similar to but more floriferous and fuller than American ipecac.

323

Helianthus spp.
Woodland sunflowers
Asteraceae (aster family)

Following are the midwestern species of woodland sunflowers. They regularly hybridize making identification sometimes challenging if not impossible. Silphium and small woodland sunflower have hybridized in my landscape creating a really cool plant.

HOW TO GROW Woodland sunflowers grow wild across the entire Midwest. Almost all spread by underground tubers and rhizomes as well as by seed; only silphium sunflower (*Helianthus silphioides*) and small woodland sunflower (*H. microcephalus*) grow in a clump that stays put (though still self-sows quite abundantly). They grow in open woodlands, savannas, and forest edges in moist, well-drained soil in partial to light shade. Clumps may be divided and woodland sunflowers easily are grown from seed. Sunflowers are known for their allelopathic chemicals that deter the growth of other plants around them.

LANDSCAPE USE Wild sunflowers make beautiful masses along the edge of woodlands or other natural landscapes where their aggressive behavior can be embraced. They all grow at least 3 feet tall (some species reach 6 feet tall or more in rich soil) and bloom in late summer or early fall. The flowers are very rich in nectar and pollen and visited by many insects so they are ideal in an insectaries garden. They also make ideal wildlife garden plants, as seeds are a favorite food of many songbirds and small mammals, just like larger sunflower seeds purchased in a bag. These plants are not for traditional gardens where they spread like wildfire, but anyone who leaves them out of a wilder natural garden is depriving themselves of spectacular beauty and an abundance of nature.

ORNAMENTAL ATTRIBUTES The glowing golden-yellow flowers atop tall, leafy stems are followed by buttonlike fruiting heads that hold into winter after the leaves have fallen. Most species have fragrant flowers producing a scent reminiscent of pure cocoa.

Helianthus divaricatus (woodland sunflower) is very similar to hairy sunflower but grows across the Eastern Midwest westward across Illinois to Iowa, eastern Missouri, and northern Arkansas. Its leaves are sessile (stalkless) opposite each other and the plant is relatively smooth. it grows about 3 feet tall.

Helianthus hirsutus (hairy sunflower) is native across most of the Midwest except for the Dakotas. It has noticeably hairy stems and leaves. The plant grows only around 3 feet tall.

Helianthus microcephalus (small woodland sunflower) often produces the truest yellow flowers of any wild sunflower.

Helianthus silphioides (silphium sunflower) in a traditional landscape shows its basal leaves and relatively leafless tall flower stalks.

Helianthus microcephalus (small woodland sunflower) grows mainly in the Ohio River drainage from the Southern Appalachian Mountains northward to eastern and southern Ohio, central and southern Indiana, southern Illinois, and southeastern Missouri. It has smaller flowers about 3 inches across and grows 3–4 feet tall. Its cultivar 'Lemon Queen' has lighter lemon yellow flowers and is a popular traditional perennial.

Helianthus silphioides (silphium sunflower) is mainly an Ozark Highlands wildflower found from Oklahoma to Kentucky. The first time I saw it in the wilds of northern Arkansas, I thought it was a silphium and couldn't identify it until I realized it was a real sunflower. It stays in a clump growing 4–6 feet tall but does self-sow abundantly.

Helianthus strumosus (pale-leaf woodland sunflower) is native east of the Great Plains and is found across all but

Pairs of leaves just below the flowers identify this large woodland sunflower as *Helianthus strumosus* (pale-leaf woodland sunflower).

Heuchera americana (woodland alumroot) usually has green basal leaves, but this Missouri-origin plant has highlights of burgundy.

The sunflowers of *Helianthus tuberosus* (Jerusalem artichoke) produce an intense chocolaty fragrance.

the westernmost Midwest. It produces 4-inch flowers and opposite leaves on short, tapered petioles; the leaves are opposite all the way to the flowers. It grows very aggressively in rich sites without competing plants.

Helianthus tuberosus (Jerusalem artichoke) is native across central and northeastern North America including across the entire Midwest. It grows huge, easily 6 feet tall (to 10 feet), and spreads by its tubers that are edible. It's an ideal plant for a food forest, but is so aggressive it should always be carefully sited and harvested to reduce its vigor. The opposite leaves have a relatively long, tapered base and petiole; at the top of the plant the leaves are no longer paired (becoming alternate) below the flowers.

Heuchera americana
Woodland alumroot

325

Saxifragaceae (saxifrage family)

HOW TO GROW Woodland alumroot is native across the Lower Midwest and northward into Michigan and southeastern Nebraska on rocky slopes usually in the company of oaks. It grows in most well-drained garden soils. It is easy to propagate from its tiny seeds but seems to be longer lived in a garden setting if it's divided every few years.

LANDSCAPE USE The late spring to early summer flowering stems can grow 2 to more than 3 feet above the plant while the basal foliage grows just 6–9 inches tall. It makes a refined woodland garden plant for winter foliage.

ORNAMENTAL ATTRIBUTES Cultivars, mainly selected from east of the Midwest, are most popular in gardens for their variegated leaves displaying darker veining or overlays, sometimes even with silvering. Most wild plants have uniformly green leaves and the airy panicles of flowers are pale green with protruding orange stamens. The flowers are smaller and more symmetrical as compared to the similar prairie alumroot.

The crinkled new leaves of *Hydrastis canadensis* (goldenseal) enhance the apetalous, tufted white flowers.

Heuchera villosa (maple-leaved alumroot) blooms on the last day of September in the native garden at Chicago Botanic Garden.

Heuchera villosa
Maple-leaved alumroot

Saxifragaceae (saxifrage family)

Maple-leaved alumroot became popular because of its purple-leaved seedling named 'Palace Purple' that began a profusion of colorful-leaved cultivars with foliage varying from yellow to brownish shades, or pinkish to purple, and dark veined to silvered. Today, hybrids are immensely popular perennials available at almost every nursery across the Midwest. They are touted for their cold, heat, and humidity tolerance.

HOW TO GROW Maple-leaved alumroot is found mainly in the southern Appalachian Mountains with outliers northward into southern Ohio, southern Indiana, and westward in local sites in the Ozark Highlands of Missouri and Arkansas. It requires moist, well-drained soil with partial (afternoon) shade to shade. It's actually quite easy to grow from its dustlike seeds.

LANDSCAPE USE Though not particularly showy, the late summer into autumn flowers are visited by many pollinating bees. The plasterer bee (*Colletes aestivalis*) is a specialist pollinator on alumroot flowers.

ORNAMENTAL ATTRIBUTES The uniquely shaped, rather coarse evergreen leaves are the best assets of the plant with the open and airy flowers set high above the foliage an added delight. The flowers can look ratty to some tastes and can be clipped off.

Hydrastis canadensis
Goldenseal

Ranunculaceae (buttercup family)

HOW TO GROW Goldenseal is native across most of the Eastern Midwest and westward to southeastern Minnesota, eastern Iowa, and barely into Kansas and Oklahoma. It grows in rich, mesic forests and may be cultivated in moist, well-drained humus-rich, loamy woodland soils in shade. Goldenseal is challenging to grow from seed and more easily propagated by division.

LANDSCAPE USE Goldenseal usually grows around 1 foot tall and makes a fitting addition to a woodland garden. Its roots are widely utilized as an herbal medicine, second only to ginseng in monetary value, so it also makes a valuable garden herb.

ORNAMENTAL ATTRIBUTES The maple-shaped foliage is subtly beautiful: marvelously corrugated when expanding, emerging bronzy, and maturing rich green. The midspring flowers are without petals but have cute tufts of white stamens tipped with yellowish anthers surrounding a button of green pistils. Fruits ripen red in summer and contrast beautifully against the foliage.

Hypericum pyramidatum
Great St. John's wort

Clusiaceae (St. John's wort family)

HOW TO GROW Great St. John's wort is found in woodland riverbanks, woodland edges, and other disturbed ground in and around woodlands. It grows well in moist, well-drained garden soil in full sun or partial shade. It is easy to grow from seed but does not self-sow wildly.

The brightly colored star-shaped flower of *Hypericum pyramidatum* (great St. John's wort) is impossible to miss in a woodland garden.

LANDSCAPE USE This 4- to 6-foot-tall perennial is suitable for the back of a traditional perennial border though it is rarely planted that way. It is usually cultivated in natural gardens in sunnier openings of a woodland garden. Pollen-collecting bees visit the flowers.

ORNAMENTAL ATTRIBUTES The up-facing flowers are 3 inches across and showy bright yellow with a dense tuft of stamens in their centers. The flowers bloom in sequence for only about two weeks in late summer but are followed by 2-inch seed capsules that are quite ornamental and hold distinctively into winter.

Iodanthus pinnatifidus
Purple rocket
Brassicaceae (mustard family)

HOW TO GROW Purple rocket's native range is centered on the Lower Midwest, growing northward to Lake Erie in Ohio, Lake Michigan in northeastern Illinois, and as far north as a couple sites in southeastern Minnesota. It is found in upland woodlands that are not densely shaded. In cultivation it prefers well-drained garden soil in full sun to light shade. It is easy to propagate by seed.

LANDSCAPE USE Purple rocket is the native equivalent of invasive biennial dame's rocket (*Hesperis matronalis*) and is an environmentally sound replacement. It grows about 3 feet tall and does well in a natural shade or woodland garden. It blooms in late spring into early summer when few other woodland plants are in flower so provides copious nectar. It is a good insectaries plant for bees and butterflies.

ORNAMENTAL ATTRIBUTES The common name is misleading, as the flowers are soft violet, sometimes white infused with a touch of violet. They are produced on tall spikes above a rosette of leaves.

Iodanthus pinnatifidus (purple rocket) flowers are soft violet and not purple.

The rich blue flowers of *Iris brevicaulis* (zigzag iris) are comprised of upright standards and uniquely upright falls with yellow bases.

Iris brevicaulis
Zigzag iris
Iridaceae (iris family)

Zigzag iris is one of the Louisiana irises and used in hybridizing with several other southern species to create a marvelous array of cultivars.

HOW TO GROW Over half of this species's native range is in the Midwest from the Kansas City region eastward to central Illinois, central Indiana, western Ohio, and southern Ontario, southward. Zigzag iris is quite localized through this range growing in moist to wet forested slopes, ravines, and floodplain terraces. In cultivation, it grows well in moist to wet garden soils in full sun to light shade yet is very drought tolerant. It's easy to divide but also grows readily from seed and is cold hardy through USDA zone 5.

LANDSCAPE USE This iris is best in a mixed perennial border, woodland garden, or rain garden. It is relatively low-growing, reaching only 18 inches tall, but goes dormant through midsummer and reappears in fall.

ORNAMENTAL ATTRIBUTES The spiky, swordlike foliage contrasts well with other spring woodland plants. It blooms in late spring after most woodland wildflowers are finished blooming, and the flowers are magnificent—4 inches across and royal blue with yellow flares on the falls—though the falls splay more outward.

Jeffersonia diphylla
Twinleaf
Berberidaceae (barberry family)

This plant was named in honor of Thomas Jefferson.

HOW TO GROW It has a rather unexplainable spotty and localized midwestern distribution from southeast Minnesota eastward across southern Wisconsin, Michigan, and into Ontario and southward across northeast Iowa and Illinois into the Kentucky Bluegrass Region. It is found in mesic forests atop limestone, usually growing in sheltered north- or east-facing slopes. This translates into planting it in sheltered shady sites with rich, calcareous soil. It is hardy throughout the Midwest.

LANDSCAPE USE This premier woodland garden perennial is grown for its springtime flowers and handsome foliage that lasts until fall. It's long-lived and self-sows a bit when planted in suitable conditions. It reaches just 18 inches tall.

ORNAMENTAL ATTRIBUTES The two-parted leaves remind me of butterfly wings; luscious all season and turning wonderfully yellow in the fall, they also show water droplets like beads of mercury. The 2-inch midspring flowers are white and very short-lived, emerging with or underneath the expanding foliage. The flowers form special capsules with a hinged top that ripen and disperse the seeds in summer.

Maianthemum racemosum
False Solomon's seal
Syn. *Smilacina racemosa*
Liliaceae (lily family)

HOW TO GROW False Solomon's seal is native across temperate North America east of the Great Plains so is found throughout the Midwest. It grows in moist forests and woodland, often forming large masses on the edge of woods. It is easy to grow and spreads slowly by rhizomes. It can be grown from seed or more easily divided.

LANDSCAPE USE The British Royal Horticultural Society has given this plant an Award of Garden Merit, but it is rarely used in traditional landscapes in its homeland, even though it far outshines popular perennials like astilbe. It makes a 24-inch-tall, spreading clump in a shady perennial border, edge of woodland mass planting, or in a natural woodland garden.

The leaves of *Jeffersonia diphylla* (twinleaf) look like pairs of butterfly wings while the flowers look similar to bloodroot.

Maianthemum racemosum (false Solomon's seal) produces plumes of creamy white flowers in late spring.

ORNAMENTAL ATTRIBUTES The plumes of frothy, creamy white flowers are a welcome sight during their early summer bloom, a strong contrast to all the fresh green of the season. The fruits are just as showy and for a long period, turning a bronzy gold with reddish spangles by midsummer and ripening ruby red by fall, often adorning a plant in golden fall color.

Mitella diphylla
Bishop's cap
Saxifragaceae (saxifrage family)

HOW TO GROW Bishop's cap is native across most of the Eastern Midwest westward to central Minnesota, Iowa, eastern Missouri, and north central Arkansas, where it is found in rocky mesic forests, usually sheltered in north- or east-facing slopes and bluffs. It thrives in woodland gardens on the east or north sides of homes or structures in moist, well-drained soil in full shade and can be propagated by dividing mature plants (it grows by very short rhizomes to create a clump) or by growing it from seed.

LANDSCAPE USE Bishop's cap is easier to grow than popular foamflowers (*Tiarella* spp.) in the western Midwest and makes a small tidy perennial for a traditional shade border or woodland garden landscape. It should be grown along

The spires of delicately fringed flowers of *Mitella diphylla* (bishop's cap) need close inspection to appreciate.

the edge of a path or on a moist, shady wall so its delicate flowers can easily be observed.

ORNAMENTAL ATTRIBUTES The evergreen basal, paired stem leaves look great all growing season. The spires of unique white flowers are exquisitely detailed upon close inspection—white cups with fine petals that make it look like a snowflake. The flowers produce yellowish fruits that look like nests holding tiny black seeds.

Monarda bradburiana
Bradbury bergamot
Lamiaceae (mint family)

HOW TO GROW This species has a rather small range centered on the Lower Midwest, throughout the Ozark Highlands and eastward across southern Illinois barely into Indiana and southward to central Tennessee, where it is found in moist to dry upland woods and savannas. It may be grown throughout the Midwest in well-drained soil in full sun or light shade. It is easy to propagate by seed or by dividing the plant.

LANDSCAPE USE The plant grows about 2 feet tall and is suitable for a traditional perennial border, mass planting, or interspersed in a woodland garden. It attracts many pollinators so is integral to insectaries and butterfly gardens as it blooms in late spring between abundant spring and midsummer blooms.

ORNAMENTAL ATTRIBUTES The showy flowers are tinted light lilac-white with darker purple spots, produced in a ring around a central disk. The ornamental fruiting head that follows holds well into winter. The foliage is clean and mildew resistant, turning rich shades of purplish and red in fall.

The delicately spotted flower lips of *Monarda bradburiana* (Bradbury bergamot) are distinctive.

Napaea dioica

Glade mallow

Malvaceae (mallow family)

HOW TO GROW Glade mallow is native throughout the Driftless Area southward into northern and central Illinois, eastward to central Indiana and Ohio. It does not grow in glades but prefers moist soils and is wild along rivers and stream terraces, including openings of mesic floodplain forest. Glade mallow is grown easily from seed off female plants.

LANDSCAPE USE This is a dramatic foliage plant for the back of a traditional perennial border but its less showy flowers make it overlooked by gardeners. It grows 6 feet tall so works well on the edge or partially sunny opening of natural woodland gardens. The flowers are nectar and pollen rich and visited by many insects.

ORNAMENTAL ATTRIBUTES The basal foliage is large and palmately divided (like the fingers from the palm of a hand), reminiscent of a huge maple leaf. It's dramatic, not as spectacular as some tropical plants, but in that category for our spirit of place. The five-petaled, small white flowers are produced in abundance atop tall stems in midsummer. Male flowers have a column supporting their tuft of yellowish to pinkish stamens that make them somewhat showier.

Oenothera fruticosa

Sundrops

Syn. *Oenothera tetragona*
Onagraceae (evening primrose family)

Oenothera fruticosa has two subspecies: subsp. *fruticosa* and subsp. *glauca*.

HOW TO GROW Sundrops are native from the Mid-Atlantic westward to Lake Michigan and the Ozark Highlands, growing in openings in forests and woodlands and persisting well in disturbed areas. They are not native to the prairie. They are cultivated across the entire Midwest, usually as a pass-along plant, and grow well in moist garden soils in full sun or partial shade. Gardeners almost always propagate sundrops by dividing clumps.

LANDSCAPE USE Sundrops have become an heirloom plant and are suitable for traditional perennial borders where they spread into masses about 18 inches tall. They are better suited to natural landscapes on the edge of woodland gardens or in open wooded gardens.

ORNAMENTAL ATTRIBUTES The brilliant yellow flowers give this species its common name and cover the plant for a short time in early summer, shining brightly against the greens of June.

The white flowers of *Napaea dioica* (glade mallow) stand above its dramatic foliage.

Oenothera fruticosa (sundrops) flowers beam intense yellow from the bright summer solstice sun.

Onoclea sensibilis (sensitive fern) perfectly fills a space between a rocky dry streambed and a north-facing foundation.

A magnificent mass of *Osmunda claytoniana* (interrupted fern) thrives at the Eloise Butler Wildflower Garden in Minneapolis.

Onoclea sensibilis

Sensitive fern

Dryopteridaceae (wood-fern family)

HOW TO GROW Sensitive fern is a widespread native found across all of temperate eastern North America and throughout the Midwest (though localized in the westernmost Midwest). It grows in moist, upland forests but also in many wetlands (where it will tolerate full sun) and is a moisture-demanding species. It can go dormant in a drought and return. It runs by underground rhizomes, but not obnoxiously so in an upland site. The rhizomes can readily be divided.

LANDSCAPE USE This fern is aggressive in wet soils though skirts underneath larger plants like Joe-Pye weeds, rose mallows, and sunflowers. It's a unique addition to a moist woodland garden where it is less aggressive, but can smother diminutive plants. At Crystal Bridges in Bentonville, Arkansas, it is beautifully used between a rocky, dry streambed and the north-facing foundation of the art museum.

ORNAMENTAL ATTRIBUTES The coarser, almost un-fernlike and triangular deeply lobed leaves make this fern a garden standout. The plant dies quickly in a frost (how it gets its name) leaving the fertile spikes of beadlike brown spore capsules to adorn the winter landscape.

Osmunda claytoniana

Interrupted fern

Osmundaceae (royal fern family)

Interrupted fern grows from a slowly creeping, ground-hugging rhizome and is often found in mature woodlands in rather large stands. In rich, moist soils it can become a huge fern (the Midwest's largest?), 6 feet tall. One spectacular stand near my hometown is lovingly called "Neanderthal City" as walking through it was like going back in time. I have since learned that such stands of interrupted ferns are probably the oldest living plants in Iowa.

HOW TO GROW Interrupted fern is native across northeastern North America and found in the Midwest mainly in the Upper Midwest westward to central Minnesota, central Iowa, and localized southward into northern Arkansas and Kentucky. It's grows in rich, mesic forests and clearings, often sheltered in ravines or on north- or east-facing slopes. In the garden it prefers moist, well-drained soil; shade is a must in drier parts of the Midwest. To easily propagate this species, simply divide growing points off the rhizome.

LANDSCAPE USE This large fern grows 3 feet tall or more and creates a fine mass planting in appropriate rich woodland locations. It is also a striking container plant.

ORNAMENTAL ATTRIBUTES The fiddleheads unfurl earlier than those of most ferns, revealing sterile leaflets interrupted by fertile leaflets comprised solely of clusters of spores about two-thirds up the frond's stalk. Above the fertile leaflets at the end of the leaf stalk there are again sterile leaflets. The fertile leaflets start dark green but become brown in late summer with spore cases, a unique contrast to the green of the sterile leaflets above and below them on the stalk.

331

including day-flying Nessus sphinx, bumblebee and hummingbird clearwings, and at dusk by the lettered sphinx.

ORNAMENTAL ATTRIBUTES The mid to late spring flowers last, are lightly fragrant, and appear in a range of colors from white to soft blue, sometimes rosy or purplish. The flower color of particular populations is often consistent, for example, those along the Pecatonica River in northern Illinois are almost all white. Each plant is not necessarily long lived, so if a particular plant is not allowed to self-sow you will lose it—you may like a particular color of flower but self-sown plants will vary. I planted many strains of native plants and cultivars but now have a blend of flower colors that are increasingly light blue.

There's nothing sweeter than the spring blooms of *Phlox divaricata* (woodland phlox).

Phlox divaricata
Woodland phlox, wild blue phlox
Polemoniaceae (phlox family)

When woodland phlox is in bloom from mid to late spring, the magic of its flower's delicate perfume fills the air. This has earned it the colloquial name sweet William.

HOW TO GROW Woodland phlox is found across the entire Midwest but localized northwestward. It grows best in mesic floodplain forests where it can carpet the ground in spring but also grows in most woodlands as long as they are moist in spring. Plant it in any well-drained woodland soil in at least a partially shaded site. It is very summer drought tolerant and is hardy across the entire Midwest. Woodland phlox is easily grown from seed and readily self-sows in between established plants, but is never smothering. It even self-sows into a shady lawn.

LANDSCAPE USE This phlox combines well with almost any woodland plant so may be integrated with shade-loving groundcovers or around larger shade perennials that will be more prominent through the growing season. Springtime flower stalks are rarely more than 18 inches tall and may look briefly unsightly but quickly disintegrate. The evergreen foliage is mainly ground hugging, sometimes disappearing after spring so it does not make a good stand-alone groundcover. Rabbits love this plant so new plantings may need to be protected until a bigger colony is established. The flower is mainly pollinated by smaller sphinx moths

Polemonium reptans
Spring polemonium, Jacob's ladder, Greek valerian
Polemoniaceae (phlox family)

HOW TO GROW Spring polemonium is found from most of Ohio and southwest Michigan westward to southeast Minnesota, southeast South Dakota, easternmost Kansas, and northeast Oklahoma. In the wild it grows in rich upland woods and is more localized in mesic forests westward. In cultivation it grows in shaded well-drained soils that are moist in spring. It is hardy throughout the Midwest. Spring polemonium is easy to grow from seed and self-sows lightly.

LANDSCAPE USE This clumping perennial grows no more than 15 inches tall and integrates well into a shady perennial border or woodland garden.

ORNAMENTAL ATTRIBUTES Spring polemonium has showy sky blue flowers and ornamental foliage comprised of pinnately compound paired leaflets (the ladder in its name "Jacob's ladder"). The light green calyxes stay on the plant after flowering, turning yellow as the capsule of seeds inside them ripens.

RELATED PLANTS The cultivar 'Stairway to Heaven' has striking white-edged, variegated leaves.

The lavender-blue flowers of *Polemonium reptans* (spring polemonium) are an iconic component of spring woodlands in the Midwest.

Polygonatum biflorum
Solomon's seal

Liliaceae (lily family)

HOW TO GROW Solomon's seal is distributed across temperate eastern North America in woodlands and forests from the Great Plains to the Atlantic and Gulf Coasts. It is common across the entire Midwest and grows in rich woodland soil in partial sun to shade. It is relatively easy to propagate from seed but slow to become a sizeable plant. It is easy to divide colonies that spread by short rhizomes.

LANDSCAPE USE This species forms architectural clumps of foliage and is suitable for a traditional garden's shady perennial border. Its variegated cousin from abroad (*Polygonatum odoratum*) is embraced, yet our native species are virtually never found at nurseries. Solomon's seal makes an important subject in a shady insectaries or bird garden. It should be planted in food forests as its rhizomes and shoots are edible; it is related to asparagus.

ORNAMENTAL ATTRIBUTES This plant is at its best in the fall when the leaves turn golden yellow while the arching stalk is adorned with pendant blue berries at each leaf axil. The late-spring flowers are pendant whitish green elongated bells that are quite nectar rich and visited by many bees and hummingbirds but are somewhat hidden by the foliage.

NOTES *Polygonatum biflorum* includes *P. commutatum* and *P. canaliculatum*. I've been enamored by the subtle beauty of this often magnificent perennial since childhood as it once lined many woodland edges and roadsides. I made a multicolored wood block print of its fall splendor in high school. Solomon's seal is a preferred browse of deer so has become rare where deer are overpopulated.

Polystichum acrostichoides
Christmas fern

Dryopteridaceae (wood-fern family)

HOW TO GROW Christmas fern is native across the Eastern and Lower Midwest northwestward to southeastern Minnesota. It grows wild in mesic forests on sandstone or cherty more acidic soils, usually on slopes and ridges. It can be cultivated in almost any moist, well-drained soil in shade with protection from hot summer winds. It's challenging to grow from spores and forms a crown that is slow to divide.

LANDSCAPE USE The plant reaches about 2 feet tall and makes a premier perennial for a shady border, mass planting, or woodland garden.

ORNAMENTAL ATTRIBUTES The fronds are dark green and evergreen. The fiddleheads unfurl exquisitely in spring, hold upright all summer, then lie flat through the winter.

Polygonatum biflorum (Solomon's seal) is most beautiful with ripe fruits and autumn foliage.

The fiddleheads of *Polystichum acrostichoides* (Christmas fern) rise above its evergreen fronds.

Rudbeckia laciniata
Green-headed coneflower, wild goldenglow

Asteraceae (aster family)

This relative of black-eyed Susan (*Rudbeckia hirta*) has flowers that look more like what most gardeners call a coneflower. Its glowing yellow ray flowers give it its other common name.

HOW TO GROW Green-headed coneflower is found across much of eastern North America east of the Rocky Mountains, an area which includes the Midwest. It inhabits sites along streams and rivers or in openings in floodplain forest or in moist woodlands. It is easy to grow in any rich soil and tolerates being inundated by water for brief periods. It is larger and more floriferous in full sun but does well in partial to light shade. It is easily divided or grown from seed.

LANDSCAPE USE This tall species grows 5–6 feet in rich soils, becoming somewhat aggressive without companion plantings. It's a good choice for a rain garden or massed along woodland edge. Many pollinating insects visit the flowers and many songbirds seek its seeds.

ORNAMENTAL ATTRIBUTES The late-summer, somewhat green-tinted, glowing yellow ray flowers droop below a cone of yellow-green disc flowers. The flower heads really lighten up the shaded leafy darkness of a woodland edge or opening. The fruiting heads persist into winter and are visited by small songbirds such as chickadees and juncos seeking a meal.

RELATED PLANTS The double-flowering cultivar 'Golden Glow' is a pass-along plant found in many farmsteads and gardens across the Midwest. The popular cultivar 'Herbstsonne' (Autumn Sun) is a hybrid green-headed coneflower with Georgia–Florida native shiny coneflower (*Rudbeckia nitida*).

Silene virginica
Fire pink

Caryophyllaceae (pink family)

HOW TO GROW Fire pink is native across the Eastern Midwest, though rare and local in Michigan. It grows westward scattered across Illinois, rare in southern Wisconsin, and common in the Ozark Highlands. It's found in rocky, mossy moist woodlands and can be tricky to cultivate. Fire pink is a short-lived perennial though it self-sows when grown in the right condition. It is best in moist, well-drained soils in light to partial shade and prefers to not mingle with other plants. It is easy to propagate from seed.

LANDSCAPE USE The plants reaches just 1–2 feet tall and is a stunning perennial for a rocky woodland garden or shaded rock garden.

ORNAMENTAL ATTRIBUTES The flowers are extremely showy, vibrant red-red, and pollinated by hummingbirds. Fire pink blooms over several weeks in mid to late spring.

Solidago caesia
Wreath goldenrod, blue-stemmed goldenrod

Asteraceae (aster family)

Goldenrods in savannas and woodlands are not as well known as their prairie and meadow counterparts yet are essential components to healthy woodland. Their late summer and early fall bloom, together with woodland asters; provide nectar for a host of pollinators and seed for songbirds.

HOW TO GROW Wreath goldenrod is mainly native across the Eastern Midwest, occasionally westward to southeastern Wisconsin, west central Illinois, and through the lower

The yellow, glowing flowers of *Rudbeckia laciniata* (green-headed coneflower) light up a woodland.

When given optimal growing conditions, *Silene virginica* (fire pink) produces an abundance of vibrant red blooms.

A blooming wand of *Solidago caesia* (wreath goldenrod) reaches into a woodland path at the Indiana Dunes National Lakeshore.

Solidago ulmifolia (elm-leaved goldenrod) in bloom is like a massive fireworks' explosion of gold.

Solidago sciaphila (shadowy goldenrod) brightens a shaded woodland garden.

Solidago rugosa 'Fireworks' (Fireworks rough-stemmed goldenrod) is a popular perennial in a traditional perennial border.

Ozark Highlands. It is found in upland forests and woodlands, usually in more acidic soils. It is propagated easily from seed.

LANDSCAPE USE The plant reaches around 2 feet tall and makes a lovely mass planting or can be integrated with other woodland goldenrods, asters, grasses, and fall berries.

ORNAMENTAL ATTRIBUTES In late summer, clusters of golden flowers adorn the leafy stems like a wreath.

RELATED PLANTS Three unique woodland goldenrods thrive in rocky, shaded woodland gardens and may be planted in a shaded rock wall. Cliff goldenrod (*Solidago drummondii*) grows wild on limestone cliffs or in rocky calcareous soils mainly in the Ozark Highlands of Missouri and Arkansas but also in adjacent Illinois. Shadowy goldenrod (*Solidago sciaphila*) is endemic to the Driftless Area where it grows in moist limestone outcrops in forests and open woodlands. Short's goldenrod (*Solidago shortii*) was once thought extinct but is found in three locations in the Shawnee Hills of Indiana and Kentucky; its cultivar 'Solar Cascade' is highly gardenworthy.

Solidago ulmifolia
Elm-leaved goldenrod
Asteraceae (aster family)

HOW TO GROW This species is native across the Eastern and Lower Midwest and northwestward to central Minnesota and southeastern Nebraska, where it is found in upland forests, woodlands, and savannas. In the garden in thrives in moist to summer-dry soils in light to partial shade. It is easy to grow from cuttings and seed.

LANDSCAPE USE Elm-leaved goldenrod is a good choice for a woodland garden and an insectaries garden in the shade. It grows about 3 feet tall and spreads just as wide.

ORNAMENTAL ATTRIBUTES The plant forms an open, multi-branched burst of classic golden-yellow flowers that lighten up a shade garden in early fall.

RELATED PLANTS Rough-stemmed goldenrod (*Solidago rugosa*) is very similar to elm-leaved goldenrod and is native across the Eastern Midwest westward to Wisconsin and the Ozark Highlands. It and its cultivar 'Fireworks' can self-sow into a dense mass under ideal conditions.

Stylophorum diphyllum
Wood poppy, celandine poppy

Papaveraceae (poppy family)

HOW TO GROW This species is native from southwestern Michigan and central Ohio westward into eastern Missouri, north central Arkansas, and much of Kentucky. It's found on mesic slope forests over limestone and is hardy and adaptable throughout the Midwest. It can self-sow aggressively in similar habitats and mesic shade gardens, and is considered invasive where not native but don't confuse it with a related nonnative poppy, the celandine (*Chelidonium majus*), which is truly invasive and usurps everything in its path. Because of the potential for confusion between the two species, I prefer not to use the name celandine poppy for our native species.

LANDSCAPE USE Wood poppy makes a lush spring perennial in a moist shade garden; it grows around 2 feet tall and can overtake less aggressive woodland garden plants. It can go dormant during summer dry spells.

ORNAMENTAL ATTRIBUTES The up to 3-inch-across, four-petaled golden-yellow springtime flowers are very showy. Each one is short-lived but they bloom in sequence and smaller flowers can be produced into summer if rainfall is consistent. The foliage is heavily lobed and attractive.

The bright yellow flowers of *Stylophorum diphyllum* (wood poppy) contrast perfectly with blooming *Mertensia virginica* (Virginia bluebells).

Symphyotrichum cordifolium
Blue wood aster

Syn. *Aster cordifolius*
Asteraceae (aster family)

A diverse group of species formerly assigned to the genus *Aster* but now dispersed in several genera thrives in the shade of woodlands. Just like the prairie asters, these species bloom in late summer or fall and add welcome seasonal color to a shade and woodland garden. Woodland asters are a last hurrah with, or just before, the dazzling fall colors of the shrubs and trees above them. Along with woodland goldenrods, they negate the notion that woodland gardens are only colorful in spring. Following is a taste of the best of the woodland aster bunch.

HOW TO GROW Blue wood aster is native to most of temperate North America east of the Great Plains and is the most widespread midwestern woodland aster found throughout except for Oklahoma. It grows wild in upland, mesic forests, and woodlands. In the garden, it is easily grown in light to partial shade in rich soils. It is easy to propagate from seed. The plant transplants well and self-sows abundantly under conditions that suit it.

The cordate (heart-shaped) leaves of *Symphyotrichum cordifolium* (blue wood aster) are distinctive and lovely.

LANDSCAPE USE Plants grow 2–3 feet tall and are integral to a woodland or natural shade garden—looking best in billowing masses or mixed with other fall-blooming asters, goldenrods, white snakeroot, and woodland grasses. The late-season flowers of all woodland asters, like the flowers of prairie asters, are rich in nectar for diverse pollinators and their seeds are tufted with a pappus and sought by many songbirds and small mammals.

ORNAMENTAL ATTRIBUTES Blue wood aster produces delightful, small whitish blue flowers in abundance that together form very showy full flower heads in fall.

RELATED PLANTS Calico aster (*Symphyotrichum lateriflorum*) is native across North America east of the Great Plains so is found in all midwestern states except North Dakota and is localized in the westernmost Midwest. It inhabits sites in upland dry to mesic forests, woodlands, and savannas and grows well in almost any rich soil in partial shade to light shade. Calico aster also forms frothy clouds of tiny but abundant flower heads in the autumn woodland garden—the disc flowers in the center change from yellow to deep pink as they age and are surrounded by white ray flowers. This lovely color variation of aged flowers is what gives the plant its common name, calico aster. Plants self-sow abundantly in disturbed soils so are often regarded as pests in traditional gardens. 'Lady in Black' is a cultivar with chocolaty-purplish leaves that fade to bronzy green by flowering time in early fall.

Crooked-stem aster (*Symphyotrichum prenanthoides*) is native to the central Appalachian Mountains and adjacent highlands westward to the Upper Midwest where it is found in northern Illinois, central and eastern Iowa, southeastern Minnesota, and southern Wisconsin. It inhabits moist forests to wet woodland seeps and forested wetlands. The plant produces an abundance of pinkish or bluish white, yellow-centered flowers that cover the plant in early fall.

Drummond's aster (*Symphyotrichum drummondii*) is another widespread midwestern woodland aster (most common west of Lake Michigan) whose ray flowers usually

Symphyotrichum shortii (Short's aster) may have the showiest flowers of the midwestern woodland asters.

337

Symphyotrichum lateriflorum (calico aster) in bloom creates a frothy mass of white flowers.

The lobed bases of the leaves (at which point the stem slightly changes direction) identify this aster as *Symphyotrichum prenanthoides* (crooked-stem aster).

open whitish and age to bluish or purplish. The disc flowers emerge yellow and age to shades of mauve.

Short's aster (*Symphyotrichum shortii*) is native throughout the Midwest east of the Mississippi River, westward to southeast Minnesota, eastern Iowa, and sparingly into the Ozark Highlands of Arkansas. It is found in mesic woodlands and is easy to grow in good garden soils in partial to light shade. It is easy to propagate from seed and transplants well from a seed-grown plug. Short's aster looks best in billowing masses and is my favorite of the woodland asters with very showy, rather full heads of possibly the truest blue flowers.

Taenidia integerrima
Yellow pimpernel

Apiaceae (carrot family)

HOW TO GROW Yellow pimpernel is native across the Eastern and Lower Midwest northwestward through southeastern Minnesota and the eastern two-thirds of Iowa. It is found in upland savannas and rocky woodlands and grows in well-drained garden soils in full sun to light shade. It is easy to propagate from seed.

LANDSCAPE USE The plant reaches 2–3 feet tall and provides an enchanting wildflower for a woodland garden. It does not self-sow as wildly as its relative golden alexander. It is a native host plant to the black swallowtail.

ORNAMENTAL ATTRIBUTES Flowering yellow pimpernel looks like a little firework in suspended animation—green-yellow umbels of flowers supported by the most delicate pedicels. Its early summer blooms light up woodlands amid the full greens of June. The delicate compound leaves are also charming through the growing season.

Taenidia integerrima (yellow pimpernel) produces the most elegant yellow flowers amid the greens of June.

Thalictrum dioicum (early meadow-rue) blooms almost discretely in a woodland at Grant Park, Milwaukee, Wisconsin.

Thalictrum dioicum
Early meadow-rue

Ranunculaceae (buttercup family)

Why do gardeners not better know this woodland perennial? What a delicate beauty from spring through fall, but hardy and long-lived.

HOW TO GROW Early meadow-rue is found throughout the Midwest so hardy everywhere but localized to mesic forests so quite infrequent westward. It requires moist, well-drained soils and shade, but is tolerant of summer dryness once established. It can be propagated by seed or by division of mature clumps.

LANDSCAPE USE This perfect woodland garden or shady perennial border plant provides springtime bloom and sturdy foliage through the growing season. It grows about 2 feet tall.

ORNAMENTAL ATTRIBUTES Male plants have showier mid-spring flowers with delightful pendant golden anthers. Female flowers are more spidery looking but produce fruiting capsules that adorn the plant through summer. The compound foliage is airy and delicate, the individual leaflets three-lobed in tripartite groupings. The leaves often turn light banana yellow in fall.

Thalictrum thalictroides
Rue-anemone

Syn. *Anemonella thalictroides*
Ranunculaceae (buttercup family)

HOW TO GROW Rue-anemone is native across the Eastern and Lower Midwest and northwestward to central Minnesota and central Iowa, where it inhabits moist upland forests and woodlands. In gardens it grows in rich, well-drained soil in light shade. It is challenging to propagate from seed, but the tiny tuberous roots of established plants may be divided.

LANDSCAPE USE This miniature plant seldom grows more than 6 inches tall. It makes an enchanting perennial for a shade or woodland garden where its small size can be easily viewed. Woodland sedges, ferns, and other companion plants that won't smother it should be planted nearby to fill in later in the season.

ORNAMENTAL ATTRIBUTES The most delightful asset is the flower, which was most commonly pink where I grew up, but is white in many areas. I have even seen a wild plant with green petals and there are several cultivars with double pink flowers. The foliage is also best described as delightful as it's often purplish or bronzy, finely textured, and compound, looking just like mini-versions of its meadow rue cousins. The foliage holds into midsummer when moisture is adequate but the plant is mainly a spring ephemeral.

This close-up of *Thalictrum thalictroides* (rue-anemone) was taken near Decorah, Iowa, where plants are most often pink-flowering.

Tradescantia ernestiana
Woodland spiderwort, Ernest's spiderwort

Commelinaceae (spiderwort family)

HOW TO GROW Woodland spiderwort is an Ozark Highland and Ouachita Mountain endemic with a few populations southward. A denizen of rocky mesic forests, it is quasi-ephemeral. The foliage emerges in the fall to about 2 inches tall, but grows and blooms in spring, going dormant by midsummer. This species is easy to grow in well-drained, rich woodland soils. Plants may be divided as they go dormant in midsummer and are easily grown from seed.

LANDSCAPE USE Growing just 12–18 inches tall, woodland spiderwort stays in a clump so is a top choice for a traditional shade garden. It self-sows sparingly in good growing conditions, but needs companion plants as it goes dormant.

ORNAMENTAL ATTRIBUTES The short foliage emerges purplish or reddish, maturing bronzy green and crowned with exceptionally showy, three-petaled flowers over arching, bladelike foliage. Flowers are often fuchsia or purple, sometimes purplish blue or white. As with all spiderworts, the flowers open in the morning, lasting through the entire day only when cloudy.

Tradescantia ernestiana (woodland spiderwort) is perhaps the most ornamental spiderwort species.

Uvularia grandiflora
Large-flowered bellwort

Liliaceae (lily family)

There is elegance to the emergence of this plant in spring, something I always look forward to in the woodland garden and have sketched and photographed. Large-flowered bellwort (often called merrybells) heralds the return of spring. The genus is named after uvula: that "little grape" that hangs down in the back of your throat.

HOW TO GROW Bellwort is native to forests and woodlands, mainly west of the Appalachians westward to the eastern edges of the Dakotas, Kansas, and Oklahoma (absent only from Nebraska). It is found in limestone-based, shady, moist, well-drained humus-rich, loamy soils and goes dormant in summer droughts. It may be grown from seed though ants quickly collect them when they're ripe as they are coated in a fatty elaiosome (as are many spring-flowering woodland plants) they seek for food—inadvertently transporting and planting the seed. Large clumps of bellwort are easily divided as the plants go dormant in late summer or fall.

LANDSCAPE USE This quintessential woodland garden plant blooms in spring and becomes a 2-foot-tall foliage plant that holds through the growing season (with adequate rainfall).

ORNAMENTAL ATTRIBUTES The plants emerge upward but become pendant as they grow, a botanical ballet producing pendant light to golden-yellow flowers, themselves shaped like a drooping pinwheel. The emerging foliage is bluish or purplish but matures true green, holds clean through the summer, and turns golden in fall or if stressed by summer drought.

RELATED PLANTS Perfoliate bellwort (*Uvularia perfoliata*) is a similar species with paler flowers and more rounded bases to its leaves. It's native mainly in the Appalachians so is common in eastern Ohio and around the southern edge of the Midwest westward to Oklahoma.

Sessile bellwort (*Uvularia sessilifolia*) is another related bellwort, though a mini-version growing just 1 foot tall with flowers that are often straw-colored like oats. It also has a more widespread but scattered range across eastern North America—even more localized in the Midwest, but absent only from Kansas and Nebraska. Sessile bellwort often grows in more acidic soils. The leaves are unstalked (sessile) and not pierced by their stem.

Viola pubescens
Downy yellow violet

Violaceae (violet family)

An abundance of violet species grows wild in forests and woodlands throughout the Midwest and they come in flower colors from white to yellow and classic blue-violet. Surprisingly, many of the native species are very difficult to cultivate and are best left in their native haunts. Look for additional violet species in the Groundcover chapter. *Viola pubescens* as described here includes *V. pennsylvanica*.

HOW TO GROW Downy yellow violet grows across most of

Look closely and you will see a tiny plant-hopper bug on the yellow sepal of this *Uvularia grandiflora* (large-flowered bellwort).

Uvularia perfoliata (perfoliate bellwort) was named for its stalked leaves that are pierced by their stem (perfoliate).

Viola pubescens (downy yellow violet) is a widespread yellow-flowering violet.

The common form of *Viola sororia* (wild blue violet) produces deep blue-violet flowers and stems covered in short woolly hairs.

The largest, most-aggressive form of *Viola sororia* (wild blue violet) was once known as *V. papilionaceae*.

central and northeastern North America and is found across the entire Midwest. In the wild it grows in moist forests from mesic floodplains to uplands and in the garden it thrives in rich woodland soils in partial shade to shade. It transplants relatively easily but self-sows only sparingly so can be expensive and difficult to procure and is not at all aggressive.

LANDSCAPE USE This delightful small clumping perennial grows just 6–10 inches tall. It's a delightful choice for a traditional shade garden border or naturalized in a woodland garden.

ORNAMENTAL ATTRIBUTES The spring flowers are true yellow with tiny black whiskers painted on the lower and lateral petals.

Confederate violet, a bi-colored form of *Viola sororia* (wild blue violet), has white flowers with blue-gray eyes. Here it grows with an all-white violet.

Viola sororia
Wild blue violet

Violaceae (violet family)

The botanist "lumpers" have prevailed over the "splitters" regarding our wild blue violets, placing a bunch of very similar forms (formerly listed as species) into this one species. As a gardener and violet lover it is important to understand the differences in the forms. *Viola sororia* is the state flower of Illinois and Wisconsin.

HOW TO GROW Almost all of Eastern North America east of the Great Plains, including the entire Midwest, is the native home to this violet. It's a plant of woodlands where it grows in humus-rich soils. Most varieties are easily cultivated in any rich soil that is moist in spring, growing in full to partial shade. Plants may go dormant in summer drought conditions. Some forms readily self-sow, often from cleistogamous (self-pollinated) flowers.

LANDSCAPE USE Violets make springtime gardens come alive with their beloved flowers and are at their best mixed among other woodland wildflowers. Some varieties can be quite aggressive growers and smother diminutive plants, even usurping grass in a shaded lawn. The flowers of these violets are edible so are a nice component in an edible landscape and food forest. They are host to several fritillary butterflies including the spectacular great spangled fritillary

whose nocturnal springtime caterpillars can actually help keep the plant in check. Note that this caterpillar emerges from its egg in fall and overwinters in the leaf litter—so raking up leaves essentially destroys it.

ORNAMENTAL ATTRIBUTES The spring flowers are usually royal blue, but various varieties can be nearly any shade of blue from navy blue to light blue and rarely white.

RELATED PLANTS Here are some strains to know. Woolly blue violet is the aptly named form with stems covered in short woolly hairs and flowers usually of deep blue-violet. This strain seems most common in undisturbed areas from mesic floodplain forest to upland oak-hickory woodland. It is not an aggressive self-sower.

Butterfly violet or dooryard violet (formerly *Viola papilionacea*) is usually the largest and most aggressive form; it's essentially a weed to many homeowners. It self-sows abundantly and with its larger sized, taller leaves (nearly 1 foot tall) can overtake some small woodland wildflowers and invade lawns. It is a major component in my garden; I simply pull any wayward plants but I welcome it as a component of lawn. The healthy population of resident great spangled fritillaries actually help keep it in check by heavily feeding on its springtime leaves.

There is a blue-gray-eyed white form known as the Confederate violet, a speckled blue-flowering form called 'Freckles', and a rosy-eyed white form called 'Rosie'.

Wetland Perennials

Wetland perennials thrive with wet feet. Some of them can be cultivated in moist gardens but virtually all of them have languished and died in upland soils without substantial irrigation in recent drought years.

PERENNIALS FOR A TRADITIONAL PERENNIAL BORDER IN WET CONDITIONS

Asclepias incarnata (swamp milkweed)

Boltonia asteroides (boltonia)

Carex stricta (tussock sedge)

Chelone spp. (turtleheads)

Eutrochium maculatum 'Gateway' (spotted Joe-Pye weed)

Helenium autumnale (sneezeweed)

Hibiscus lasiocarpos (rose mallow)

Juncus effusus (soft rush)

Oligoneuron ohioense (Ohio goldenrod)

Oligoneuron riddellii (Riddell's goldenrod)

Osmunda cinnamomea (cinnamon fern)

Osmunda regalis (royal fern)

Phlox glaberrima (smooth phlox)

Phlox maculata (marsh phlox)

The fragrant pink flowers of *Asclepias incarnata* (swamp milkweed) host a wealth of pollinators.

Boltonia asteroides 'Snowbank' creates a snowbank of late-summer flowers in the Perennial Garden at Powell Gardens, Missouri.

Asclepias incarnata

Swamp milkweed, marsh milkweed

Apocynaceae (dogbane family)

HOW TO GROW Swamp milkweed grows wild in swamps, marshes, riverbanks, fens, sedge meadows, and wet prairies and is best cultivated in moist to wet organic soils in full sun to partial shade. The plant is often rather short-lived, living just three years in upland gardens, but is easy to grow from cuttings or seed.

LANDSCAPE USE This species grows 3–4 feet tall and is best in water gardens, rain gardens, and pond or stream-side wetland gardens. It is tolerant of clay soils and moist, well-drained sites and is sometimes planted in traditional perennial borders. It attracts abundant insect pollinators so should be included in insectaries gardens and is host plant to the monarch.

ORNAMENTAL ATTRIBUTES The up-facing, flat clusters of fragrant pink and white bi-colored flowers (occasionally pure white) bloom in late summer into early fall. The dried plant, with its pod fruit remnants, is quite attractive in the winter landscape.

Boltonia asteroides

Boltonia

Asteraceae (aster family)

HOW TO GROW This species is native from Canada to the Gulf through the heart of North America and is found throughout the Midwest, though very localized in the Eastern Midwest. In the wild it grows in wet floodplain prairies and

345

open floodplain woodlands and the edges of wetlands. It thrives in average to wet garden soil in full sun to partial shade. It is easy to grow from seed.

LANDSCAPE USE Boltonia grows about 4 feet tall and is suitable for the back of a traditional perennial border. It makes a colorful late-season addition to a wetland garden. I have indelible memories of it growing wild with sneezeweed and cardinal flower lobelia that bloom at the same time.

ORNAMENTAL ATTRIBUTES The foliage is attractive bluish green through the season but its flowers are spectacular in mass, creating a snowbank of white flowers in late summer and early fall. Individual flowers are small and aster-like with white ray flowers around a center of yellow disc flowers.

RELATED PLANTS Cultivars 'Pink Beauty' and 'Snowbank' were once popular in perennial borders.

A maturing flower of *Calla palustris* (wild calla) clearly shows its developing fruits beside the persistent milky white spathe.

Calla palustris

Wild calla

Araceae (arum family)

HOW TO GROW Wild calla ranges from central Minnesota eastward across Wisconsin into northeastern Illinois, northern Indiana, Michigan, and northern Ohio, in bogs, fens, and lakeshore muck. It can easily be grown in a water garden or water-garden container provided it's planted in half peat, half sand. It must have a constant supply of moisture and light shade and will grow into the Lower Midwest in such conditions.

LANDSCAPE USE The plant, which grows just 1 foot tall, spreads by a rhizome that can be in water as much as 2 inches deep.

ORNAMENTAL ATTRIBUTES Wild calla has lustrous heart-shaped leaves. The flowering stem consists of a knob-shaped spadix of yellow flowers surrounded by a showy milky white spathe. The fertile flowers produce a cluster of red berries and the spathe can remain intact through fruiting.

Caltha palustris (marsh-marigold) gleams at many of the springs, seeps, and fens at Seed Savers Heritage Farm near Decorah, Iowa.

HOW TO GROW This species is found mainly across the Upper Midwest from Minnesota eastward and sparingly westward into the Dakotas and Nebraska and southward in isolated habitats into Kentucky and one fen in Missouri. It prefers sites in springs, seeps, and fens in forests, sedge meadows, or prairies. It needs constant moisture to survive, even while dormant. Only one plant has survived at Powell Gardens in a wetland habitat. Oddly we have found no fruits on the sole Missouri population that grows just 30 miles away, though we had hoped that the local population might be more adapted to our hotter climate. Marsh-marigold may be grown from seed or division.

LANDSCAPE USE In the right wetland habitat marsh-marigold makes a remarkable spring perennial but is considered invasive in cool, wet climate areas where not native. It grows to 2 feet tall and often in the company of skunk cabbage, cinnamon and royal ferns, or tussock sedge—creating a memorable composition.

Caltha palustris

Marsh-marigold

Ranunculaceae (buttercup family)

Marsh-marigold was the "cowslips" of my mom's childhood, growing in the pasture's slough, the brilliant flowers bursting from the melt waters of spring. It is a spring ephemeral, emerging quickly as soon as the ground thaws, blooming, setting seed, and often dying by midsummer.

ORNAMENTAL ATTRIBUTES The early emergence and vivacious, molten golden-yellow flowers in spring make it the most dazzling spring wetland wildflower, often forming mystical rivers or masses of gold in wetlands.

Carex muskingumensis
Palm sedge

Cyperaceae (sedge family)

HOW TO GROW Palm sedge is native in the heart of North America from Ontario and Ohio southward to Kentucky and Arkansas and westward to eastern Kansas and central Minnesota. It's found in openings in swamps and floodplain forests and occasionally in sedge meadows. It grows in rich, continually moist to wet soils in full sun to light shade. Propagate it by cuttings or seed.

LANDSCAPE USE This sedge grows about 3 feet tall and is an excellent choice for a rain garden, retention basin, or wetland garden. It produces copious seed fruits for water and songbirds.

ORNAMENTAL ATTRIBUTES The interesting tufts of stems are comprised of whorls of three arching leaves—the overall look is reminiscent of a palm. The arching fruiting heads turn golden brown when ripe and are pretty in summer.

The palmlike foliage of *Carex muskingumensis* (palm sedge) make it one of the easier sedges to identify.

Carex stricta
Tussock sedge

Cyperaceae (sedge family)

Tussock sedge has been a beloved plant all my life; it defined the slough at my grandparents' farm, its tussocks a memorable part of many "swamp stomps" to find wonderful wildflowers like the small white lady's slipper or wetland birds like rails. It is shocking how little known this great plant is in traditional landscaping.

HOW TO GROW The species is most common across the Upper Midwest but grows more localized southward into the Lower Midwest. It's a major component in sedge meadows but is also found in wet prairies, marshes, fens, and openings in swamps. It requires moist to wet organic soils in full sun to partial shade. Propagate it by division of clumps or by seed.

LANDSCAPE USE This perfect grasslike plant stays in a clump (the proverbial tussock) and grows about 2 feet tall and wide. It's a nice structural plant for the lowest part of a rain garden or retention basin as well as for a wetland garden and is an important component of grassland wetland reconstructions. Tussock sedge produces seeds that are sought by many ducks and songbirds, and it is a host plant to several wetland-dependent butterflies including several skippers and satyrs.

ORNAMENTAL ATTRIBUTES Tussock sedge is early to emerge in spring and makes a beautiful tufted mound of arching fine-textured foliage that looks great through the entire growing season as well as when dormant. It blooms in late spring with little-noticed spikelets.

347

Carex stricta (tussock sedge) emerges and blooms early creating a crown of golden inflorescences amid fresh upright growth.

Chelone glabra
White turtlehead
Plantaginaceae (plantain family)

HOW TO GROW White turtlehead grows across much of eastern North America but is more localized southward. It's found across the Eastern Midwest and is common westward to central Minnesota, though more localized westward to central Iowa and western Missouri. It grows wild in marshy ground including sedge meadows, seeps, and fens, but will grow in continually moist garden soil—languishing in dry conditions. Propagate turtleheads by dividing clumps or by sowing seed.

LANDSCAPE USE Turtleheads do best on the edge of pond, stream, or in a wetland garden. They spread slowly by rhizomes forming clumps so are suitable for traditional perennial borders that remain moist. Plants usually grow around 3 feet tall in a garden setting. White turtlehead is the sole host plant for the dazzling Baltimore checkerspot butterfly across most of the Upper Midwest, though the butterfly is usually restricted to remnants of natural areas as it overwinters in leaf litter as a caterpillar. You may see the butterfly at the Minnesota Landscape Arboretum's gardens and wild wetlands in midsummer. Hummingbirds and large bumblebees that are able to open and enter the flowers are responsible for pollination.

ORNAMENTAL ATTRIBUTES The showy spikes of milky white flowers are shaped like the heads of turtles and protrude from the stem. They bloom from late summer into early fall and the seedpod fruits ripen brown and stand tall for winter interest.

RELATED PLANTS Rose turtlehead (*Chelone obliqua*) requires the same growing conditions as white turtlehead and can be used similarly in the landscape. Its flowers are rosy pink with the same flower shape. Rose turtlehead has a range centered on the lower Wabash River and mid-Mississippi River so is mainly found in Illinois, Indiana, and Kentucky with populations westward into Missouri, Iowa, and one record from Minnesota.

Cypripedium reginae
Showy lady's slipper
Orchidaceae (orchid family)

Cypripedium reginae is the state flower of Minnesota.

HOW TO GROW The species is found across the Upper Midwest from North Dakota to central Iowa, central Illinois and eastward to central Ohio. A few relict populations can be found in the Ozark Highlands of Missouri and Arkansas. This slipper grows in sites that have a constant supply of moisture but are never flooded, often on hammocks of fens, wet prairies, algific talus slopes, and cliff seeps in full sun to partial and light shade in neutral to alkaline substrate. Showy lady's slipper requires continually moist, but specific landscape conditions—if it were easy to cultivate it would be in every garden. Consult *Growing Hardy Orchids* by John Tullock for how to build a bed suitable to grow this slipper.

LANDSCAPE USE Showy lady's slipper can grow 3 feet tall and should be cultivated in a landscape only with suitable conditions or specifically constructed beds. These beds may be a very special addition to water or wetland gardens if creatively integrated and not left looking like an elevated box.

348

Chelone glabra (white turtlehead) flowers are reminiscent of their relative the snapdragon.

The flowers of *Chelone obliqua* (rose turtlehead) are nearly identical to those of white turtlehead but flushed with lavender-rose.

Cypripedium reginae (showy lady's slipper) with buds and spent flowers growing among last year's stalks shows how this wildflower is not always picture perfect.

My hand shows how diminutive Cypripedium candidum (small white lady's slipper) really is.

Cypripedium parviflorum var. parviflorum (small yellow lady's slippers) flowers hide amid taller foliage on a fen near Minneapolis.

ORNAMENTAL ATTRIBUTES This is the largest North American lady's slipper whose slipper is white and flushed with pink and surrounded by two white sepals and two white petals. Its magnificent blooms can be enjoyed around the summer solstice.

RELATED PLANTS Small white lady's slipper (*Cypripedium candidum*) is native across the Upper Midwest with just a few outliers southward in Kentucky, Illinois, and Missouri, where it's now rare throughout its range. It is found mainly in wet prairies and fens and requires a constant supply of moisture in neutral or alkaline conditions. I have found it once on a rocky, south-facing hill prairie in northeast Iowa but most certainly its roots went into moist, cool crevices in the bedrock. This aptly named plant grows around 1 foot tall with a tiny milky white slipper flower that blooms around Memorial Day. Wherever it grows with small yellow lady's slipper you may find hybrid Andrew's lady's slipper (*Cypripedium* ×*andrewsii*) with larger, creamier white slippers.

Small yellow lady's slipper (*Cypripedium parviflorum* var. *parviflorum*) is native across northeastern North America and usually found in swamps and wetlands. It is tolerant of a wider range of pH and is a smaller plant than large yellow lady's slipper with smaller yellow flowers surrounded by more purplish sepals and petals.

NOTES The lady's slipper species described here are now grown commercially and available to gardeners. Along with the woodland denizen large yellow and pink lady's slippers, they should only be purchased from reputable sources. As much as I love lady's slippers, I do not have them in my current garden as I feel I don't have the conditions for them to thrive. I prefer to experience them in nature's gardens.

Eutrochium maculatum
Spotted Joe-Pye weed

Syn. *Eupatorium maculatum*
Asteraceae (aster family)

Joe-Pye was allegedly a Native American healer or medicine man who used this plant to cure ailments and that is how the plant got its name. The dome of flowers forms an ideal butterfly-landing pad, and when the plant is in bloom in summer, I immerse myself in it to look for butterflies. Slogging through hummocky sedge meadows on sultry summer days also comes to mind when I think of this flower.

HOW TO GROW Spotted Joe-Pye weed is native to New England, westward across the Upper Midwest to locations

In late summer, *Eutrochium maculatum* (spotted Joe-Pye weed) produces dramatic domes of pink flower heads above beautiful whorls of highly textured leaves.

Filipendula rubra (queen-of-the-prairie) provides a glorious pink flower in a prairie restoration at Fernwood, Michigan.

in the Rocky Mountains. It is very localized in the Lower Midwest and absent from Oklahoma and Arkansas. In the wild it is found in wet prairies, sedge meadows, fens, and disturbed moist sites, and in cultivation it grows well in good garden soil but performs best with adequate moisture and wet conditions in full sun to partial shade. Divide large plants to propagate them or grow new plants from seed.

LANDSCAPE USE The wild species can be used in traditional perennial borders where it grows 4–5 feet tall. It is a solution to a problematic wet site, pondside, or rain garden. It is a must for a butterfly or insectaries garden and completely neglected in the night garden as it's a magnet for nocturnal moths.

ORNAMENTAL ATTRIBUTES The domes of rosy pink flower clusters atop this plant are its crowning glory. The leaves also are notable, whorled in threesomes or foursomes along the purplish or purple-spotted stem. The plant also is attractive when covered with its delicate tufted fruits in autumn.

RELATED PLANTS Cultivar 'Gateway' is commonly used in traditional perennial borders where it grows 5–6 feet tall and has larger domes of flowers.

Filipendula rubra
Queen-of-the-prairie
Rosaceae (rose family)

HOW TO GROW Never a common plant, queen-of-the-prairie grows wild in a few sites in the Ozark Highlands of Missouri eastward to Ohio and northward to Michigan. It has naturalized or persists from cultivation northward into Minnesota and Wisconsin. The wild plant is found in open

wetlands from seeps to fens in the prairie states, but is only occasionally a wet prairie plant. It will grow well in moist, rich garden soil but is best with extra moisture. It flowers robustly in full sun but will survive in partial shade. It is usually grown from division and virtually all plants in cultivation are the cultivar 'Venusta'.

LANDSCAPE USE This species makes a fine back-of-the-border perennial in a traditional garden where it reaches 4–5 feet in height. It makes a nice mass and spreads slowly into a clump, never running amok or self-sowing. It makes a magnificent plant for rain garden, wetland garden, or pondside.

ORNAMENTAL ATTRIBUTES Cotton candy on a stick best describes the midsummer flowers, though they consist of frothy masses of pearly round buds that open to carmine pink blossoms with a darker center and stamens. The flowers produce seed capsules that tarnish reddish before ripening brown and persisting into winter so that the tall fruiting head is quite showy in the winter landscape. The foliage is also beautiful, mainly basal (smaller up the stem) and deeply lobed, looking very maplelike on the end of the leaf's stem.

Helenium autumnale
Sneezeweed
Asteraceae (aster family)

HOW TO GROW Sneezeweed grows wild across much of temperate North America though is most common in the central and northeastern parts of the continent including the entire Midwest. It is found in wetlands from riverbanks and floodplains to marshes, fens, and sedge meadows. It grows in moist garden soil but is at its best in gardens with extra

Helenium autumnale (sneezeweed) flowers don't make you sneeze. Native Americans allegedly made a snuff from them to rid the body of evil spirits.

Although not as large as other rose mallows, the flower of *Hibiscus laevis* (Halberd-leaved rose mallow) features a distinctive red eye.

moisture in full sun or partial shade. Sneezeweed is easy to propagate from seed.

LANDSCAPE USE This plant is suitable for a traditional perennial border, rain garden, pondside, or streamside setting. It grows around 4 feet tall and attracts a wealth of small pollinators.

ORNAMENTAL ATTRIBUTES The memorable yellow flower head is comprised of a button of disc flowers surrounded by a whirling skirt of yellow ray flower notched "petals." It blooms in late summer into fall.

NOTES The species has been hybridized into a wealth of readily available cultivars of more compact stature and colors ranging from yellow to orange and burnt reds, and combinations thereof.

Hibiscus laevis

Halberd-leaved rose mallow

Syn. *Hibiscus militaris*
Malvaceae (mallow family)

Rose mallows are hardy, perennial versions of the tropical hibiscus so much beloved as a container plant for its vibrant colors. The American native species have been hybridized into a vast array of very popular traditional perennials. The hybrids have huge (12-inch-diameter) flowers in colors varying from white to cream, pink to red, and even purplish. Purple-leaved and divided-leaved cultivars are currently most popular.

HOW TO GROW The species is native mainly in the Mississippi River drainage northward to southeastern Minnesota, up the Missouri River to eastern Nebraska and up the Ohio River to central Ohio. It also grows northward to southern Lake Michigan and western Lake Erie and is found sparingly on the Eastern Seaboard.

LANDSCAPE USE Halberd-leaved rose mallow, like rose mallow, grows to 5 feet tall and makes a unique natural wetland garden subject or insectaries garden plant.

ORNAMENTAL ATTRIBUTES The 3- to 4-inch white or pinkish flowers have a distinctive red eye but are not as large as other rose mallows. The foliage is three-pointed and smooth.

Hibiscus lasiocarpos / Hibiscus moscheutos

Rose mallow

Malvaceae (mallow family)

These two species are essentially the same, differing only in hairs on their leaf surfaces and fruits. Some botanists lump them into one species but gardeners need not split hairs here.

HOW TO GROW Rose mallow is native across most of the Lower Midwest and northward to areas near southern Lake Michigan and near Lake Erie. It is found along rivers and larger streams, in backwaters, swamps, pondsides, and even wet ditches where it grows in standing water. In cultivation, it does just fine in good garden soil but prefers wet conditions. Propagate plants from seed or cuttings but beware they often hybridize and wild-sourced plants growing near showy cultivars may cross-pollinate.

LANDSCAPE USE Plants of wild origin are best grown in natural landscapes and are ideal subjects for the shores of rivers, lakes, ponds, or other wet locales. They grow about 5 feet tall and can be planted in containers or in water gardens. The flowers are very nectar- and pollen-rich so are excellent choices for insectaries gardens.

ORNAMENTAL ATTRIBUTES The very showy flowers are 5 inches or more across with a red center called an "eye" from which the column of stamens tipped with a five-parted pistil protrudes. The velvety, pleated looking petals may be white to varying shades of pink. The ornamental seedpod

The brief appearance of *Iris virginica* var. *shrevei* (Shreve's blueflag iris) flowers creates an indelible early summer experience.

Hibiscus lasiocarpos (rose mallow) blossoms are usually white with a red eye.

Two clumps of *Juncus effusus* (soft rush) stand out along this stream garden with their arching blue-green tubular leaves.

fruits look like short okra (its relative), turning blackish brown in late fall, opening to show the round seeds, and holding well on the plant through winter.

Iris virginica var. shrevei
Shreve's blueflag iris, southern blueflag iris
Iridaceae (iris family)

HOW TO GROW Shreve's blueflag iris ranges across most of the Midwest except for the Dakotas and is native in a diversity of wetlands including swamps, lakeshores, wet prairies, sedge meadows, and fens. It tolerates dryness better than many wetland perennials, but still blooms best in moist to wet soils in full sun to partial shade. It spreads by rhizomes into clumps that can be readily divided though it also grows well from seed.

LANDSCAPE USE This iris grows 2–3 feet tall and is a good choice for mass plantings to control erosion as its spreads moderately from sturdy, well-anchored rhizomes. Its foliage holds well through the growing season and it's a good choice for rain gardens and the margins of lakes and ponds. Shreve's blueflag iris may be grown in a traditional, moist and rich-soiled perennial border as long as it is occasionally divided.

ORNAMENTAL ATTRIBUTES The swordlike leaves are attractive from emergence to autumn, but the plant is beloved for its sweet fragrant, lavender-blue to violet-blue flowers whose falls are splotched with yellow and striated with black nectar guides. It blooms in early summer and flowers produce interesting seedpod fruits that ripen dark brown in fall.

Juncus effusus
Soft rush
Juncaceae (rush family)

HOW TO GROW Soft rush is native across most of the Midwest except is absent or localized in much of the western prairie portion of the region. It is found in swamps, marshes, sedge meadows, wet prairies, and disturbed wetlands. In the garden it may be grown in moist soils but is best in wet conditions in full sun to partial shade. Clumps may be divided easily or new plants grown from its tiny seeds.

LANDSCAPE USE The plant grows at least 2 feet tall and provides a tidy, tufted structure in a wetland, rain garden, or water garden and looks almost formal when planted in masses. It also makes a great container plant.

ORNAMENTAL ATTRIBUTES The arching spiky tufts of green foliage look spiffy and are ornamental from summer through fall and into winter.

Justicia americana (American water-willow) is a premier pollinator plant for a wetland garden.

White pollen-covered stigmas above brilliant red-petaled flowers define *Lobelia cardinalis* (cardinal flower).

Justicia americana
American water-willow

Acanthaceae (acanthus family)

HOW TO GROW American water-willow is native across the Lower Midwest northward to southeastern Iowa, the Chicago region, southeastern Michigan, and Lake Erie. It is found on muddy shores and in shallow waters of slow streams, lakes, and ponds. This easy-to-grow species may be propagated by cuttings or by seed.

LANDSCAPE USE The plant grows 2–3 feet tall and should be utilized in water gardens, along shorelines, and in water garden containers. It has very nectar-rich flowers that attract many insects including skipper butterflies as well as hummingbirds. The species is a hardy version of shrimp plants commonly cultivated as a summer tropical.

ORNAMENTAL ATTRIBUTES The flowers are small but in whorled clusters atop spikes that emerge from the base of the plant's upper leaves. Each flower has four petals with wine purple flares at their base; the upper petal is often fused with lavender pink. The flowers are produced for an extended period from mid to late summer.

Lobelia cardinalis
Cardinal flower

Campanulaceae (bellflower family)

A profile of cardinal flower's flower fits the head of ruby-throated hummingbirds, its primary pollinator. Hummingbirds develop white racing stripes of pollen over the crown of their heads while the cardinal flower is in bloom.

HOW TO GROW This species is found across the Eastern and Lower Midwest, westward to southeast Minnesota along the Mississippi and St. Croix Rivers, eastern Iowa, and southern Nebraska. It grows in moist to wet floodplain soils (deep silty, sandy, or cherty based but not limestone gravel of calcareous areas) in full sun to light shade. This short-lived perennial produces basal offshoots that should be separated each May and planted singly to keep a planting thriving indefinitely. The plant self-sows in bare, moist soils and is easy to grow from seed—adaptations to flooding in its ever-changing riparian habitat. It does not like mulch except as a protective covering through the winter.

LANDSCAPE USE Cardinal flower grows 3- to 5-feet-tall spires and is best in natural gardens along a stream or river. It's often planted in rain gardens and perennial borders where it is short-lived unless siltation or disturbance creates bare soil for seeds to germinate in. It's a quintessential plant for a hummingbird garden. A few butterflies with long proboscises such as the spicebush swallowtail and cloudless sulphur also will nectar on the flowers.

ORNAMENTAL ATTRIBUTES I'm not sure any flower is redder than cardinal flower. The flowers are produced in long spikes and bloom from the bottom up. The flowers are in bloom for six weeks in late summer into early fall during migration of hummingbirds. The fruiting heads add interest to the late-fall landscape but don't hold well through winter.

NOTES Where I grew up, a rendezvous to the Mississippi River floodplain over Labor Day ensured an experience with this exquisite flower. It did not grow "inland" along the calcareous, Upper Iowa River.

The bright fuchsia purple flowers of *Lythrum alatum* (winged loose-strife) occur for a long period through summer.

Lobelia siphilitica
Great blue lobelia

Campanulaceae (bellflower family)

HOW TO GROW Great blue lobelia is native throughout central and northeastern North America and found throughout the Midwest in moist to wet soils from prairies and meadows to openings in floodplain forests. It is tolerant of a wide range of soils as long as they are not dry. Propagation is by basal offshoots or by seed.

LANDSCAPE USE This lobelia grows about 3 feet tall but is a rather short-lived perennial and like cardinal flower needs to be divided or allowed to self-sow to persist. This adaptation makes it best suited to a natural garden or a site where gardener will give it extra attention.

ORNAMENTAL ATTRIBUTES Tall showy spires of light to rich blue flowers are produced in late summer. This species blooms at the same time as cardinal flower but is pollinated by bees—mainly bumblebees.

Lythrum alatum
Winged loosestrife

Lythraceae (lythrum family)

HOW TO GROW Winged loosestrife is found throughout the Midwest, which defines the core of its range across eastern North America. It is weakly rhizomatous and grows in wet prairies, sedge meadows, fens, marshes, and swamps. In

cultivation it prefers full sun in moist to wet soils including clay. Propagation is seed.

LANDSCAPE USE This loosestrife grows 2 to as much as 3 feet tall and is the native equivalent of the invasive exotic, noxious weed purple loosestrife (*Lythrum salicaria*). It makes a subtle addition to a wetland restoration or natural wetland garden.

ORNAMENTAL ATTRIBUTES Winged loosestrife displays a bushy form of finely textured leaves through the growing season. It is studded with light fuchsia purple flowers in mid to late summer—not as showy as the invasive species. In fall the whole plant turns dark burgundy purple and is a real standout among other wetland plants.

Lobelia siphilitica (great blue lobelia) blooms on a roadside near Decorah, Iowa.

Nelumbo lutea
American lotus

Nymphaeaceae (water lily family)

HOW TO GROW This aquatic plant is native along major rivers, especially the Mississippi and Missouri Rivers, and is found across the Midwest except for the Dakotas. It needs shallow water to grow in, at least 18 inches deep. It aggressively runs by rhizomatous roots to create dense colonies, growing so quickly that it is not recommended for water gardens even when planted in containers. It is easy to grow from seeds or divisions of its roots.

LANDSCAPE USE This wildflower is best where native but may be allowed in a natural-themed pond where it can periodically take over, occasionally being reduced by muskrats

Nelumbo lutea (American lotus) is the most spectacular wildflower in the Midwest.

Nuphar advena (yellow pond-lily) flowers are succulent and rich yellow, forming an exquisite rounded cup.

or other calamities like drought, but returning. It has substantial edible seeds and roots, which makes it a choice plant for an edible landscape.

ORNAMENTAL ATTRIBUTES American lotus is most certainly the showiest flower native to the Midwest with large (9-inch) creamy yellow flowers consisting of cupped petals and a tuft of stamens surrounding a central doorknob-shaped cone. The magnificent, blue-green peltate leaves (like plates on a stick) have a surface that repels water, stunning when adorned with beads of water. The seedpod fruits are highly ornamental and sought by florists; gardeners often cut them for indoor decor.

NOTES The water-repellent leaves have inspired biophilic design, whereby nano-technology has recreated the leaf's surface to make synthetic water-repellent surfaces.

Nuphar advena
Yellow pond-lily, spadderdock

> Syn. *Nuphar lutea* subsp. *advena*
> Nymphaeaceae (water lily family)

Nuphar advena is very similar to *N. variegata* (see following). Both species were once thought synonymous with the Eurasian species *N. lutea* but now are proven to be distinct. Although they look alike, they have different growth habits: one is an emergent aquatic, the other a floating aquatic.

HOW TO GROW Yellow pond-lily is native to the Eastern and Lower Midwest northward only across northern Illinois into southern Wisconsin. It grows more readily in seeps, fens, marshes, and swamps, and even open water but is an emergent aquatic. Propagation is by division or seed.

LANDSCAPE USE This species is suitable for a wetland garden, pondside, or lakeshore in shallow water. Its leaves may emerge 2 feet from the water's surface.

Nuphar variegata (variegated yellow pond-lily) floats on a Mississippi River backwater swamp in the northeast corner of Iowa.

ORNAMENTAL ATTRIBUTES The shiny, rich green, heart-shaped leaves may float on the water's surface but are more apt to rise above the water. The flowers are round and cup-shaped about 2 inches across.

Nuphar variegata
Variegated yellow pond-lily

> Syn. *Nuphar lutea* subsp. *variegata*
> Nymphaeaceae (water lily family)

HOW TO GROW Variegated yellow pond-lily is native mainly in the Great Lakes region and across the Upper Midwest from the Dakotas, across northern Iowa and eastward to the southern shores of the Great Lakes. It is a floating aquatic found in ponds and lakes. Plant it in water 18 inches to several feet deep. In the wild it grows in water to 7 feet deep. Propagation is by division or seed.

LANDSCAPE USE This is a distinctive plant for floating on the surface of a pond or water garden.

ORNAMENTAL ATTRIBUTES Variegated yellow pond-lily forms heart-shaped leaves that float on the water. The 1- to 2-inch flowers are cup-shaped and yellow, variegated with reddish inside the sepals.

Nymphaea odorata
Fragrant white water lily

Nymphaeaceae (waterlily family)

HOW TO GROW Fragrant white waterlily is native across the entire Midwest but is most common in the Great Lakes Region and along the major rivers with large backwater lakes. It grows in water as shallow as 18 inches to several feet deep. Plant it in low containers of heavy soil—half sand, half clay—to keep it anchored in a water garden. Propagation is by divisions of its tubers or from seed.

LANDSCAPE USE This waterlily holds its own with other hybrid waterlilies and makes a unique choice for a water garden as a floating aquatic. If planted in a large water container, its flowers are easier to view and their lovely aroma is easier to small. Many pollinators visit the pollen-rich flowers including bees, flies, and beetles.

ORNAMENTAL ATTRIBUTES The pristine white flowers (rarely pink) are 4–6 inches across with yellow stamens in the center and last about four days but only open in the morning. The shiny floating, circular leaves have a slit to the leaf stem and are quite lovely too.

RELATED PLANTS American white waterlily (*Nymphaea odorata* subsp. *tuberosa*) is a larger plant with nonfragrant larger, more blunt-petaled white flowers to 8 inches across. Intermediate plants occur in the wild.

Oligoneuron ohioense
Ohio goldenrod

Syn. *Solidago ohioensis*
Asteraceae (aster family)

HOW TO GROW Ohio goldenrod is a Great Lakes endemic scattered across Ohio, Ontario, Michigan, and westward to southeastern Wisconsin and central Illinois. It is found in wet prairies, sedge meadows, fens, and Great Lake interdunal wetlands. It grows best in full sun in rich, moist soil and can be propagated from seed.

LANDSCAPE USE This magnificent goldenrod grows 3–4 feet tall and is suitable for a water garden, wetland, or rain garden. It can also be used in a traditional perennial border as

Nymphaea odorata (fragrant white waterlily) is aptly named for its flowers.

Oligoneuron ohioense (Ohio goldenrod) glows in the early autumn sun at Illinois Beach State Park.

long as the soil is rich in clay or remains moist.

ORNAMENTAL ATTRIBUTES The yellow flowers are among the largest of the goldenrods in flat-topped clusters in late summer into early fall. The foliage is handsome and upward pointing, lance-shaped but with a blunt upper end. The basal foliage is wider and makes a showy tuft even without a flower stalk. Fall color is often a rich burgundy red.

The pointed, arching leaves beneath flat-topped flower clusters identify this as *Oligoneuron riddellii* (Riddell's goldenrod).

The orange-brown fertile fronds standing at the center of arching green sterile fronds identify this fern as *Osmunda cinnamomea* (cinnamon fern).

Maturing brown fertile fronds adorn the center *Osmunda regalis* (royal fern).

Oligoneuron riddellii
Riddell's goldenrod

Syn. *Solidago riddellii*
Asteraceae (aster family)

HOW TO GROW Riddell's goldenrod is native across much of the Upper Midwest (absent from Nebraska) and sparingly in the Lower Midwest from the eastern Ozark Highlands eastward to Ohio. It is found in calcareous seeps, fens, wet prairies, and sedge meadows and requires moist to wet alkaline soil in full sun. Propagate it by seed.

LANDSCAPE USE The plant grows 3 feet tall and is suitable for water gardens, wet prairie reconstructions, and rain gardens.

ORNAMENTAL ATTRIBUTES The rich yellow flowers are condensed in flat-topped clusters that tower above the plant. The leaves are lance-shaped, curve outward and downward, and turn rich burgundy in the fall. Flowers produce clouds of tufted white fruits.

Osmunda cinnamomea
Cinnamon fern

Osmundaceae (royal fern family)

HOW TO GROW This species grows wild across eastern North America east of the Great Plains but is absent from much of the Corn Belt. It is found mainly in at least seasonally wet forests in sandy or other acidic-based, organic or humus-rich soils in partial to full shade. The rhizomes of mature plants may be divided to propagate this fern.

LANDSCAPE USE Cinnamon fern makes a magnificent vase-shaped foliage perennial, 3 feet tall. It is readily available and planted in woodland gardens where it usually languishes without extra moisture or regular irrigation. It's better in waterside gardens or sites with extra moisture, including containers.

ORNAMENTAL ATTRIBUTES The unfurling fiddleheads are cloaked in rusty woolly hairs that are memorably showy and give this plant its name. In a mature plant, the fertile fronds in the center are like flames of orange-brown in late spring to early summer surrounded by arching green sterile fronds. The fertile fronds disintegrate by late summer.

Osmunda regalis
Royal fern

Osmundaceae (royal fern family)

HOW TO GROW This species is native across much of eastern North America and is found across the Eastern Midwest westward to central Minnesota, eastern Iowa, eastern Kansas, and eastern Oklahoma. It thrives in organic sandy or

Phlox glaberrima (smooth phlox) blooms at the outlet of my home's downspout, where it receives extra moisture.

Phlox maculata (marsh phlox) has distinctive upright clusters of flowers.

acidic wetlands in full sun to light shade but is slow growing in neutral or alkaline conditions. Mature clumps may be divided.

LANDSCAPE USE Plants grow around 3 feet tall in moist soils but mature plants in ideal conditions may be 5 feet tall. This is a magnificent fern for lining a pond or streamside garden and actually makes a gorgeous container plant. A mature plant is almost shrublike or palmlike creating a hummocky base.

ORNAMENTAL ATTRIBUTES Royal fern is beautiful beginning with its coppery to purplish fiddleheads followed by rich green fronds of pinnately, doubly or triply compound leaves, finishing the season with golden fall color. The fertile fronds topped with golden spores mature brown and persist into winter.

Phlox glaberrima
Smooth phlox
Polemoniaceae (phlox family)

HOW TO GROW Smooth phlox is native from southeast Wisconsin southward across Illinois into southeast Missouri and Arkansas; and southeastward across Indiana into southern Ohio and Kentucky. It grows in continually moist to wet prairies, edges of marshlands, and openings in floodplain forests. In the garden it requires continually moist,

organic or humus-rich soils in full sun to partial shade. It is easy to propagate from cuttings and seed, but only self-sows sparingly.

LANDSCAPE USE The plant grows 18–30 inches tall and is suitable for any perennial border that remains moist to wet where it will form a multistemmed clump without running roots. Butterflies, moths, and hummingbirds pollinate its midsummer flowers.

ORNAMENTAL ATTRIBUTES Smooth phlox is quite similar to marsh phlox but with flowers on flattish umbels colored acid fuchsia pink—really pink with a stunning blue overtone that reminds me of redbuds in springtime. Gensburg-Markham Prairie in Illinois is where this flower shines like no other place I've seen and in stunning combination with marsh blazingstars and rattlesnake masters. Marsh phlox's leaves are narrow and show a fine texture in lovely contrast to most other perennials even while not in flower.

Phlox maculata

Marsh phlox

Polemoniaceae (phlox family)

HOW TO GROW Marsh phlox is native from southeast Minnesota, across eastern Iowa, northern and central Illinois, into Indiana, southwest Michigan, Ohio, and disjunct in southeast Missouri. In the Midwest it is found mainly in wet prairies, sedge meadows, fens, and openings in floodplain forest. It may be cultivated in moist organic or humus-rich soils, but it thrives in wet soils. It is easy to grow from cuttings.

LANDSCAPE USE The plants grows 2–3 feet tall and forms an upright clump suitable to a traditional perennial border where irrigation is available to keep the soil moist. It's an ideal plant for a rain garden, downspout, or other landscape site that receives extra moisture including a water garden or wet prairie reconstruction. The early summer flowers are nectar rich and visited by long-tongued insects including butterflies and sphinx moths as well as by hummingbirds. Marsh and smooth phlox are a favorite fodder of herbivores like rabbits, woodchucks, and deer and usually need protection in a garden.

ORNAMENTAL ATTRIBUTES The elongated flower clusters are somewhat reminiscent an ear of corn but in striking vibrant fuchsia-pink (rarely white). The stems are spotted and streaked with purple.

RELATED PLANTS There are many cultivars as well as some hybrids with garden phlox (*Phlox paniculata*).

Pontederia cordata

Pickerelweed

Pontederiaceae (pickerelweed family)

HOW TO GROW This species is native mainly in lakes, ponds, and swamps in and around the Great Lakes and along major river backwaters below the Great Plains. It grows in shallow water as an emergent aquatic and requires full sun to partial shade. Propagation is by division or seed.

LANDSCAPE USE Pickerelweed grows about 3 feet tall and makes a premier shoreline, water garden, or water container plant. Its flowers are nectar rich and visited by many bees, butterflies, and other insects. The seeds are eaten by waterfowl and are edible by humans.

ORNAMENTAL ATTRIBUTES The flowers are simply stunning—violet blue with a yellow blotch on the upper petal—and are produced on full upright spikes that bloom over a long period from summer into fall. The leaves also are handsome, a polished green in the shape of an elongated heart. Why this spectacular native wildflower is not in every garden is beyond me.

Pontederia cordata (pickerelweed) blooms in front of Victoria waterlilies on the Island Garden at Powell Gardens, Missouri.

Sanguisorba canadensis

American burnet

Rosaceae (rose family)

American burnet in full bloom takes your breath away, so why has this magnificent and gorgeous perennial not hit mainstream gardens?

HOW TO GROW The species is native mainly in the Northeast but is found westward into Michigan and Illinois in isolated populations. It's also found in the Northwest. In the wild it grows mainly in wet prairies, seeps, and fens in the

The fluffy white flowers of *Sanguisorba canadensis* (American burnet) are a standout at the Boerner Botanical Gardens near Milwaukee.

The aptly named flowers of *Saururus cernuus* (lizard's tail) rise above Pinhook Bog at the Indiana Dunes National Lakeshore.

Midwest. In cultivation it requires wet organic soils to persist but can grow in full sun or partial shade. It is best propagated by division.

LANDSCAPE USE American burnet grows 3–5 feet tall and is a dramatic plant for a garden that remains moist to wet. It's a great plant for honeybees and other small nectar- and pollen-collecting native bees.

ORNAMENTAL ATTRIBUTES The flowers are simply "wow" held high above the plant and looking like fluffy white wands (4–8 inches long) abuzz with bees. Prairie Moon Nursery's description of the flowers looking like "Mr. Snuffleupagus" bears repeating.

Saururus cernuus
Lizard's-tail

Saururaceae (lizard's-tail family)

HOW TO GROW This species is native across the Eastern and Lower Midwest and northwestward into the Chicago region and western Illinois. It is found in swamps, marshes, and ponds and is easily cultivated in water-saturated soil or in shallow water in full sun to light shade. The plant spreads slowly and can be divided or rooted from cuttings.

The huge spring leaves of *Symplocarpus foetidus* (skunk cabbage) provide a contrasting coarse texture to a wetland garden.

LANDSCAPE USE Lizard's-tail grows 18–36 inches tall and forms a nonaggressive clump so is an ideal water garden or water garden container plant. I will warn that this plant is to squirrels what catnip is to cats. Squirrels often wildly decimate it when it is growing in water containers they can access. Go figure as its foliage is considered toxic. Do I dare say squirrel catnip?

ORNAMENTAL ATTRIBUTES The plant produces lush and attractive heart-shaped foliage, but its crowning glory is its creamy white, tail-like flower spikes that droop or nod at their ends. The flowers are slightly aromatic and bloom for more than a month from midsummer into late summer.

The "flower" of *Symplocarpus foetidus* (skunk cabbage) consists of a yellow spadix sheltered by a succulent madder red hoodlike spathe.

Symplocarpus foetidus

Skunk cabbage

Araceae (arum family)

After a long winter in the Upper Midwest, skunk cabbage is one of the first wildflowers that blooms from the mud of March along with hepaticas and snow trillium—members of my most beloved trifecta of wildflowers that welcome spring.

HOW TO GROW This species is native across northeastern North America, mainly in the Great Lakes region—westward to central Minnesota and northeastern Iowa, southward to central Illinois, central Indiana, and northernmost Kentucky. It is found in seeps, along springs, edges of swamps, and in low wet forests and requires constant moisture in organic soils. Propagation is by seed.

LANDSCAPE USE Skunk cabbage is rarely cultivated but may be grown along wetlands and in seeps. It is a spring ephemeral whose 2-foot foliage is coarse and tropical looking but quickly fades with the dense shade of full leaf out in the tree canopy above. Companion plantings to hold its place through summer include royal and cinnamon ferns.

ORNAMENTAL ATTRIBUTES The madder red, succulent spathes of this flower emerge in early spring, often creating their own heat to melt snow. The flowers look like little 4-inch-tall fairy teepees enclosing the spadix of flowers. The foliage is large and heart-shaped, magnificent in the springtime.

Groundcovers

One of the ways gardeners may readily propagate groundcovers is by dividing them. This can be accomplished in several ways. One way is to cut off daughter plants that have spread by stolons (aboveground stems or runners) and transplanting them to a new site. Another was is by cutting up sections of rhizomes (belowground stems) and replanting them. A third way is by transplanting their abundant seedlings.

The whole reason to plant groundcovers is to create an often aggressive or smothering mass to fill a large space. Groundcovers are perfect plants to control areas of soil erosion. They create relatively uniform masses that are more acceptable in traditional landscapes.

GROUNDCOVERS FOR SHADY TRADITIONAL LANDSCAPES

Antennaria plantaginea (plantain-leaved pussytoes)
Asarum canadense (wild ginger)
Carex pensylvanica (Pennsylvania sedge)
Chasmanthium latifolium (river oats)
Diarrhena obovata (American beakgrain)
Erigeron pulchellus (Robin's plantain)
Eurybia macrophylla (big-leaf aster)
Iris cristata (crested iris)
Packera aurea (golden groundsel)
Packera obovata (round-leaved groundsel)
Solidago flexicaulis (zigzag goldenrod)
Viola striata (cream violet)
Waldsteinia fragarioides (Appalachian barren strawberry)

GROUNDCOVERS FOR SUNNY TRADITIONAL LANDSCAPES

Antennaria neglecta (lesser pussytoes)
Artemisia ludoviciana (Louisiana sage)
Chasmanthium latifolium (river oats)
Conoclinium coelestinum (mistflower)
Fragaria virginiana (wild strawberry)
Phlox bifida (cleft phlox)
Phlox paniculata (garden phlox)
Phlox subulata (moss phlox)
Pycnanthemum muticum (short-toothed mountain-mint)
Rudbeckia fulgida varieties (orange coneflower)
Ruellia humilis (prairie petunia)
Spartina pectinata (prairie cordgrass)
Symphyotrichum ericoides (heath aster)

Anemone canadensis (Canada anemone) lights up a prairie with a milky way of white, starlike flowers.

Ammophila breviligulata (dunegrass) gleams as a dense ground-cover on the sandy shores of Lake Michigan at Muskegon State Park, Michigan.

Ammophila breviligulata
Dunegrass
Poaceae (grass family)

HOW TO GROW Dunegrass grows on the sand dunes of the Great Lakes and along the dunes of the Atlantic Ocean in New England and the Mid-Atlantic states. It's the grass that first colonizes and stabilizes sand dunes, initiating the first wave of flora colonization that leads to a steady succession of plants that may climax in rich moist forests. It requires shifting pure sand and full sunshine to be fruitful and aggressively multiply.

LANDSCAPE USE Dunegrass grows 2–3 feet tall with slightly taller spikes of sandy blond inflorescences and is an iconic groundcover planted to stabilize sand dunes or sandy development sites on the shores of the Great Lakes. It apparently requires mycorrhizal fungi found in such environments to extract nutrients and protect it from other soil pathogens and so it will languish under cultivation in seemingly appropriate situations. Established plants have been documented to grow an amazing 6 feet with 100 stems in a single season.

ORNAMENTAL ATTRIBUTES The gleam of this stiff grass stitching together dunes of sand coupled with the sparkling reflection of the sun on the waters of the glorious Great Lakes make it the ideal zipper between land and inland freshwater sea.

Anemone canadensis
Canada anemone
Ranunculaceae (buttercup family)

HOW TO GROW Canada anemone is native across the entire Upper Midwest and only barely and sparingly into the Lower Midwest. It grows in moist prairies, meadows, and disturbed habitats such as roadsides.

LANDSCAPE USE This species reaches 12–18 inches tall and spreads aggressively in all directions by underground rhizomes. It is best utilized in a natural landscape in conjunction with established grasses and robust perennials that it can weave through and fill in between and create a plant tapestry.

ORNAMENTAL ATTRIBUTES The flush of bright white flowers in late spring into early summer creates a spectacular sparkling sight. The foliage is palmately divided and quite distinctive too.

Antennaria neglecta
Lesser pussytoes

Asteraceae (aster family)

HOW TO GROW Lesser pussytoes is native in all midwestern states but localized or absent in some regions including the Ozark Highlands, Shawnee Hills, and Appalachian Plateau. It grows in dry prairies, savannas, and open woodland in challenging, often eroded spots where little else will compete with its small stature. It spreads by rhizomes into a large mass that may be several feet across. Plants are dioecious (either male or female) and may be propagated by division or from seed.

LANDSCAPE USE This species is a great groundcover for exceptionally dry locations in both sun and shade working well in "hell strips" or underneath a red cedar. The basal leaves reach just 3 inches tall with flower spikes growing up to 8 inches tall.

ORNAMENTAL ATTRIBUTES The fine-textured, small basal leaves with a single noticeable vein are silvery, contrasting nicely with companion plants. The flowers are silky white tight clusters like a little toe. Female flowers produce a tuft of awned fruits that look frothy when ripe whereas the male flowers quickly disintegrate after flowering.

RELATED PLANTS Plantain-leaved pussytoes (*Antennaria plantaginifolia* including *A. parlinii*) has a strange midwestern range from Minnesota and Wisconsin southward through Iowa, Illinois, and Missouri into Arkansas and Kentucky—just sparingly eastward into Indiana. It is very similar to lesser pussytoes but has wider basal leaves like a plantain that show three noticeable veins. Plantain-leaved pussytoes grows better with more shade than lesser pussytoes and its leaves may actually burn if it's in a location that is too sunny, too hot, or too dry.

366

Aralia nudicaulis
Wild sarsaparilla

Araliaceae (aralia family)

HOW TO GROW Wild sarsaparilla is native across northern North America southward in mountains to Colorado and Georgia. Its Midwest range includes most of the Upper Midwest southward rarely into eastern Kentucky and east central Missouri. It grows mainly in mesic upland forests—occasionally in low or swampy woods, but always on well-drained lenses of soil above saturation. It spreads by rhizomes into large patches and thrives in humus-rich, moist, well-drained soil in shade.

Close up of a female *Antennaria neglecta* (lesser pussytoes) depicts how this plant gets its name.

The basal leaves of *Antennaria plantaginifolia* (plantain-leaved pussytoes) look remarkably like a lawn plantain (*Plantago* spp.).

The leaves of *Aralia nudicaulis* (wild sarsaparilla) rise above the naked stem of springtime flower umbels.

LANDSCAPE USE The species grows 12–18 inches tall and makes a lovely groundcover in a woodland setting. The plant's root has been used as a substitute for sarsaparilla flavoring in root beer.

ORNAMENTAL ATTRIBUTES The leaves usually emerge bronze along with the round golf-ball-sized flower umbels that open greenish white in mid to late spring. Mature plants produce three leaves that are usually comprised of five leaflets: three at the end of the leaf rachis and two along the side; sometimes there are just three leaflets and occasionally up to seven. Flowers produce blackish fruits by late summer.

Arctostaphylos uvaursi
Bearberry
Ericaceae (heath family)

HOW TO GROW Bearberry grows mainly in sandy- or sandstone-based soils near the Great Lakes but southward along the Illinois River into central Illinois and westward

Arctostaphylos uvaursi (bearberry) creates a lush groundcover with a bit of thyme at the Matthaei Botanical Gardens in Ann Arbor, Michigan.

This glorious patch of *Artemisia ludoviciana* (Louisiana sage) was photographed at St. Croix Savanna, Minnesota.

into the Driftless Area of northeastern Iowa and southeastern Minnesota. It may be cultivated only in sandy acidic soils in full sun or light shade. Plants may be grown from cuttings, layering, or from seed.

LANDSCAPE USE The plant grows up to 6 inches tall and is an exquisite evergreen groundcover, but requires sandy soils to thrive. It is perfect for sandy lakeshores and combines well with common and creeping junipers, little bluestem, and dunegrass.

ORNAMENTAL ATTRIBUTES The foliage is rounded, glossy, and evergreen. The white urn-shaped flowers form showy red fruits from fall into winter.

Artemisia ludoviciana
Louisiana sage
Asteraceae (aster family)

HOW TO GROW Louisiana sage is native in dry upland prairies in well-drained soil in full sun. It is a rampant spreader by underground rhizomes and can be shockingly aggressive when planted without competition in a perennial border. It should be grown in poor, dry soil in sites with good air circulation. Foliar issues can afflict it when grown in soil that is too rich, especially in areas of high humidity in the Lower Midwest.

LANDSCAPE USE The plant grows 2–3 feet tall and makes a marvelous foliage contrast filler between established short prairie grasses such as little bluestem, prairie dropseed, or sideoats grama in challenging droughty soils.

ORNAMENTAL ATTRIBUTES The foliage and stems are wonderfully silvery all growing season.

RELATED PLANTS The cultivar 'Valerie Finnis' is a British selection of the midwestern native plant that grows more densely and does not spread as vigorously.

Asarum canadense
Wild ginger
Aristolochiaceae (birthwort family)

HOW TO GROW This species is native to all midwestern states except Nebraska. It grows in mesic woodlands and forests in well-drained soils. Because the species is somewhat variable, gardeners should choose regional plants whenever possible. Upper Midwest strains may have larger, thinner leaves while Lower Midwest strains have thicker leaves that are more at home in summer heat.

LANDSCAPE USE The plant grows 6–9 inches tall and spreads by rhizomes on or near the surface of the soil. It makes

The heart-shaped leaves of *Asarum canadense* (wild ginger) have a taffeta-like sheen to them.

Lifelong friend Ruth Little succumbs to the inviting soft texture of a groundcover of *Calamagrostis canadensis* (bluejoint grass).

a welcome woodland groundcover in almost any moist, woodland site.

ORNAMENTAL ATTRIBUTES Wild ginger produces two heart-shaped leaves whose marvelous sheen inspires me to describe them as living taffeta. Jug-shaped flowers are produced in spring between the two leaves and hug the ground—opening before the leaves are fully formed. The outermost whorl of the flower is comprised of a three-parted calyx that opens to reveal madder red inside. The calyx is highly variable depending on the population of plant, so wild ginger is considered a polymorphic species. There are plants whose flowers have long calyxes that open from the beak-shaped flower bud while in others the calyx lobes are folded inward on the "beak" of the bud.

Calamagrostis canadensis
Bluejoint grass

Poaceae (grass family)

As with tussock sedge, bluejoint grass brings back fond memories exploring nature with lifelong friends and discovering wondrous flora and fauna that find sustenance and companionship among it.

HOW TO GROW Bluejoint grass is a cool-season grass that grows across the entire Upper Midwest and sparingly southward into the northern part of the Lower Midwest. It is found in wet soils of fens, sedge meadows, and wet prairies, where it aggressively spreads into a uniform, though tall, groundcover. Propagation is by division or seed.

LANDSCAPE USE The species reaches around 3 feet tall with taller arching inflorescences and is an important component of wet prairie and sedge meadow restorations. It works well in large bioswales where it controls erosion and absorbs runoff.

ORNAMENTAL ATTRIBUTES This thin-bladed grass is an icon of Upper Midwest wet grasslands. Its arching form and fine texture are unmatched in all seasons whether green in summer or blonde in winter. It always invites one to lie

Carex pensylvanica (Pennsylvania sedge) creates a wild lawn in a sand savanna at Sugar River Forest Preserve outside Rockford, Illinois.

in its graces, as soft as a comfortable bed. Bluejoint grass produces pinkish tinged, light sage green inflorescences in midsummer that ripen tan.

Carex pensylvanica
Pennsylvania sedge

Cyperaceae (sedge family)

If you have never been to a native remnant savanna carpeted by a natural, no-mow lawn of Pennsylvania sedge, you are missing out on an experience that proves a lawn need not be nonnative bluegrass.

HOW TO GROW This species is common across most of the Upper Midwest from central Ohio, central Indiana, central Illinois, northern Missouri, Iowa, Minnesota, and the eastern Dakotas northward. There are localized populations southward through Kentucky and sparingly into Arkansas. Pennsylvania sedge is found in savannas and open woodlands. In cultivation it requires moist to dry, well-drained soil in partial shade to light shade. It's most easily propagated by division.

Carex jamesii (James's sedge) makes rich tufts of green in a ground-cover trial bed at Shaw Nature Reserve in Franklin County, Missouri.

The lovely seed heads of *Chasmanthium latifolium* (river oats) make it one of the most ornamental of woodland grasses.

LANDSCAPE USE The plant grows 4–6 inches tall and spreads about 6 inches a year to create a low turflike groundcover that never needs mowing.

ORNAMENTAL ATTRIBUTES Pennsylvania sedge is a cool-season plant that greens up early in spring with fine-textured leaf blades just like bluegrass but it never sets a tall inflorescence, rather a shorter golden torchlike flower. The blade color is more yellow or golden green and becomes straw colored in winter.

RELATED PLANTS Two other sedges recommended by Grow Native! "Top 10 List" for use as native groundcovers under 6 inches include cedar sedge and James's sedge.

Cedar sedge (*Carex eburnea*), also known as ivory sedge or bristle-leaf sedge, is native in all midwestern states except Kansas and Oklahoma, but is most widespread in the Driftless Area and Ozarks where it is often associated with red cedar trees. It is more finely textured than Pennsylvania sedge with more ivory colored anthers when in bloom. It's a great choice for planting in dry shade beneath red cedar trees.

James's sedge (*Carex jamesii*) is native from the Mid-Atlantic westward into the Midwest where it's found throughout the Eastern and Lower Midwest westward to southeastern Minnesota, southeastern Nebraska, and eastern Kansas. It forms tidy tufted clumps no more than 1 foot

tall and wide, but self-sows into a handsome groundcover. I find it as a marvelous no-maintenance turf under trees in historic cemeteries in Kansas City (places that don't use lawn herbicides). It's our favorite sedge in trials at Powell Gardens and its flowering and fruiting remain inside the foliage for neatnik gardeners.

Short-beak sedge or plains oval sedge (*Carex brevior*) is a similar groundcover and grows 12–24 inches tall while spreading about 12 inches through a couple of growing seasons.

Chasmanthium latifolium
River oats
Poaceae (grass family)

HOW TO GROW This species is native across the Lower Midwest and up the Mississippi River valley to northeastern Missouri. It has naturalized northward into Iowa, Wisconsin, and Michigan. It is found mainly in mesic floodplain forests or other wooded sites with rich loamy soils—often on terraces beneath bluffs. It does not self-sow much in heavy, poor clay soils. This cool-season grass languishes in drought-prone soils. With moisture, it will grow in full sun but it prefers partial to light shade.

LANDSCAPE USE This native grass is commonly sold as an ornamental grass for shade, and gardeners have labeled it a "thug" for its abundant seedlings in perennial borders in humus-rich, loamy soil. River oats is better suited as a tall (2–3 feet) groundcover where it can seed into a large mass without interfering with prized perennials. The grain is edible and the foliage is a host plant to several woodland butterflies including the northern pearly-eye and little wood-satyr.

ORNAMENTAL ATTRIBUTES The flattened, pendant spikelets in lovely arching infructescences produced by midsummer make it one of the most ornamental of grasses for shade. The fruiting heads may be cut and dried for bouquets and garden decor though they will hold well into winter so that the grass looks lovely for at least six months.

Conoclinium coelestinum
Mistflower
Syn. *Eupatorium coelestinum*
Asteraceae (aster family)

HOW TO GROW Mistflower is native across the Lower Midwest and has escaped northward into northeastern Illinois and Michigan. It is found along wooded streams and

369

The abundant blue flowers of *Conoclinium coelestinum* (mistflower) look nearly identical to the annual ageratum.

in floodplain forests but thrives in almost any soil that is not dry in full sun to light shade. Propagation is by cuttings, division, or seed.

LANDSCAPE USE Plants grow 24–30 inches tall and spread aggressively by underground rhizomes. I have had many gardeners complain what a garden thug this plant is and ask me how to get rid of it. The underground rhizomes are nigh impossible to remove though pulling the aboveground part of the plant in mid to late summer sets it back for a season. Mistflower is best in natural landscapes where it must compete with other aggressive species and should not be planted in a traditional mixed perennial border where it is not contained. It has very nectar-rich flowers and is visited by many pollinators including butterflies. It's a great plant to control erosion.

ORNAMENTAL ATTRIBUTES Mistflower is sometimes called hardy "ageratum" as it looks so much like that annual plant. It produces equally beautiful light to medium blue (sometimes white) flowers capping the plant in late summer

Diarrhena obovata
American beakgrain

Poaceae (grass family)

HOW TO GROW This species is distributed in a scattered pattern across most of the Midwest, often in mesic floodplain forests to dry upland woodlands in humus-rich loamy soils. It grows best in partial to full shade in well-drained soils and is easily divided or grown from seed.

LANDSCAPE USE American beakgrain reaches 18–24 inches tall and spreads by rhizomes, making a great groundcover in humus-rich soils of woodland gardens or the upper edges of rain gardens.

ORNAMENTAL ATTRIBUTES The arching foliage consists of glossy green blades that hold well into late fall or early winter (Lower Midwest) before turning golden yellow and bleaching a blond tan through the winter. The inflorescences are lovely arching stems of pendant beaked grains.

NOTES *Diarrhena obovata* here includes *D. americana*.

Enemion biternatum
False rue-anemone

Syn. *Isopyrum biternatum*
Ranunculaceae (buttercup family)

When in bloom, *Enemion biternatum* (false rue-anemone) is one of our finest woodland groundcovers.

HOW TO GROW False rue-anemone is native across most of the Midwest, absent only from Nebraska, and is localized or rare in western Iowa, western Minnesota, and eastern South Dakota. It grows in mesic floodplain to mesic upland forests in moist, humus-rich, well-drained soils in shade.

LANDSCAPE USE The plant reaches 6–9 inches tall and spreads into marvelous carpets of vegetation. The plant is slow to propagate which limits its availability and makes it expensive.

ORNAMENTAL ATTRIBUTES The delicate compound leaves look darling in spring into summer and again in the fall. The brilliant white flowers are difficult to photograph and spangle the plant through spring.

NOTES It always pains me to see this plant destroyed for development knowing it probably has a value of at least $40 per square foot. What I wouldn't give to have the moist parts of my woods carpeted with this premier wildflower.

Diarrhena obovata (American beakgrain) fills the end of a parking lot island at Missouri Botanical Garden.

Erigeron pulchellus
Robin's-plantain
Asteraceae (aster family)

HOW TO GROW This species is found across the Eastern Midwest, westward to eastern Minnesota, eastern Iowa, northeastern Missouri, and throughout the Ozark Highlands. It grows in savannas and open woodlands usually in thin humus-rich soils over bedrock. It may be cultivated in almost any well-drained garden soil in partial to light shade. It is readily divided and quite easy to grow from seed. Remember that you need two different plants (not cloned divisions) for cross-pollination and seed set.

LANDSCAPE USE The basal leaves grow less than 6 inches tall but the flower spikes reach 18 inches tall while flowering and fruiting in late spring into early summer. The plant makes a plantainlike ground-hugging groundcover of rather coarse leaves the rest of the year and is a good choice for rocky open woodlands or locations where the soil is rather lean so that other plants won't outcompete it. The very nectar rich flowers attract many pollinators especially little butterflies such as the pearl crescent.

ORNAMENTAL ATTRIBUTES Gardeners grow this species for its cute mini-daisy flower heads comprised of white fringy ray flowers and yellow disc flowers.

Euonymus obovatus
Running strawberry-bush
Celastraceae (staff tree family)

If this prostrate shrub were evergreen like its invasive Asian relative the purpleleaf wintercreeper (*Euonymus fortunei* 'Coloratus'), it would be a very popular groundcover.

HOW TO GROW Running strawberry-bush's native range is centered on the Eastern Midwest but grows westward sparingly to northeastern and east central Illinois and across southern Illinois westward into the Ozark Highlands. It grows in mesic forest with good drainage, occasionally in wet woods but always on better-drained lenses of soil above saturation. In cultivation it prefers moist soils in partial to full shade and will root where its stems touch the ground. It can easily be propagated by division and by replanting rooted stems.

LANDSCAPE USE This prostrate shrub sprawls approximately 4 feet and occasionally clambers 18 inches up the base of a tree. It makes a fine groundcover in moist but well-drained woodlands and forests.

ORNAMENTAL ATTRIBUTES The species is most showy in spring with bright green new foliage. The spring to early summer flowers are dull purplish green and not showy but form striking fruits in fall comprised of rosy to pinkish warty capsules that open to display vermillion to scarlet red arils inside. In fall the foliage is pale yellow and in winter the decumbent stems are strikingly green.

Euonymus obovatus (running strawberry-bush) bears tiny purplish green flowers in midspring.

Finely fringed, white daisylike flowers adorn *Erigeron pulchellus* (Robin's-plantain) in late spring.

Eurybia macrophylla (big-leaf aster) forms a dense carpet of heart-shaped leaves by midspring.

Eurybia macrophylla
Big-leaf aster

Syn. *Aster macrophyllus*
Asteraceae (aster family)

HOW TO GROW This species is native across northeastern North America from central Minnesota through Wisconsin, northeastern Illinois, northern and southeastern Indiana, northern and eastern Ohio, and into Kentucky. It prefers well-drained, more acidic soils but will grow in most humus-rich, moist soils. It spreads by underground rhizomes and can be readily divided to create more plants. As with most asters, it requires cross-pollination to produce seed and most plant divisions are clones so there is no fruit set. This aster survives but behaves more like a nonspreading perennial in the heavy clays and heat of my western Lower Midwest garden.

LANDSCAPE USE Big-leaf aster creates a full 12-inch-thick carpet of larger leaves for a coarser texture than many groundcovers and is suitable for planting under trees like pines that may be rather challenging locations for plantings. Flowering stems are formed sparingly and reach towards the light 24–30 inches above the foliage.

ORNAMENTAL ATTRIBUTES The large, heart-shaped toothed basal leaves create a rich carpet of foliage from spring through fall when it turns a calico of various warm-colored autumn shades. Mature plants produce white to light blue flowers in late summer into fall, followed by tufts of fruits if flowers are cross-pollinated.

RELATED PLANTS Forked aster (*Eurybia furcata*, syn. *Aster furcatus*) is similar looking and behaving, but is a midwestern endemic, considered rare or threatened across its entire native range from the eastern Ozark Highlands northward to southeastern Iowa, southeastern Wisconsin, northern Illinois, northwestern Indiana, and two locations in Michigan. It usually is found in calcareous, mesic wooded slopes but is easy to grow in well-drained, humus-rich woodland soils in shade.

Eurybia macrophylla (big-leaf aster) produces flat-topped clusters of white to bluish flowers in late summer into early fall.

Fragaria virginiana
Wild strawberry

Rosaceae (rose family)

HOW TO GROW Wild strawberry is native throughout the Midwest and all but the hottest and driest parts of North America. It is found in moist to dry prairies, savannas, open woodlands, old fields, and other disturbed areas. In cultivation it thrives in almost any well-drained soil. Plants are

Eurybia furcata (forked aster) has flowers similar to big-leaf aster.

Fragaria virginiana (wild strawberry) produces showy white flowers in midspring.

The maple-shaped, water-spotted leaves of *Hydrophyllum canadense* (Canada waterleaf) are showy in both spring and fall.

usually male or female, so fruit is only produced on female plants. Gardeners often ask me why their wild strawberries don't set fruit. The answer is, because they are male clones of plants.

LANDSCAPE USE This groundcovering species is as aggressive as traditional vinca or English ivy, filling any space quickly with its stoloniferous "daughter" plants. It makes a tremendous stand-alone groundcover in full sun or even dry shade. It is perfect for a natural landscape where plants may run between exiting plants, and it should be plugged into established prairie reconstructions. Wild strawberries are integral plants for edible landscapes and food forests; they have a more intense flavor than the cultivated strawberries hybridized from them.

ORNAMENTAL ATTRIBUTES The spring flowers are shockingly white with a yellow center. Fruits are small and red, as ornamental as they are delicious. The three-parted leaves often turn rich red shades in autumn and persist through winter hugging the ground.

Hydrophyllum canadense
Canada waterleaf, maple-leaf waterleaf

Hydrophyllaceae (waterleaf family)

HOW TO GROW This species is native across the Eastern Midwest and westward to eastern Missouri, where it is found in moist forests in rich, calcareous soils. It is easy to grow in rich shady woodland gardens in limestone. Propagation is by division or seed.

LANDSCAPE USE Canada waterleaf makes a massing perennial for a moist woodland garden. Its flowers are rich in nectar and pollen and attract flies and bees as pollinators.

ORNAMENTAL ATTRIBUTES The mature foliage looks just like maple leaves with the classic silvery grayish spots on fresh newly emerging leaves that give this plant the name waterleaf. The flowers are round clusters of white or blushed with pinkish or lavender purple and are held above the foliage (they bloom solely white in Michigan). The foliage holds well through summer with adequate rainfall but may

The compound, water spangled spring leaves of *Hydrophyllum virginianum* (Virginia waterleaf) make a lovely groundcover.

go dormant in dry summers, only to reemerge in the fall.

RELATED PLANTS Appendaged or great waterleaf (*Hydrophyllum appendiculatum*) is a biennial so look for its profile in that chapter.

Hydrophyllum virginianum
Virginia waterleaf

Hydrophyllaceae (waterleaf family)

HOW TO GROW Virginia waterleaf is native throughout central and northeastern North America including the entire Midwest. It grows in mesic floodplain and mesic upland forests in moist, well-drained soils and can be quite aggressive in moist, shady gardens.

LANDSCAPE USE This species makes a fantastic groundcover in moist woodland gardens where it is best integrated with sturdy perennials.

ORNAMENTAL ATTRIBUTES Virginia waterleaf is most lovely in the spring with its watermark spangled leaves. Its foliage is not so special later in the season so grow it with ferns such as maidenhair or ostrich that will help overshadow it later in the season. One of my favorite compositions is with maidenhair fern as its flowers are tall enough to shoot up through the foliage.

The bladelike leaves *of Iris cristata* (crested iris) are adorned with classic blue flowers for a brief time in midspring.

The purple-leaved cultivar of *Lysimachia ciliata* (fringed loosestrife) emerges dark purple and creates a dramatic groundcover that contrasts with the greens of spring.

Iris cristata
Crested iris

Iridaceae (iris family)

HOW TO GROW Crested iris is native in the Ozark Highlands, Shawnee Hills, Appalachian Plateau, and Lake Erie Drift Plains. It is widely cultivated across the entire Midwest where it thrives in humus-rich, moist, well-drained soils in shade. Propagation is by division of the rhizomes.

LANDSCAPE USE This plant grows about 6 inches tall and makes a low groundcover especially good for rocky border edges.

ORNAMENTAL ATTRIBUTES The spiky short leaf blades create a distinct texture and look through the entire growing season. This iris is beloved for its blue to rarely white flowers which are stunningly beautiful but briefly adorn the plant in midspring.

Maianthemum stellatum (starry false Solomon's seal) produces starry white flowers in spring.

RELATED PLANTS Several purple-leaved cultivars are very showy early in the season though they all fade to bronzy green by late summer.

Lysimachia ciliata
Fringed loosestrife

Primulaceae (primrose family)

HOW TO GROW Fringed loosestrife is native across the entire Midwest and all but the hottest and driest parts of North America. It is found in prairies, seeps, open woodlands, savannas, and adjacent ditches where it thrives in full sun to light shade in moist, well-drained soils. It requires moisture and afternoon shade on my western, Lower Midwest property. Propagation is by cuttings, division, or seed.

LANDSCAPE USE This species aggressively spreads by rhizomes so it makes a 3-foot-tall filler between larger sturdy perennials in partial shade. It is a good companion to purple coneflower, Culver's root, and American ipecac.

ORNAMENTAL ATTRIBUTES The midsummer flowers are true yellow and make a handsome contrast to the greens of the season.

Maianthemum stellatum
Starry false Solomon's seal

Syn. *Smilacina stellata*
Liliaceae (lily family)

HOW TO GROW Starry false Solomon's seal is native across the entire Upper Midwest and very sparingly southward to Kentucky and Arkansas. It grows in a wide variety of habitats from sand dunes to fens though is found mainly in moist, well-drained upland savannas and woodlands in all types of soil from sand to humus-rich loam. It can be cultivated in full sun with available moisture but is best in partial to full shade in drier areas.

LANDSCAPE USE The plant spreads by underground rhizomes into a large patch over time. It is an aggressive spreader in light sandy soils.

ORNAMENTAL ATTRIBUTES My favorite aspect of starry false Solomon's seal are the fruits in mid to late summer which become golden green marked with dark burgundy longitudinal lines, maturing to dark red. The flowers are produced in late spring and are starry white in short 1-inch-long clusters atop the plant.

Matteuccia struthiopteris
Ostrich fern

Dryopteridaceae (wood fern family)

HOW TO GROW This fern is native across the Upper Midwest with just a few outlying populations southward in Missouri, Indiana, and Ohio. It is found mainly in mesic floodplain forests, occasionally in mesic forests or as relicts from cultivation. It requires moist, humus-rich soils and shade; some sun is tolerated in more northern and eastern locations as long as they stay moist. Ostrich ferns spreads quickly by ground-hugging rhizomes to create extensive masses, -growing twelve times as fast as interrupted fern. Propagation is by division.

LANDSCAPE USE The plant reaches 3–4 feet tall and spreads so vigorously it is best used as a tall groundcover in moist, shady sites.

ORNAMENTAL ATTRIBUTES Ostrich fern is most beautiful when its fiddleheads emerge in spring and form a vase-shaped plant comprised of fresh fronds of vibrant green that are widest towards the top and come to an abrupt point. In winter the stiff, brown plumelike fertile fronds of sporangia stand around 2 feet tall after the surrounding sterile fronds have died.

Packera aurea
Golden groundsel, ragwort

Syn. *Senecio aureus*
Asteraceae (aster family)

HOW TO GROW Golden groundsel is native across most of the Midwest but absent from the Dakotas, Nebraska, and Kansas. It grows in low wet woodlands or forested seeps and thrives in moist to wet soils in partial to full shade. It may be cultivated from seed but you must have a couple of seed-grown strains for cross-pollination to produce fruits.

LANDSCAPE USE This is a showy spring groundcover for moisture-rich woodland gardens. The nearly evergreen basal leaves grow under 12 inches tall but the late-spring flower scapes reach 18 to sometimes 30 inches tall in perfect wet, rich-soil conditions. The plant is a fine choice for the lower portions of rain gardens or to infill problematic wet areas in any landscape. The flowers are very nectar rich and attract many pollinating insects including butterflies.

ORNAMENTAL ATTRIBUTES Golden groundsel forms large masses of golden flowers in mid to late spring. It has evergreen basal foliage that holds its place through all seasons and resists deer browse.

RELATED PLANTS Round-leaved groundsel (*Packera obovata*, syn. *Senecio obovatus*) is fairly similar to golden groundsel, but native across the Lower Midwest northward into northeastern Missouri, west central Illinois, and southern Michigan. It is more of an upland species thriving in moist to dry upland woodlands and forests. Its evergreen basal foliage is

A May groundcover of *Matteuccia struthiopteris* (ostrich fern) fiddleheads will soon be a solid stand of delicate fronds.

Packera aurea (golden groundsel) creates a golden mass of springtime bloom in a wet woodland garden.

Packera obovata (round-leaved groundsel) is a great choice for a groundcover in a dry woodland garden setting.

A mass of wild *Phlox paniculata* (garden phlox) blooms in various shades between pink and purple.

smaller and more round than that of golden groundsel and it is more loosely flowered. The flowers are equally nectar rich for pollinators and its foliage is the only confirmed host plant for the diminutive, rare and local northern metalmark (*Calephelis borealis*).

Phlox paniculata
Garden phlox
Polemoniaceae (phlox family)

This perennial is one of the most popular midwestern garden flowers and so is appropriately named garden phlox. It has been hybridized into a wide array of colors from white, shades of pink to almost embarrassing bright coral pinks, purples, and almost blue. Over time, plantings of cultivars self-sow and revert back to the wild form, which is bright pink with a lovely aroma I can't find words for. I've known garden phlox virtually all my life as my aunt's farm had a large mass of it beside the milk house and that's where I would watch for treasured eastern tiger swallowtails nectaring on its blossoms. Every farmstead and older home had a patch of garden phlox or so it seemed.

HOW TO GROW This species is native across the Lower Midwest from the Ozark Highlands to the Appalachian Plateau of Ohio. It has naturalized from plantings northward across the entire Midwest, often an indicator of a long-forgotten homestead. In the wild, the species is found in moist open woodland, woodland edges, and savannas including adjacent disturbed ground such as roadsides and meadows. It self-sows wildly from seed and is easy to grow in average garden soil in full sun to partial shade.

LANDSCAPE USE True wild garden phlox self-sows so abundantly and is hard to remove; it is best grown as a groundcover mass with spring ephemerals or other robust herbaceous perennials that can hold their ground against it. New England aster, Joe Pye-weed, woodland sunflowers, native thistles, sedges, woodland delphinium, and violets are good companions. Garden phlox grows 3–4 feet tall and when in flower in late summer it is a butterfly and moth magnet—the narrow floral tube fits the thin proboscis of lepidopteron pollinators.

ORNAMENTAL ATTRIBUTES Garden phlox produces a large panicle of showy flowers atop its leafy stem and blooms for a long period from late July into September and sometimes later. The foliage is prone to mildew that looks unsightly, but it appears true wild strains are less susceptible. In fall, the fertile flowers make tiny tan, round fruit capsules that audibly snap open on a dry day, flinging the black seeds a good distance.

RELATED PLANTS This phlox has been hybridized with several other species of phlox and I have yet to find a cultivated form as attractive to pollinators. I must say some of its cultivars have beautiful flowers, my favorite being 'Blue Paradise'.

Phlox subulata
Moss phlox
Polemoniaceae (phlox family)

HOW TO GROW This species is native only in the Lake Erie Drift Plains, Appalachian Plateau of Ohio into Kentucky, and the southern Ozark Highlands in Arkansas. It

One can identify true *Phlox subulata* (moss phlox) by its orange stamens visible at the center of its flowers.

The fresh spring umbrellas of *Podophyllum peltatum* (mayapple) foliage create a dramatic groundcover in a woodland garden.

A groundcover of blooming *Pycnanthemum muticum* (short-toothed mountain-mint) creates a cloak of silvery white.

is commonly planted throughout the Midwest and has naturalized northward into Michigan, Wisconsin, and Minnesota.

LANDSCAPE USE Moss phlox grows about 6 inches tall and may spread by stems that root as they sprawl more than 3 feet across over many years. This species is suitable for a traditional landscape where a low evergreen groundcover is needed in a harsh sunny site.

ORNAMENTAL ATTRIBUTES The plant is smothered in blooms in early to midspring so that it becomes a mass of color—usually soft blue. There are many cultivars with variably "eyed" and shaded white, pink, blue, or purplish flowers.

RELATED PLANTS Cleft phlox (*Phlox bifida*) is a lesser known, though very similar, midwestern endemic phlox found solely from southwestern Lower Michigan and southern Wisconsin southward through Missouri, Illinois, and Indiana to northwestern Arkansas and central Tennessee. It usually grows in sand savannas in the Upper Midwest and more often in glades or rocky slopes in the Lower Midwest. It is similarly easy to cultivate as moss phlox. The flowers are white to light blue or lavender with narrower corolla lobes (not true petals) that split into two tips.

Podophyllum peltatum
Mayapple
Berberidaceae (barberry family)

HOW TO GROW This species is native across the entire Eastern and Lower Midwest, northwestward to southeastern Minnesota, central Iowa, and southeastern Nebraska. Found in upland woodlands and forest, it grows well in humus-rich moist to summer-dry soils in partial to full shade. Plants may be successfully transplanted when actively growing, but are set back for a season or two, so division after plants have gone dormant is the best way to propagate them.

LANDSCAPE USE Mayapple grows 2 feet tall and spreads by underground rhizomes into large patches. It's an ideal part of a natural woodland groundcover looking spiffy in spring into early summer, often declining by midsummer and gone by late summer.

ORNAMENTAL ATTRIBUTES The fresh new umbrellas of leaves emerge from the woodland soil like mushrooms. Their coarse texture contrasts well with most upland woodland plants of springtime and the peltate leaves are bright green, sometimes marked by darker olive shades. Mayapple produces a white showy flower in the Y where its two leaves join. The flower may be creamy to pinkish, especially in cooler spring weather. Pollinated flowers form a rounded fruit that ripens yellow and is aromatic. The flesh is edible only when fully ripe; the seeds are poisonous and should not be eaten. Box turtles eat the fruits and help spread the seeds.

Pycnanthemum muticum
Short-toothed mountain-mint
Lamiaceae (mint family)

HOW TO GROW Short-toothed mountain-mint is native in the Lake Erie Drift Plains and lower Ozark Highlands with scattered, possibly escaped, locations in between. It is found in low woodlands and moist meadows and grows best in full sun in moist to wet soils. Propagation is mainly by division of the stolons.

LANDSCAPE USE The plant grows around 3 feet tall and spreads aggressively by aboveground stolons into large masses so it's a prime erosion control plant. It also may be the top insectaries garden plant as its flowers attract a full range of insect pollinators. A mass of short-toothed mountain-mint in full sun is abuzz with bees, butterflies, flies, and wasps. The foliage is strongly aromatic of mint and made into teas and used as a flavoring.

ORNAMENTAL ATTRIBUTES This mountain-mint is the show-iest in flower because the upper foliage acts like a beacon of silvered bracts surrounding and enhancing the flower heads. It blooms over a long period from midsummer into fall. The fruiting heads ripen dark gray and hold well into the winter.

Ranunculus hispidus
Hispid buttercup, swamp buttercup
Ranunculaceae (buttercup family)

HOW TO GROW Buttercups are native across almost the entire Midwest, absent only from the far western fringe of the tall-grass prairie region. They are found mainly in mesic flood-plain forests but often in mesic forests as well. The plants thrive in moist to wet humus-rich woodland soils in shade and should be propagated by division.

LANDSCAPE USE Buttercups grow less than 1 foot tall and make a reliable groundcover on moist woodland sites where they spread by stolons. They make good filler between larger, sturdier woodland plants in a natural landscape but may smother more diminutive woodland plants that are not well established.

ORNAMENTAL ATTRIBUTES The glowing yellow flowers have a unique glossy sheen as if lacquered and are challenging to photograph.

NOTES Once considered a separate species, the swamp but-tercup (*Ranunculus septentrionalis*) is now considered a subspecies of hisbid buttercup.

Rubus flagellaris
Northern dewberry
Rosaceae (rose family)

HOW TO GROW This species is a decumbent shrub native to central and northeastern North America including all of the Eastern and Lower Midwest and westward to eastern Min-nesota, Iowa, and southeastern Nebraska. It grows in prai-ries, savannas, woodlands, and adjacent disturbed sites in full sun to light shade in moist to dry well-drained soils. It spreads by stoloniferous stems and should be propagated by division.

LANDSCAPE USE Northern dewberry grows prostrate or sometimes taller when climbing over companion plantings. It makes a unique tapestry groundcover in a natural gar-den when it is allowed to spread among established grasses, forbs, and smaller shrubs. When planted as a sole ground-cover mass, it must be periodically cut back and rejuve-nated which is never a fun task because of the thorny stems. Weeding a mass planting also requires leather gloves, long sleeves, and sturdy footwear.

ORNAMENTAL ATTRIBUTES Dewberry is underappreciated shrub with showy white flowers in late spring and with rich

Ranunculus hispidus (hispid buttercup) is known for its lacquered-looking, intensely yellow flower.

Rubus flagellaris (northern dewberry) produces abundant crops of beautiful, delicious, and seedy fruits.

green foliage studded with fruits that change from green to red and then black when ripe. In late fall, this groundcover turns various shades of red to burgundy purple, often holding some colorful leaves into early winter.

Rudbeckia fulgida
Orange coneflower
Asteraceae (aster family)

Orange coneflower, a perennial species of black-eyed Susan, is relegated to this chapter because it self-sows so abundantly that it is considered a nuisance by many gardeners (including Jennifer Barnes, senior gardener in the Perennial Garden at Powell Gardens'). This species is best left to create exuberant flowering masses of summer gold. Its seed strain 'Goldsturm' was selected in the Czech Republic and is a very popular plant earning the title of Perennial Plant of the Year in 1999. 'Goldsturm' is no longer recommended in many areas because of foliar septoria leaf spot, but several natural-occurring varieties of wild plants still make outstanding perennials suitable for traditional and natural gardens.

HOW TO GROW Orange coneflower is native across the Eastern Midwest and westward across Illinois to southeastern Wisconsin and the Ozark Highlands. It grows in savannas and open woodland in soils that are moist to wet—usually near waterways, edges of marshes, seeps, and fens. Growers propagate the species by seed or division, and different varieties usually grow true-to-type from seed.

LANDSCAPE USE The plant grows 2–3 feet tall and makes dramatic sweeping masses in larger landscapes. It is often planted in a traditional perennial border, but its prolific seedlings may become a nuisance. Wetland tolerant varieties make colorful additions to rain gardens and bioswales. Orange coneflower is a favorite host plant for the silvery checkerspot.

ORNAMENTAL ATTRIBUTES This species is a popular perennial because of its long-lasting orange-golden-yellow flowers that cover the plants like frosting in late summer into early fall. The botanical name's specific epithet *fulgida* means shining, gleaming, or glittering, which describes the glowing nature of the flower's color. The foliage comprises attractive basal leaves of variable shapes depending on the variety. The fruiting heads of all orange coneflower varieties mature dark brown and hold well through the winter landscape.

RELATED PLANTS Several varieties of this species are native in the Midwest. **VARIETY** *fulgida* has the narrowest basal leaves (at least three times longer than wide), is native mainly in

The black-eyed, golden flowers of *Rudbeckia fulgida* (orange coneflower) pair well with the purple flowers *Phlox paniculata* (garden phlox).

Wide, fuzzy leaves are distinctive on *Rudbeckia fulgida* var. *deamii* (Deam's orange coneflower).

the Ohio River drainage, and is found in dry to mesic condition in full sun to partial shade.

Deam's orange coneflower (var. *deamii*) is native in Indiana and adjacent Illinois and Ohio and is in cultivation. It has wider and hairier leaves of a more gray-green appearance, is fairly tolerant of drier conditions, and self-sows abundantly.

Showy coneflower (var. *speciosa*) is listed as native from the Ozark Highlands northeastward to Michigan with two collections from southeastern Wisconsin but should probably be "lumped" with Sullivant's orange coneflower (var. *sullivantii*) based on updated regional flora books from Missouri to Michigan. Sullivant's orange coneflower produces relatively narrow leaves and is more tolerant of drier conditions; Missouri's wild origin forms of this plant are the least invasive of orange coneflowers in my garden. The cultivar 'Goldsturm' is supposedly a selection from this variety.

A patch of blooming *Rudbeckia fulgida* var. *umbrosa* contrasts beautifully with the blue flowers of *Salvia azurea* 'Grandiflora' (pitcher's sage).

Wild strains of *Rudbeckia fulgida* var. *sullivantii* (Sullivant's orange coneflower) grow taller with less profuse and larger flowers than other varieties of orange coneflower.

The most moisture-demanding variety of orange coneflower (var. *umbrosa*) spreads by underground rhizomes into dense patches and is ideal for the bottom of a rain garden or any place water collects.

Ruellia humilis (prairie-petunia) flower is marvelously showy upon close inspection.

Ruellia humilis
Prairie-petunia
acanthaceae (acanthus family)

HOW TO GROW This species is native from central Kansas and Oklahoma eastward to central and southwestern Ohio and northward to southern Iowa, southern Wisconsin, and southwestern Michigan. It grows most often in dry upland prairies and savannas in full sun to partial shade. It readily self-sows so is easy to propagate by seed or transplanting seedlings.

LANDSCAPE USE Prairie-petunia grows about 12–18 inches tall and may aggressively self-sow into a mass without competition from sturdy perennials. It makes a good edging perennial for borders of natural landscaping. It will grow well in harsh conditions and so is suitable for "hell strips" or other dry, rocky challenging locations.

ORNAMENTAL ATTRIBUTES The foliage is hairy, giving it a distinct gray-green cast and the 2-inch lavender-blue flowers are reminiscent of a petunia but it is unrelated.

Sedum ternatum
Widow's cross sedum
Crassulaceae (orpine family)

HOW TO GROW This species is native from Ohio, Indiana, Illinois, Missouri, and points southward—possibly native northward into Michigan where it also has naturalized. It usually is found on or below rock outcrops in woodlands or forests where little else will grow. Plant it in well-drained rocky soil in partial to full shade. It requires protection from rabbits in a garden setting.

LANDSCAPE USE The plant reaches 3–6 inches tall and makes a luscious succulent groundcover for a shady or sheltered rock garden or rock wall.

ORNAMENTAL ATTRIBUTES The fleshy rounded leaves are rich green, holding in part into winter. The delightful midspring inflorescence is a spreading spike of starry white flowers with stamens that emerge plump and red and mature purplish black. The whole inflorescence reminds me of giant snowflakes over the plant. The flowers produce dried brown fruits comprised of four follicles, each follicle containing up to a dozen tiny seeds.

An inflorescence of *Sedum ternatum* 'Larinem Park' (widow's cross sedum) creates the effect of a giant snowflake.

Spartina pectinata (prairie cordgrass) is particularly showy when in bloom.

The stem of *Solidago flexicaulis* (zigzag goldenrod) is not straight-growing, as the common name implies.

Spartina pectinata
Prairie cordgrass
Poaceae (grass family)

HOW TO GROW This warm-season grass is native across much of central and northeastern North America including the entire Midwest. It grows wild in wet prairies, sedge meadows, and fens. It can be propagated by division of its rhizomes or by seed.

LANDSCAPE USE The foliage attains 3–4 feet tall with much taller inflorescences that may reach 4, 6, or even 8 feet tall. It is best grown in sweeping masses in a traditional landscape though better suited to wet prairie reconstructions, rain garden basins, and shorelines. Its leaf edges are sharp and can cause cuts, while its cordlike rhizomes make it an ideal plant for erosion control.

ORNAMENTAL ATTRIBUTES The gleaming green, arching blades bleach straw-colored in winter. The late-summer inflorescences are tall and also arching with distinctive flat-topped spikelets that are gorgeous in bloom with pendant creamy to pinkish stamens. The infructescences hold well into winter.

Solidago flexicaulis
Zigzag goldenrod
Asteraceae (aster family)

HOW TO GROW Zigzag goldenrod is native over most of the Midwest except for the western edge, absent only from Oklahoma. It grows in moist to summer-dry upland forests and woodlands and is easy to propagate from seed or division.

LANDSCAPE USE This goldenrod reaches around 2 feet tall, spreads by rhizomes, and makes an ideal mass planting in a shade or natural woodland garden. It also combines well with other woodland species blooming at the same time as woodland asters and white snakeroot.

ORNAMENTAL ATTRIBUTES This groundcover blooms in mid-fall, producing wands of classic goldenrod yellow flowers in small tufted clusters at the bases of the leaves.

Symphyotrichum ericoides
Heath aster
Syn. *Aster ericoides*
Asteraceae (aster family)

HOW TO GROW Heath aster is native across most of the Midwest, absent only from much of the Ozark Highlands and Kentucky. It is found in dry, upland sandy or gravelly prairies as well as in moist loamy "black soil" prairies in well-drained soils. It grows in most well-drained soils in full sun and may be propagated by division or seed.

LANDSCAPE USE This aster reaches 1–2 feet tall and spreads by rhizomes. It makes a fine groundcover for the edge of a

Symphyotrichum ericoides (heath aster) has heathlike tiny leaves adorned with snowy white flowers in autumn.

Tiarella cordifolia (foamflower) creates a foamy tapestry of white flowers when in bloom in late spring. The purple flowers are Phlox stolonifera.

perennial border, a filler groundcover between plantings of stout perennials or the shorter native grasses, or an edging for a natural landscape of the Midwest's tall grasses. It works as a stand-alone groundcover in harsh conditions of dry, rocky soil in full sun and exposure to the elements. As with most asters, it makes an ideal insectaries garden plant that attracts a wide diversity of pollinators when it's in bloom.

ORNAMENTAL ATTRIBUTES Heath aster is aptly named because the tiny leaves make it look like those of heath shrubs (*Erica* spp.) creating a very fine texture. In early autumn, the plant is covered with tiny and starry white flowers with yellow centers.

Tiarella cordifolia
Foamflower

Saxifragaceae (saxifrage family)

HOW TO GROW Foamflower is native sparingly in the Midwest from northeastern Wisconsin, northern and eastern Lower Michigan, Lake Erie Drift Plains, and the Appalachian Plateau in southeastern Ohio into Kentucky. It grows wild in mesic forested ravines or "hollows" or on the edges of nonfloodplain swamps including the organic soils of arborvitae and hemlocks "swamps." It's not a wetland species but thrives with constant moisture in well-drained soils in shade. It is easily propagated by seed, but is mainly propagated by dividing colonies of the plant.

LANDSCAPE USE This species grows less than 1 foot tall and makes a memorable shade garden or woodland groundcover in appropriate moist soils or in irrigated landscapes throughout the Midwest.

Viola canadensis (Canada violet) is the largest and showiest of the white-flowering violets.

ORNAMENTAL ATTRIBUTES The foliage is lobed and somewhat maple shaped as well as evergreen so it looks good year-round. It blooms in mid to late spring with spires of white foamy flowers that give the plant its name.

Viola canadensis
Canada violet

Violaceae (violet family)

HOW TO GROW This species grows mainly in the Upper and Eastern Midwest with disjunct populations in the Ozark Highlands of northern Arkansas. It is found in moist woodlands and forests in well-drained soils. I remember tall carpets of this wildflower in bloom on moist forested slopes along the Big Sioux River—the river that divides Iowa from South Dakota—so certain strains of the plant are very adaptable to a wide range of temperatures and moisture.

Viola striata (cream violet) in bloom carpets the Shade Native beds of Powell Gardens' Perennial Garden, Missouri.

LANDSCAPE USE Canada violet reaches 12–15 inches tall and spreads into an aggressive solid groundcover in moist shade. Flowers and fresh leaves are edible so it makes a good groundcover along with other native violets in a food forest.

ORNAMENTAL ATTRIBUTES The flowers are showy and white with purple "whisker" marks, produced from upright stems in spring into early summer above the heart-shaped foliage. This violet is larger and coarser textured than any other midwestern native violet.

Viola striata
Cream violet
Violaceae (violet family)

HOW TO GROW Cream violet is native across the Eastern Midwest and westward across Illinois to southeastern Iowa, the Ozark Highlands, and rarely to Wisconsin. It's found in moist woodlands and forests in well-drained soils and grows in about any good garden soil in partial to full shade. It self-sows rather abundantly and seedlings readily transplant as a way to propagate the plant.

LANDSCAPE USE The plant reaches only 6–9 inches tall and also forms a rather aggressive and solid groundcover in moist shade. It does allow for sturdy woodland perennials to grow through it, so is a good companion groundcover that "ties the room together" so to speak. Cream violet has edible flowers and leaves and is often host to caterpillars of the great spangled fritillary.

ORNAMENTAL ATTRIBUTES Cream violet is somewhat similar to Canada violet but with smaller, creamier white flowers that have black streaks. The stems are shorter and the leaves smaller creating a more refined look in a garden.

Waldsteinia fragarioides (Appalachian barren strawberry) produces an abundance of yellow flowers in spring.

Waldsteinia fragarioides
Appalachian barren strawberry
Rosaceae (rose family)

American-native Appalachian barren strawberry should not be confused with Eurasian barren strawberry (*Waldsteinia ternata*) or with the invasive lawn weed Indian strawberry (*Duchesnea indica*).

HOW TO GROW This species is native mainly in the northern and eastern Great Lakes and Appalachians, rarely westward across the Lower Midwest to the Ozark Highlands. Northern populations of the plant are not heat and humidity tolerant so be sure to grow Ozark Highlands or appropriate local, heat-tolerant strains in the Lower Midwest. Appalachian barren strawberry is found in mesic upland, rocky woodlands and forests (often containing pines) and will grow in moist, well-drained garden soils in partial to light shade.

LANDSCAPE USE The plant reaches 4–6 inches tall and makes a gem of a groundcover in moist, well-drained raised beds, rock gardens, or rock walls.

ORNAMENTAL ATTRIBUTES Appalachian barren strawberry looks a lot like wild strawberry but with yellow flowers and a bit more refined trifoliate leaves that are semievergreen. The fruits are dry achenes; consequently, the plant is "barren" compared to the fleshy red fruits of wild strawberry.

383

Bulbs

Bulbs (including other geophytes such as corms, tubers, and rhizomes) don't hold their place in a landscape or garden for very long, but without them the seasonal delights of our flora would be missing. Most bulbs need to be planted with companions in a garden setting and may be tucked into appropriate groundcovers or between perennials that remain summer through fall. Observing where bulbs grow in the wild will give ideas for companion plantings that fill the space while the bulbs are dormant during the rest of the growing season.

Most bulbs in this chapter are spring ephemerals, a marvelous design to emerge early and capture the spring sunshine before trees or large perennials steal the light and extra moisture later in the season. Many of the woodland species have a unique relationship with ants to disperse their seeds. Each seed is covered in a fatty elaiosome that ants seek. Ants take the seeds to their nest, eat the elaiosome, and discard the seed into their rubbish—the perfect site for a seed to germinate away from the mother plant and with fertilizer to boot.

Some traditional bulb catalogs are now selling many of these native plants, which are more expensive than tulips, daffodils, and crocus because most of our native bulbs take a long time to mature and they are not grown in quantity. If they are not requested or specified by designers, that won't change either. I am thrilled that Prairie Moon Nursery, Minnesota, sells a bunch of the native bulbs in late summer after they have gone dormant and hope one day this will become a common practice.

BULBS FOR A TRADITIONAL LANDSCAPE

Allium stellatum (prairie onion)
Arisaema triphyllum (Jack-in-the-pulpit)
Camassia scilloides (wild hyacinth)
Delphinium tricorne (woodland delphinium)
Dodecatheon meadia (Midland shooting stars)
Hypoxis hirsuta (yellow stargrass)
Lilium michiganense (Michigan lily)
Mertensia virginica (Virginia bluebells)
Oxalis violacea (violet wood-sorrel)
Sanguinaria canadensis (bloodroot)
Trillium grandiflorum (large-flowered trillium)
Trillium recurvatum (prairie trillium)

The lavender-pink flowers of *Allium canadense* var. *lavandulare* (showy wild garlic) produce no bulblets in the inflorescence unlike typical wild garlic.

Allium canadense var. *lavandulare*
Showy wild garlic

Syn. *Allium muticum*
Liliaceae (lily family)

HOW TO GROW This is the showiest flowering variety (it's a distinct species in my mind) of wild garlic native only to the western Midwest in eastern South Dakota, eastern Nebraska, eastern Kansas, northeastern Oklahoma, the northwestern corner of Arkansas, and western and central Missouri. It is found in open woodland, savannas, and prairie in soils that are moist in spring. It is a spring ephemeral so it grows in full sun or light shade in almost any soil that is not wet. Propagation is by division of bulbs or by seed.

LANDSCAPE USE Showy wild garlic grows about 18 inches tall and is one of only a handful of wildflowers in my backyard woods spared by the deer. It provides late spring color in a prairie planting or combines well with woodland wildflower plantings. It does not produce flower head bulblets like weedy true wild garlic (*Allium canadense* var. *canadense*) or self-sow abundantly like nodding wild onion. It makes a fine addition to an edible landscape and can be used as a garlic substitute.

ORNAMENTAL ATTRIBUTES Muffin-shaped umbels of light pink flowers occur in late spring into early summer when few showy flowers are in bloom in prairies, savannas, and open woodlands.

Allium cernuum
Nodding wild onion

Liliaceae (lily family)

HOW TO GROW This species is native across much of North America from the Appalachian Mountains westward sparingly across the Midwest to the Rocky Mountains. It is found in moist prairies to moist rocky forests and so is tolerant of full sun or light shade. It self-sows abundantly and also may be propagated by dividing bulbs.

LANDSCAPE USE Nodding wild onion grows about 2 feet tall. Because it self-sows aggressively, it is best in natural landscapes where it must compete with dense established plantings. The bulb is edible and may be used as an onion substitute.

ORNAMENTAL ATTRIBUTES The upright foliage has a characteristic oniony aroma. The flower scapes are lovely from bud to bloom as they elegantly bend just below the umbel of showy light to deep pink flowers. I like the plant for

Allium cernuum (nodding wild onion) blooming at Chiwaukee Prairie in the southeastern corner of Wisconsin shows its characteristic nodding umbel of flowers.

This mass of *Allium stellatum* (prairie onion) at Powell Gardens in Missouri shows how this bulb can become a spectacular ornamental onion when grown in a garden setting.

providing color in late summer to woodland plantings, blooming when little else does and before the main autumn hurrah of asters and goldenrods.

Allium stellatum
Prairie onion, fall glade onion

Liliaceae (lily family)

HOW TO GROW This species is native mainly in the northern Great Plains of the Dakotas, eastward into northwestern and central Wisconsin and southward into northwestern Iowa. In this northern range, it is found in upland prairies. It has a separate southern range from the Flint Hills of Kansas eastward across the Ozark Highlands into southwestern Illinois. In its southern range, it is often called fall glade onion as it grows on rocky prairies and glades and blooms in early fall. The plant requires full sun to partial shade in well-drained soils.

LANDSCAPE USE Prairie onion grows about 2 feet tall and does not self-sow as aggressively as nodding onion. Prairie onion is tolerant of extremely hot and dry conditions so is suitable for rock gardens and other challenging "hell strips." It will grow just fine in a traditional perennial border as well.

ORNAMENTAL ATTRIBUTES This onion produces a showy pink "golf ball" of flowers in late summer (northern populations) into early fall that form lovely fruiting heads adorned with black fruits.

Allium tricoccum
Wild leek, ramps
Liliaceae (lily family)

HOW TO GROW This species is native across northeastern North America in most of the Eastern and Upper Midwest, but absent from Nebraska, Kansas, Oklahoma, and Arkansas. It is found in mesic forests or sites with rich soils from floodplain to upland, most abundant on floodplain terraces or north- and east-facing slopes in humus-rich, well-drained soils. Propagation is by division of dormant bulbs or by seed.

LANDSCAPE USE Wild leek is a great woodland garden plant where its early spring, coarser textured, 6- to 8-inch foliage contrasts well with other spring ephemeral wildflowers. It is a delicious edible and so is an important component of a food forest.

ORNAMENTAL ATTRIBUTES The bold foliage emerges in early spring and is a rich glossy green with a red base. The umbels of whitish flowers arise after the summer solstice, appearing naked as they bloom after the foliage has disappeared. The flowers produce fruiting heads that dry well and sport black seeds by early fall.

RELATED PLANTS A very similar species of wild leek (*Allium burdickii*) has a nearly identical native range, bears narrower, paler (almost silvery green) leaves, and blooms earlier around the summer solstice with fewer flowers in a pale greenish white umbel. The two species require the same cultivation and landscape use, though Burdick's wild leek shows a finer texture in a spring garden or landscape. Burdick's wild leek is even more common than wild leek in the Chicago region and is Chicago's namesake—the place was once known as Chicagoua because of the quality of wild garlic growing in the woods there.

Arisaema triphyllum
Jack-in-the-pulpit
Araceae (arum family)

HOW TO GROW This species is native in all but the driest woodlands across all of eastern North America including the entire Midwest. It grows in humus-rich, well-drained soils in partial to full shade. Propagation is by division of dormant clumps or by seed.

LANDSCAPE USE The plant grows 1–3 feet tall and is an integral addition to a shady perennial border, shade garden, or natural woodland garden.

ORNAMENTAL ATTRIBUTES Jack-in-the-pulpit is beloved for its inflorescence: the "pulpit" is a striated spathe that shelters "Jack," the flower's spadix, inside. The spathe may be green or purplish yet always striated. No one can resist lifting up the pointed hood of the flower to see Jack inside. As the spathe withers, the developing fruit looks like a drumstick of fleshy green berries that turn glossy lipstick red when ripe. The stems are often flecked with purple or dark green and the foliage comprises lush three-parted leaves

The striated spathes of *Arisaema triphyllum* (jack-in-the-pulpit) flowers are distinctive.

The bold spring foliage of *Allium tricoccum* (wild leek) contrasts with the finer textures of other vernal wildflowers.

The tropical-looking foliage of *Arisaema dracontium* (green dragon) towers above the dragonlike flower.

The midspring flowers of *Camassia scilloides* (wild hyacinth) are usually pale lavender-blue with contrasting yellow stamens.

The month-later flowers of *Camassia angusta* (prairie wild hyacinth) are nearly identical to those of wild hyacinth.

that hold into mid or late summer depending on available moisture and rainfall.

RELATED PLANTS Green dragon (*Arisaema dracontium*) is closely related and native across the Eastern and Lower Midwest and northwestward to southeastern Minnesota and southeastern Nebraska. It requires the same cultural conditions and grows to the same size as Jack-in-the-pulpit, but its striking leaves are deeply divided into multiple lobes. Green dragon is named after its flowers that are comprised of a spathe that looks like a green hood with the spadix bending outward then upward like a dragon's tongue. It develops bright red fruits just like Jack-in-the-pulpit in late summer into fall.

Camassia scilloides
Wild hyacinth
Asparagaceae (asparagus family)

Three western species of *Camassia* known as quamashes (*C. cusickii, C. leichtlinii, C. quamash*) have become popular traditional bulbs while our two midwestern species (*C. angusta, C. scilloides*) known as wild hyacinths remain native plant specialties.

HOW TO GROW Wild hyacinth is native across the Lower Midwest and northward to southeastern Michigan, southern Wisconsin, and southern Iowa. It grows in moist prairies, savannas, and woodlands where it thrives in humus-rich soil in full sun or partial shade. It is slow to propagate or reach flowering size when grown from seed which is why

it is not readily available—though worth the wait as a colony does multiply over time.

LANDSCAPE USE The flower spike grows 15–30 inches tall and makes a wonderful addition to both prairie and woodland gardens for its midspring flowers. It should be included in food forests for its edible bulbs.

ORNAMENTAL ATTRIBUTES The bladelike foliage appears first in early spring followed by the lovely spike of bluish white to soft lavender-blue flowers with contrasting yellow stamens. Fruits quickly mature and disintegrate.

RELATED PLANTS Prairie wild hyacinth (*Camassia angusta*) is virtually identical to wild hyacinth but blooms about a month later in late spring or the beginning of summer. It's native in the Flint Hills and Osage Plains northeastward rarely to central Illinois and central Indiana. It grows in mesic prairies and savannas in full sun or partial shade. Plant prairie wild hyacinth with wild hyacinth to enjoy an encore performance of lovely soft blue flowers spikes.

Chamaelirium luteum (fairy wand) in bloom lightens up any woodland garden.

An early spring carpet of blooming *Claytonia virginica* (spring beauty) enhances shaded turf at Powell Gardens, Missouri.

Chamaelirium luteum
Fairy wand

Liliaceae (lily family)

HOW TO GROW Fairy wand is native in the East from southern New England to the Gulf Coast and is found only in the Eastern Midwest including southern Ontario, possibly southeastern Michigan (one historic collection from Detroit), eastern and southern Ohio, southern Indiana, and southernmost Illinois. It mainly grows in moist woodlands, savannas, or meadows and requires humus-rich moist soils in full sun to shade. Propagation is by division or by seed from female plants. The species is probably hardy only through USDA zone 6. I have had success cultivating it after taking advice to "grow it like a maidenhair fern."

LANDSCAPE USE This bulb lightens up a shady nook or makes a unique component in a traditional perennial border. The flower stalk of female plants may grow 1–3 feet or slightly taller while the male plant's inflorescence is usually shorter.

ORNAMENTAL ATTRIBUTES Fairy wand produces wands of creamy white, frothy flowers that rise above the basal rosette of leaves in late spring. The flowers of female plants produce spikes of fruits. The basal foliage is semievergreen and often tarnishes purplish in autumn.

Claytonia virginica
Spring beauty

Portulacaceae (purslane family)

HOW TO GROW This species is native in eastern North America from the Mid-Atlantic westward through most of the Midwest (absent from western Minnesota and the Dakotas) and southward to the western Gulf states. It grows in upland mesic, drier oak-hickory forests, and woodlands that are at least moist in spring. Propagation is by division of dormant clumps or by seed.

LANDSCAPE USE Spring beauty is less than 6 inches tall and has naturalized into lawns beneath native trees at Powell Gardens. Naturalizing it in a lawn is one of my favorite landscape uses of this wildflower but it also makes a fine addition to any shade or woodland garden. Spring beauty is deer candy and I've lost it to them in my own woods, forgetting deer deterrents and not fencing off the few remaining remnant clumps. Where superabundant it can handle some browse but starting new populations require deer protection. The foliage and tubers also are edible so it is a plant for a food forest. It has a specialist pollinator—the spring beauty bee (*Andrena erigeniae*)—and is an important spring source of nectar and pollen to a wide range of pollinators.

ORNAMENTAL ATTRIBUTES The plant produces just a pair of straplike leaves from which a flower cluster shines in early through midspring. The flowers have showy pink stamens and petals that are white striated with varying intensities of pink veins from faint to bold.

Delphinium tricorne
Woodland delphinium, dwarf larkspur

Ranunculaceae (buttercup family)

I hope one day that this wonderful spring ephemeral will become as popular as nonnative delphiniums and larkspurs. I was thrilled to find it native on my property and have easily expanded it throughout my woodland gardens.

HOW TO GROW This delphinium is found across the entire Lower Midwest and northward a bit, into central Illinois, throughout Missouri, and up the Missouri River drainage into southern Iowa and southeastern Nebraska. It grows in mesic to summer-dry upland woodlands and forest where it is a true spring ephemeral emerging early, blooming by midspring, quickly setting seed, and gone by midsummer. It is easy to propagate from seed or to transplant. It self-sows mainly in my paths and I move these seedlings to the garden with ease—a good reminder that many woodland wildflowers germinate best in leaf-free, bare or disturbed soil sites—and why I forgive my wild turkeys from scratching around in the woods and woodland garden beds. Woodland delphinium is hardy throughout the Midwest.

LANDSCAPE USE This premier wildflower can be used like any bulb for naturalizing in a woodland garden or shaded groundcover. It needs companion plants to fill in after it goes dormant and combines well with woodland sedges, roundleaf groundsel, woodland phlox, and violets. It usually grows about 15 inches tall but some robust plants are 2 feet tall or more.

ORNAMENTAL ATTRIBUTES The deep blue flowers of woodland delphinium are as richly colored as the gorgeous spires of garden delphiniums found in cooler summer climates. This spring ephemeral packs springtime punch to woodlands, unlike the color of any other in bloom at that time. The palmately divided leaves are good-looking in spring and through the plant's bloom time. I also don't mind the fast-maturing infructescence that turns yellow when ripe and salmon-colored when dry—quickly disintegrating and not requiring any cutting or removal.

Dicentra cucullaria
Dutchman's breeches

Papaveraceae (poppy family)

HOW TO GROW Dutchman's breeches is native throughout temperate eastern North America and grows wild in mesic forests throughout the Midwest. It needs humus-rich, well-drained soil and is usually found on north- or east-facing wooded slopes over limestone. Propagation is by seed or by division of dormant tuberous roots.

LANDSCAPE USE The plant reaches about 1 foot tall and is a classic addition to shade or woodland rock gardens, woodland slopes, or in shaded raised beds with improved drainage.

ORNAMENTAL ATTRIBUTES The fernlike, highly divided foliage is beautiful from emergence into late spring. The flowers are well named as they look like pendant, upside-down white britches. The flowers are inflated and translucent which adds to their luminous white character.

Dicentra cucullaria (Dutchman's breeches) flowers look like inflated pantaloons.

The creamy white heart-shaped flowers of *Dicentra canadensis* (squirrel corn) show it's related to the "fringe-leaf" bleeding heart.

Delphinium tricorne (woodland delphinium) sports flowers as blue as any hybridized delphinium.

RELATED PLANTS Squirrel corn (*Dicentra canadensis*) has a more restricted range than Dutchman's breeches growing across the Eastern Midwest and westward to southeastern Minnesota, eastern Iowa, and scattered through Missouri as far west as bluffs along the Missouri River near Kansas City. It is found in similar habitat to Dutchman's breeches and requires the same care. Its tubers really do look like golden kernels of corn and are what give the plant its name. The flowers are white and heart-shaped above ferny spring ephemeral foliage.

Dodecatheon meadia
Midland shooting stars

> Syn. *Primula meadia*
> Primulaceae (primrose family)

The downward-pointed flowers of *Dodecatheon meadia* (Midland shooting stars) are designed for bumblebees to collect pollen and cross-pollinate them.

Midland shooting stars are my favorite prairie wildflower and as a young man the reason to trek over a county to Hayden Prairie in late spring. After a burn, Hayden Prairie has the most spectacular floral display of shooting stars I have ever laid eyes on (Chiwaukee Prairie a close second). The flowers bloom white, light lavender, pink, rose, and every hue in between. Together with their pollinating bumblebees, the other wildflowers, and the sounds of prairie songbirds, shooting stars make me want to lie back down in those moments—laughingly remembering watching a northern harrier through my binoculars and suddenly realizing through its focused eyes it was aiming right for me as I was apparently lying close to its nest. From the wolf whistle of the upland sandpiper to the winnowing of Wilson's snipe at dusk, the multiple harmonics from the bobolink's bill, and the soft insectlike call of Henslow's sparrow, shooting stars exemplify (to me) the last tiny remnants of rich black soil tallgrass prairie and all the now-rare life that goes with them. When I wish upon a star . . . I wish I were among blooming shooting stars.

Blooming *Dodecatheon amethystinum* (jeweled shooting star) carpets the ground above North Bear Creek, Winneshiek County, Iowa.

HOW TO GROW The species is native to central North America and found in the Midwest from Ohio and southwestern Michigan westward to southern Wisconsin, barely into southeastern Minnesota, eastern Iowa, eastern Kansas, and Oklahoma. It grows in prairies, savannas, and woodlands in humus-rich, well-drained soils that are moist in spring. The plant may go dormant by midsummer, well adapted for a summer drought. Propagation is by division of the dormant plant; a bare-root plant looks like a massive nerve cell or octopus. Plants are slow and challenging to grow from seed.

LANDSCAPE USE Shooting stars makes a glamorous addition to any landscape from a traditional perennial border to a natural landscape and is a must for native plant community reconstructions where it is native. The flowers are produced atop 12- to 30-inch scapes and do not contain nectar but are pollinated by queen bumblebees that buzz while hanging upside down from the flowers, thus releasing their pollen (there is no nectar).

ORNAMENTAL ATTRIBUTES This wildflower is beloved for its downward-pointed "beaklike" pistil with surrounding stamens and its exquisite corolla of upswept petals that may be white, various shades of pink, or deep rose. The fruiting capsule ripens brown and holds well into winter. The basal leaves emerge in early spring and go dormant by midsummer, occasionally lingering much later in moist conditions.

RELATED PLANTS Jeweled shooting star (*Dodecatheon amethystinum*) is native in the Midwest from the western

Driftless Area southward into western Illinois and north-eastern and central Missouri. It's smaller than Midland shooting stars with flowers usually more deeply colored (and the source of the specific epithet "amethyst"), and it also goes dormant in summer. In the wild, it is found in unique sites, usually below cliff faces of limestone, dolomite, or St. Peter sandstone that face north or east creating a cooler microclimate shaded from sun and sheltered from hot, dry summer winds. I grow the species at my home sheltered on the north side of the house above a limestone rock wall so it has perfect drainage and a cool, sheltered site.

Erythronium albidum
White trout-lily, dogtooth violet

Liliaceae (lily family)

Trout-lilies are America's closest relative to the tulip. May they one day be as popular to gardeners.

HOW TO GROW White trout-lily is native mainly in central North America throughout the Midwest in mesic floodplain forests and mesic wooded slopes. It spreads into extensive colonies consisting of carpets of single leaves where only the oldest and most mature bulbs produce a pair of leaves and a flower. It grows well in humus-rich, well-drained soils and is best propagated by division of bulb masses when the plant is dormant.

LANDSCAPE USE This beautiful spring foliage and flowering plant reaches about 6 inches tall in a woodland garden. It can be allowed to naturalize in lawns or in masses under trees and it combines well with almost any woodland

wildflower. Its only drawback is it takes years to establish large masses of the plant and 30 years to flower.

ORNAMENTAL ATTRIBUTES The mottled burgundy or reddish and medium green leaves are highly ornamental in spring. Flowers are delightful as they nod with pendant yellow stamens while the white tepals open only in sunshine and curve skyward.

RELATED PLANTS Prairie trout-lily (*Erythronium mesochoreum*) is native in the western Midwest from southeastern Nebraska, southern Iowa, and west central Illinois southward through Missouri, eastern Kansas, Arkansas, and Oklahoma as far south as northern Texas. It grows in prairies, or savannas, and open woodlands that were probably once prairie. It does not form dense colonies; its leaves are thinner than those of white trout-lily and not mottled.

Erythronium mesochoreum (prairie trout-lily) has thin, unmarked leaves but nearly identical flowers to white trout-lily.

The flowers of *Erythronium albidum* (white trout-lily) remind me of the flying nun's hat.

Erythronium americanum (yellow trout-lily) is part of the prevernal floral display at the Eloise Butler Wildflower Garden in Minneapolis.

Erythronium americanum
Yellow trout-lily

Liliaceae (lily family)

HOW TO GROW This species is native mainly in northeastern North America and is found across the Eastern Midwest (where it's more prevalent than the white trout-lily) westward to eastern Minnesota, eastern Iowa, and eastern Missouri. It grows in rich mesic forests in humus-rich, well-drained soils and is more apt to form clumps rather than colonies in northeastern Iowa but that is not the case in most of its range from Michigan to Missouri. Propagation is best by division of dormant bulbs.

LANDSCAPE USE Yellow trout-lily may be used just like white trout-lily in a landscape.

ORNAMENTAL ATTRIBUTES The flowers remain nodding with pendant yellow or contrasting rusty-colored stamens. The flower's tepals open only in sunshine and curve skyward just as with white trout-lily.

RELATED PLANTS Beaked yellow trout-lily or golden-star trout-lily (*Erythronium rostratum*) is a Lower Midwest species from the Ozark Highlands (south to Louisiana) and southern Ohio and adjacent Kentucky (with a third population in southern Tennessee and Alabama). It has yellow flowers that are similar to yellow trout-lily but are up-facing, flatter "golden stars" in sunshine and the fruits retain a beaklike projection from the flower's pistil. Beaked yellow trout-lily requires the same cultural requirements and is found in the same habitats as yellow trout-lily.

The up-facing, golden trumpetlike flowers of *Erythronium rostratum* (beaked yellow trout-lily) make it the showiest species.

Airy, grasslike leaves float above starlike flowers in the aptly named *Hypoxis hirsuta* (yellow stargrass).

Hypoxis hirsuta
Yellow stargrass

Liliaceae (lily family)

HOW TO GROW Yellow stargrass is native across temperate eastern North America including the entire Midwest. Surprisingly, it is also native in China. It is found in prairies and savannas, fens, and edges of wetlands in humus-rich or organic soils. It may be grown in rich garden soil in full sun or partial shade. Propagation is by seed.

LANDSCAPE USE This 6-inch-tall, diminutive bulb is slow to propagate and as yet still expensive so it is rarely cultivated. It would make a fine addition to the edge of a traditional flower border and naturalized in a lawn. It is an important component of a prairie or savanna restoration but almost never included because it's not "cheap and easy."

ORNAMENTAL ATTRIBUTES The leaves look like folded blades of grass and this tuft of foliage is adorned with a flat crown of bright golden-yellow flowers in late spring.

Lilium michiganense
Michigan lily

Liliaceae (lily family)

HOW TO GROW This lily is native nearly throughout the Midwest, absent only from North Dakota and eastern Ohio, and is found in moist to wet prairies, savannas, and open woodlands, or on the edges of sedge meadows and in fens, colonizing adjacent ditches. It requires moist to wet humus-rich soil in full sun to light shade. Propagation is easiest from small bulb offshoots. I once assisted in transplanting (rescuing?) some offshoots from a pasture where cattle cropped

The brilliant red-orange flowers of *Lilium michiganense* (Michigan lily) stand out at Searls Prairie in Rockford, Illinois.

Even without flowers, the beautifully tiered whorls of leaves make *Lilium superbum* (Turk's cap lily) a spectacular ornamental plant.

them so that they never bloomed. They flowered the very next year in their new garden and are thriving to this day.

LANDSCAPE USE Michigan lilies grow 3–6 feet tall and make tremendous plants for the back of a traditional perennial border, edge of a wetland garden, or other natural garden with moist to wet, rich soil conditions. Their flowers are deer candy and because of that, this native plant is now rare to see in flower and even exterminated in areas with over-populations of deer.

ORNAMENTAL ATTRIBUTES The plant produces a spectacular candelabrum of pendulous buds whose sepals and petals curl back to their pedicel when the flower opens. The flowers are orange to brilliant vermillion, sometimes seeped in scarlet red in full sun. The insides of the sepals and petals are freckled with dark purple or brown. The flowers produce capsules that dry brown and are quite ornamental in the winter landscape.

RELATED PLANTS Canada lily (*Lilium canadense*) is a similar species with more bell-shaped flowers and is found widely along the eastern fringe of the Midwest; it is the most common native lily at Holden Arboretum outside Cleveland, Ohio.

Prairie lily (*Lilium philadelphicum* var. *andinum*) produces up-facing flowers like the Asiatic garden lilies and they are a delicious and sumptuous orange to vermillion unlike any other. This wood lily is native from New England and the Appalachians westward across the Upper Midwest to the Northern Great Plains and Rocky Mountains. There's a growing sadness in my soul as I watch this glorious element of Midwestern flora fade away into oblivion. So far it defies captivity (we haven't figured out how to grow it in gardens) and the remnant natural areas where it once persisted are overrun with deer, eating every plant. The biggest population of prairie lily that I ever experienced is now a vacation home development.

Turk's-cap lily (*Lilium superbum*) is fairly similar to Michigan lily and found mainly in the Appalachian Mountains eastward to southern New England and westward into

Manfreda virginica (false agave) inflorescences may tower as tall as 6 feet.

eastern Ohio and the southern Lower Midwest in southeastern Missouri (one known site) and Arkansas. Turk's-cap lily grows in similar conditions although in acidic soils and may become a superb plant 6 to even 8 feet tall. Its flower color is usually more orange and the center of the flower is marked by a green star created by a nectary on the midrib base of the sepals and petals.

Manfreda virginica
False agave

Asparagaceae (asparagus family)

HOW TO GROW False agave is native across the Lower Midwest from the Ozark Highlands and eastward across southern Illinois, southern Indiana, to southern Ohio. It grows wild in glades and barrens in thin soil over bedrock in full sun or only a bit of shade. It may be cultivated in rocky soils or even droughty clays where it self-sows readily. It may be hardy only across the Lower Midwest. Propagation is easy by seed.

LANDSCAPE USE The basal leaves grow less than 1 foot tall but the flower spikes may reach 3–6 feet tall. This plant makes a good choice for a "hell strip" or rock garden, thriving in challenging conditions without irrigation. It actually will grow quite well in a traditional perennial border where its unique flowers can be paired with other species and its fruiting spikes left for winter interest.

ORNAMENTAL ATTRIBUTES The basal rosettes of succulent foliage are handsome; the tall spikes of yellowish and greenish flowers are unique; but the most ornamental aspect is the rounded fruits that begin green but mature almost black and remain on the infructescence through winter.

Melanthium virginicum
Bunchflower

Liliaceae (lily family)

HOW TO GROW Bunchflower has a spotty range from the Mid-Atlantic states westward to the core of the Midwest and southward to the Gulf Coast. It is found mainly in the Lower Midwest but northward into southern Iowa and northern Indiana. It grows in open savannas or the edges of prairies, usually in moist swales that are wet in spring. It flowers best in full sun but is fine in partial shade too. Propagation is best accomplished by dividing bulbs after the plant has gone dormant in summer.

LANDSCAPE USE This wildflower is suitable for the back of

The 1- to 2-foot-tall inflorescence of *Melanthium virginicum* (bunchberry) is comprised of hundreds of flowers that make it glow like a beacon while in bloom.

a traditional perennial border in rich, moist soils where its spectacular, 4- to 5-foot towering bunch of flowers in late spring fills an early-season void and where its infructescence may be left for interest into fall. It works ideally in moist meadows or prairie swales in natural gardens.

ORNAMENTAL ATTRIBUTES The bright green basal leaves emerge in spring as vibrant and showy tufts of foliage. The flowers are produced much later and in abundance on a tall, large cone-shaped inflorescence; they open creamy white with green highlights and age to a very special light creamy green. The fruits are little pods that are also attractive.

Mertensia virginica
Virginia bluebells

Boraginaceae (borage family)

There is nothing as beautiful in springtime as a forested floodplain terrace carpeted by this wildflower. I wish I could share experiences along the Turkey River in northeastern Iowa, the Pecatonica River in northern Illinois, the Blackwater River in western Missouri, or the Meramec River at Shaw Nature Reserve outside Saint Louis. No photograph can do Virginia bluebells justice; one must ensconce oneself in such a place when they are at peak bloom.

HOW TO GROW The species is native from the Mid-Atlantic states westward to the core of the Midwest and is mainly found in the Ohio River basin, middle Mississippi River basin (southeastern Minnesota southward to northern

Sedges and early meadow-rue provide good companions to this clump of *Mertensia virginica* (Virginia bluebells) at the Chicago Botanic Garden.

Northoscordum bivalve (false garlic) provides nectar for insects early in the season but is not fit for human consumption.

Arkansas), and Lower Missouri River basin barely reaching Kansas where it is extirpated in the wild. It has naturalized almost anywhere near where planted in the Midwest. Virginia bluebells is mainly found in mesic floodplain forests where it thrives in deep loamy soils. It languishes in poorly drained clays or poor dry soils. Propagation is by division of the fleshy rhizome after the plant goes dormant or by seed. In the garden, this species readily self-sows.

LANDSCAPE USE Virginia bluebells grows 1–2 feet tall and may be used like any traditional bulb that has spectacular spring flowers and then goes dormant by early summer. It is perfect naturalized in a lawn or included for seasonal color in a traditional perennial border. It is an essential component of natural woodland and shade gardens.

ORNAMENTAL ATTRIBUTES The flowers mature from pink buds yet are mainly shades of blue, from soft smoky blue to medium porcelain blue, but one may always find a pink- or white-flowering plant growing amid large populations.

Nothoscordum bivalve
False garlic, crowpoison
Liliaceae (lily family)

HOW TO GROW False garlic is native from Kansas eastward across Missouri to central and southern Illinois, southern Indiana, southern Ohio, and all points south. It will grow in a wide variety of soils and in prairie or woodland sites, thriving in challenging dry upland sites. Propagation is by division of clusters of bulbs or by seed.

LANDSCAPE USE The plant grows less than 1-foot tall and works well in "hell strips" or other dry and rocky, challenging situations. It is very rich in nectar and important for many pollinators including butterflies. The bulb is poisonous so keep it out of food forests or edible landscapes so there is no chance of mistaking it with wild garlic or other edible onions and wild hyacinths.

ORNAMENTAL ATTRIBUTES This delightful little bulb is appreciated for its delicate white flowers in early to midspring when few other plants are blooming in open habitats.

Oxalis violacea
Violet wood-sorrel
Oxalidaceae (wood-sorrel family)

HOW TO GROW Violet wood-sorrel is native throughout the Midwest though barely reaching southern Michigan where it may now be extirpated. It grows in upland prairies, savannas, and open woodlands, thriving in well-drained soils.

The pale flowers of *Oxalis violacea* (violet wood-sorrel) above shamrock-shaped foliage contrast with dark blue violas.

Propagation is by division of plants right as their leaves are going dormant or by seed.

LANDSCAPE USE Various tender nonnative "shamrock" oxalis can be found in summer bulb catalogs, but our native species is rarely utilized though it can be naturalized in lawns and is an important spring wildflower in prairie restorations, savannas, or woodland gardens. The developing fruits are edible and contain oxalic acid, which creates the same tartness found in rhubarb.

ORNAMENTAL ATTRIBUTES This wildflower has lovely pink to violet-pink flowers in mid to late spring and beautiful three-parted foliage typical of all oxalis "shamrocks" though the leaves are bronzy or purplish-tinged beneath. Some plants have delightful dark burgundy markings on their leaf surfaces that make them highly ornamental foliage plants. Violet wood-sorrel does go dormant in summer, sometimes producing a few new leaves in fall.

The luminous white flower with egg yolk yellow stamens makes *Sanguinaria canadensis* (bloodroot) one of the showiest of early spring woodland wildflowers.

Stenanthium gramineum (eastern feather bells) produces large inflorescences of dainty white flowers.

Sanguinaria canadensis
Bloodroot
Papaveraceae (poppy family)

I remember this flower from when I was a small child—what a gimmick it has with its bloody red (actually more orange-red) sap. Now I can't bear for it to bleed, as it is such a treasured wildflower I am slowly reestablishing it in my landscape and woodland.

HOW TO GROW The species is native in the temperate deciduous forests of eastern North America including the entire Midwest. It prefers mesic upland forests in humus-rich, loamy soils. Propagation is by division of the fleshy root after the plant goes dormant or by seed.

LANDSCAPE USE Bloodroot is a quintessential wildflower for a woodland or shade garden. It provides copious pollen for spring active bees.

ORNAMENTAL ATTRIBUTES The showy spring flowers are comprised of pristine white petals encircling a bunch of yellow stamens that surround a central green pistil. The flower buds are often tinged pink in cool weather and equally as beautiful. The flowers are very short-lived and open only on sunny days but the leaves are exquisitely shaped with deep undulating lobes forming a unique shieldlike form around the flower and developing fruit.

Stenanthium gramineum
Eastern feather bells
Liliaceae (lily family)

HOW TO GROW Eastern feather bells is native from Ohio and Indiana westward to southern Illinois, Missouri, and Oklahoma and points southward. It is an uncommon to rare species and in decline, usually found in open woodlands in moist, acidic soils at the base of a bluff, in mesic floodplain terraces of streams and rivers, and occasionally in disturbed habitat along roads and railroads. It blooms heaviest in full sun but tolerates partial shade. The species has naturalized in Michigan's Upper Peninsula and is probably hardy throughout the Upper Midwest where insulating snow cover is reliable in winter.

LANDSCAPE USE The flower spike grows 2–6 feet tall and makes a stunning addition to a natural garden where moist, acidic soils are present.

ORNAMENTAL ATTRIBUTES The basal leaves emerge lushly in spring, looking grassy like the specific epithet *gramineum* suggests The flower spike emerges in late spring and produces a magnificent panicle of white starry, lightly fragrant flowers (actually comprised of six tepals as sepals and petals are undifferentiated) in midsummer. The fruit has a hooked beaklike capsule and matures almost black.

Trillium spp.
Trilliums
Liliaceae (lily family)

Trilliums are beloved woodland wildflowers with leaves of three, and flowers with three white, yellow, or maroon petals. They don't grow from true bulbs but actually short, thickened rhizomes. I'll admit I'm a trilliumophile. What is it about trilliums that makes so many gardeners, including me, worldwide want to collect them? The genus is at its most diverse in the American Southeast, but many species are at their best or endemic to the Midwest.

All trilliums may be propagated by dividing their rhizomes (best done as the plant is going dormant in late summer or fall) or from seed. Seedlings take several years to reach flowering size, making trilliums very expensive. A few growers have created mass beds of trilliums from which they divide out plants at a more economical cost. May that practice become more prevalent along with these iconic wildflowers in gardens. In the wild, all species grow in moist, humus-rich woodland soils and all are in decline because they are a favorite food of deer. Whether cropped by deer or picked (with their leaves included) by humans, trilliums are usually destroyed or set back severely. I am old enough to remember wonderful masses of this spring wild-flower where now there are virtually none.

All trilliums make distinctive additions to woodland or shade gardens. Most grow 12–15 inches tall. They bloom in midspring, except for the snow trillium that blooms in early spring. They come in two flowering formats: ped-iceled trilliums have flowers setting on pedicels (flowering stems) with open, outward sepals and petals; sessile tril-liums have stemless flowers set atop the plant's whorl of three leaves with petals that grow upward and create a sort of teepee-shaped flower.

Trillium erectum
Purple trillium

Liliaceae (lily family)

HOW TO GROW Purple trillium is native to northeastern North America westward into eastern Kentucky, eastern Ohio, and Lower Michigan, with a few disjunct populations westward into northern Illinois. It grows in both acidic and neutral soil so it is fairly easy to cultivate in woodland gar-dens across the entire Midwest.

LANDSCAPE USE This species is a premier addition to a wood-land wildflower garden where it grows 1–2 feet tall. (I have never once noticed its brief odor from afar.)

ORNAMENTAL ATTRIBUTES The pedicellate showy reddish maroon-purple flowers have three petals facing outward. The flowers may occasionally be albinistic white (forma *alba*). This species readily hybridizes with nodding tril-lium to produce intermediate plants that flower in blends of white and purple.

Trillium flexipes
Drooping trillium

Liliaceae (lily family)

HOW TO GROW Drooping trillium is native from southern Michigan, Ohio, and the Bluegrass Region of Kentucky westward to northeastern South Dakota, eastern Iowa, east-ern Missouri, and rare in northern Arkansas. It grows in woodland soils that are neutral and limestone based.

LANDSCAPE USE The plant reaches 18–24 in. tall and is a choice addition to a woodland wildflower garden.

ORNAMENTAL ATTRIBUTES The flowers, produced on pedi-cels 4 inches long, have creamy white petals and may open facing outward but soon droop beside or below the plant's umbrella of three leaves so that one has to hold them up to

Trillium erectum (purple trillium) produces purplish red flowers with contrasting yellow stamens.

The white flower of *Trillium flexipes* (drooping trillium) opens on a drooping pedicel beneath the classic whorl of three leaves.

see the flower. This species may produce a showy red fruit (with a fruity aroma) when ripe—contrasting strikingly against the green foliage of the forest's ground flora.

RELATED PLANTS Nodding trillium (*Trillium cernuum*) is a very similar species restricted to the Upper Midwest from northernmost Ohio, Indiana, northeastern Illinois, and northern Iowa northward. It has flowers that nod on a shorter (less than 1¼ inch) pedicel beneath the leaves and the flower's stamens are pinkish on longer filaments.

Trillium grandiflorum
Large-flowered trillium
Liliaceae (lily family)

HOW TO GROW This species is native to the Appalachian Mountains and Great Lakes region in neutral to alkaline soils that are limestone based. It is found in the Midwest only in the states that touch the Great Lakes, but grows southward in the Appalachian Highlands to Kentucky and southward in the Appalachian Mountains to northernmost Georgia. It is readily cultivated across the entire Midwest and naturalized plantings persist, for example, as far west as Fontenelle Forest near Omaha, Nebraska.

LANDSCAPE USE The plant reaches 12–15 in. tall and is frequently grown in wildflower gardens because of its attractive, conspicuous flowers.

ORNAMENTAL ATTRIBUTES Large-flowered trillium has the largest, showiest milky white petals on outward-facing flowers with a yellow eye of stamens. The flower sits on a short pedicel, lasts a long time, and ages gracefully to rosy pink. Large populations usually contain aberrant plants with multiple petals including some with stacked or fully double petals. Double-flowering trilliums are highly prized and sought after showy flowers by shade gardeners, but such flowers are sterile producing no pollen or fruits.

Trillium grandiflorum (large-flowered trillium) is the showiest trillium producing white flowers with contrasting yellow stamens.

Mature flowers of *Trillium grandiflorum* (large-flowered trillium) age gracefully to rosy or violet pink.

Trillium recurvatum (prairie trillium) is a robust grower and quickly forms a colony of flowering stems.

RELATED PLANTS Snow trillium (*Trillium nivale*) was the most abundant native trillium where I grew up in northeastern Iowa and one of three species of herbaceous plants that were the first to bloom in early spring. It's the Midwest's smallest species reaching just 4–6 inches tall with showy white flowers like miniature large-flowering trilliums. It may be very challenging to cultivate and rarely persists for long so is best enjoyed and protected in its wild haunts.

Trillium recurvatum
Prairie trillium

Liliaceae (lily family)

Prairie trillium is probably the trillium easiest to grow in a garden and quickly multiplies into a clump of many flowering stems.

HOW TO GROW Prairie trillium is found throughout Illinois and northward into southern Wisconsin and southwestern Michigan, into most of Indiana, southeastern and east central Iowa, Missouri, western Kentucky and Arkansas. Despite its common name, it does not grow on prairies but forests, woodlands and savannas.

LANDSCAPE USE This trillium grows about 1 foot tall and can be used for naturalizing in woodland gardens.

ORNAMENTAL ATTRIBUTES The petals curve upward like a bud unlike those of toadstool trillium that spread outward or slightly upward. The sepals fall downward. The leaves are beautifully mottled green with burgundy splotches.

Trillium sessile
Toadshade trillium

Liliaceae (lily family)

Most trilliums sold as *Trillium sessile* in cultivation are actually another species—*T. cuneatum*—native to the American Southeast, now naturalized from plantings as far north as New York and Michigan.

Classic *Trillium sessile* (toadshade trillium) has rich maroon-purple flowers and beautifully mottled leaves.

Trillium sessile (toadshade trillium) may occasionally have yellowish flowers and unmottled leaves

Trillium viridescens (Ozark trillium) shows its svelte form of narrow leaves and sleek flower petals.

HOW TO GROW Toadshade trillium is native throughout Ohio, southern Michigan, the Bluegrass Region and points westward through Indiana, Illinois, the Ozark Highlands, and Osage Plains as far west as eastern Kansas. It's most commonly found in mesic floodplain forests.

LANDSCAPE USE This trillium grows about 1 foot tall and can be used for naturalizing in woodland gardens.

ORNAMENTAL ATTRIBUTES This highly variable species usually has madder red-purple sessile flowers and mottled leaves, occasionally with yellow flowers and plain green leaves. The sepals spread outward or slightly upward unlike prairie trillium's downward-pointing sepals,

RELATED PLANTS Ozark trillium (*Trillium viridescens*) is one of two western Lower Midwest specialty species. It is native in the Ozark Highlands southward to the northeastern corner of Texas. Ozark trillium has slender green to maroon-purple sessile flowers and slender, lightly mottled leaves while green trillium has sessile flowers comprised of slender green petals and sepals. Both species grow in moist woodlands in rich soils. Ozark trillium is occasionally cultivated and grows well across the entire Midwest.

Green trillium (*Trillium viride*) is also native to the western Lower Midwest, but is restricted to a range centered on St. Louis in east central Missouri and southwestern Illinois. It has sessile flowers comprised of slender green petals and sepals. Like toadshade trillium, it grows in moist woodlands in rich soils.

Annuals
and
Biennials

Traditional varieties of annuals are the most popular garden flowers worldwide. No other plant group can bloom with such color over such a long period. Annuals are often sold simply as "color," and gardeners and designers create landscapes with them, as an artist would utilize paint for a painting.

Midwestern native annuals do not bloom as long as most traditional varieties (petunias, marigolds, and zinnias) but still add amazing seasonal color to any landscape. Biennials and short-lived perennials also are included here. All must be propagated by seed each growing season; it just takes an extra year for biennials to bloom.

Most annuals and biennials do not like to be transplanted except those grown in containers. Robust biennials that set a full rosette of leaves may be transplanted bare root while they are dormant in early fall or early spring. Biennials that germinate in the fall and complete their life cycle in spring are more accurately called "winter" annuals.

Go to any nursery or garden center and you will see seed packets of wildflowers; some are sold in cylinders marketed as "meadow in a can." Most of the seeds in these mixes are from quick-to-flower annuals, biennials, and short-lived perennials yet most of the seeds included are not midwestern native species.

Most midwestern annuals are weeds that are the bane of many gardeners, farmers, and prairie reconstructionists; they emerge in disturbed soils, filling their role as the first wave of plant succession to quickly heal a scar on the earth. Because most of these plants are considered rather unsightly and compete with our ornamentals and crops, they are considered that ugly "scab" one must go through to heal.

Most annuals germinate when the soil is disturbed via erosion, animals, or cultivation. It's amazing how long the seed of many annuals may remain viable in the soil "seed bank"—the germination of long-dormant seeds initiated by sunlight. When I worked for the Winnebago County Forest Preserve District, we actually seeded one native prairie planting with a planter that was shaded so that sun did not hit the earth while we planted. We observed much reduced annual weed competition in that planting.

Ragweeds (*Ambrosia* spp.) exemplify some of the most hated annuals, especially if you suffer allergies because of their abundant windborne pollen. They are, however, great producers of nutritious seeds sought by many wintering song and game birds. I am fond of their obligate moth: the ragweed flower moth (*Schinia rivulosa*). Ragweed's botanical name *Ambrosia* may say it all. I'm for cures to allergies not eradication of species—though ragweed must be controlled in crops and plantings.

Cocklebur (*Xanthium strumarium*) epitomizes a human-hated weed, but it was the favorite food of the now-extinct Carolina parakeet.

ANNUALS FOR A TRADITIONAL LANDSCAPE

Bidens aristosa (bur-marigold)
Centaurea americana (American basketflower)
Chamaecrista fasciculata (partridge pea)
Coreopsis tinctoria (Plains coreopsis)
Euphorbia marginata (snow-on-the-mountain)
Glandularia canadensis (rose verbena)
Helenium amarum (Plains sneezeweed)
Helianthus annuus (common sunflower)
Monarda citriodora (lemon bergamot)
Monarda punctata (dotted horsemint)
Palafoxia callosa (palafoxia)
Rudbeckia hirta (black-eyed Susan)

Agalinis tenuifolia
Slender gerardia, false foxglove

Syn. *Gerardia tenuifolia*
Orobanchaceae (broomrape family)

HOW TO GROW Native throughout the Midwest, this species may be the easiest to grow and widespread of a colorful bunch of similar species. It's found in seasonally wet to dry prairies, savannas, and wetlands and also may be found in disturbed meadows and roadsides. It is at its best in natural landscapes planted between established grasses like little bluestem, sideoats grama, and prairie dropseed. Slender gerardia is probably partially parasitic on neighboring plants so it won't survive in a prepared, bare soil planting bed.

407

Agalinis tenuifolia (slender gerardia) flowers have added charm (speckled throats and fringed edges) upon close inspection.

The inflorescence of *Angelica atropurpurea* (purple-stemmed angelica) is an angelic fireworks of white flowers.

LANDSCAPE USE The plant reaches about 2 feet tall and may be integrated into a natural grass planting. Smaller species of bumblebees are the main visitors to the flowers.
ORNAMENTAL ATTRIBUTES Slender gerardia is aptly named with long slender stems and leaves. In late summer into fall, its light pink-purple flowers look like short *foxglove* blooms and are beautiful upon close inspection. Plants often are rather open with sporadic blooms so look best intertwined with shorter grasses and other forbs.

Angelica atropurpurea
Purple-stem angelica

Apiaceae (carrot family)

HOW TO GROW Native across northeastern North America, this species is found mainly in the Upper Midwest from central and northern Ohio westward across Indiana to northern Illinois (disjunct in southernmost Illinois), east central Iowa, eastern Minnesota, and points north. It's a biennial growing wild in seeps, springs, and the edges of wetlands including floodplains. It prefers moist to wet rich soils that are calcareous or with a high pH (sometimes mildly acidic) in full sun to partial shade. It is easy to propagate from seed and it grows successfully in the Lower Midwest.
LANDSCAPE USE Plants reach 4–6 feet tall, sometimes taller and are a fitting addition to a wetland garden or wet swale.
ORNAMENTAL ATTRIBUTES Purple-stem angelica forms gorgeous spherical heads of cream to greenish flowers. The stems are usually purplish, especially in full sun and in cooler northern climates. The seed head stalks hold well into winter and are highly ornamental.

The filamentous silvery, basal leaves of *Artemisia campestris* subsp. *caudata* (beach wormwood) make it a striking floral component in its native haunts.

Artemisia campestris subsp. *caudata*
Beach wormwood

Syn. *Artemisia caudata*
Asteraceae (aster family)

HOW TO GROW This wormwood is native mainly to the Upper Midwest from Ohio's Lake Erie border counties westward to northern Indiana, southward down the Illinois River's valley in Illinois, across Iowa, and Nebraska and points north and west. It also is present in the Lower Midwest a bit in east central Missouri and disjunct in the central Ozarks. It is found in sandy shores and dunes; sand prairies and savannas; and gravel, hill, and other prairies. It is usually a biennial, producing a basal rosette of foliage, flowering, setting fruits, and then dying.

LANDSCAPE USE The basal foliage reaches 6 inches tall, while flowering stems are 1–4 feet tall. Wormwood is a neat plant for a sandy or dry prairie reconstruction in the Upper Midwest. It should be considered in an annual border where its finely textured foliage and flower stems create such a unique statement that it looks exotic.

ORNAMENTAL ATTRIBUTES Beach wormwood is most loved for its finely divided, silvery basal foliage rosette. It bolts a spectacular spike of lime green flowers set atop feathery green leaves its second or third year.

Bidens aristosa
Bur-marigold, swamp-marigold, tickseed sunflower, bearded beggar-ticks

> Syn. *Bidens polylepis*
> Asteraceae (aster family)

The genus name, composed of *bi* (two) and *dens* (teeth), refers to the two spines present on the fruits of most species. These teeth help the fruit attach to fur and clothing as a way for the fruit to travel and gain the plant its common names "tickseed" and "beggar-ticks." *Bidens* species are related to marigolds (*Tagetes* spp.).

HOW TO GROW Bur-marigold is native across much of the Lower Midwest northward to northeastern Ohio, southern Wisconsin, southern Iowa, and southeastern Nebraska. It has naturalized northward into Michigan. It grows in moist sunny swales and roadside ditches and is easy to cultivate in moist to wet garden soils.

LANDSCAPE USE The plant reaches 24–30 inches tall and is best added to natural landscapes and rain gardens where it usually will self-sow.

Bidens aristosa (bur-marigold) creates stunning drifts of yellow flowers in early autumn across the Lower Midwest.

ORNAMENTAL ATTRIBUTES The species blooms profusely in early fall, covering itself with golden-yellow flowers. It's a seasonal splash of solid color I look forward to each fall when it's as showy as any traditional garden annual.

RELATED PLANTS Tall swamp marigold or northern tickseed sunflower (*Bidens coronata*) is a very similar species native across the Eastern Midwest and west to central Minnesota, southern South Dakota, central Nebraska, and Illinois. The species is naturalized in eastern Missouri. I think of it as more restricted to natural habitats like sedge meadows, fens, and the edges of marshes compared to the more southern bur-marigold. Tall swamp marigold reaches 4 feet tall with nearly identical flowers but the flower bracts are curly and sparsely fringed or bristled.

Campanulastrum americanum
American bellflower

> Syn. *Campanula americana*
> Campanulaceae (bellflower family)

HOW TO GROW American bellflower is native throughout the Midwest in mesic to dry upland woodlands and savannas. It grows in almost any rich, well-drained soil in partial to light shade. It is a true biennial and must be propagated by seed.

LANDSCAPE USE This plant is found in almost every woodland, surviving human disturbance well. It makes a great addition to a natural woodland garden where it will self-sow—rarely becoming a nuisance.

Bidens coronata (tall swamp marigold) blooms in a sedge meadow at Anna Page Park in Rockford, Illinois.

Campanulastrum americanum (American bellflower) is often the most colorful wildflower in summer woodlands.

The ferny leaves of *Chaerophyllum procumbens* (wild chervil) fill in nicely between other spring woodland wildflowers.

Centaurea americana (American basketflower) is the native bachelor's button that looks like it's on steroids and always elicits an astonished "what is that?" by those who have not seen it before.

ORNAMENTAL ATTRIBUTES The first-year tufts of foliage look like violets and the second-year blooming plant rockets skyward studded with very showy blue starry flowers in late summer and into fall.

Centaurea americana
American basketflower

Asteraceae (aster family)

HOW TO GROW American basketflower is native in the southern Great Plains northward into central Kansas, southwestern Missouri, and naturalized as far north as Wisconsin. It's found in dry, rocky prairies and glades but grows well in almost any well-drained soil in full sun. As an annual, it must be propagated from seed each season.

LANDSCAPE USE This magnificent annual reaches 4 feet

and occasionally taller so should be used at the back of a flower border or in large containers. It is showy for about a month so should be placed with other plants such as later-blooming native thistles that can fill in for it as it goes to seed. A plethora of pollinators visit it, and songbirds seek its seeds as soon as they are ripe.

ORNAMENTAL ATTRIBUTES The mid- to late-summer flowers reach 4 inches across with pinkish lavender ray flowers surrounding creamy disc flowers. The large egg-shaped flower buds are a study in architecture too as they are covered with bristled scales creating a woven-looking pattern.

Chaerophyllum procumbens
Wild chervil

Apiaceae (carrot family)

HOW TO GROW Wild chervil is native to the Eastern and Lower Midwest and northwestward to southern Wisconsin, Iowa, and southeastern Nebraska. It usually is found in mesic floodplain forests and along moist stream terraces, ravines in woodlands, and disturbed sites. It is a winter annual that germinates in the fall, flowers, sets seed, and dies by early summer. I inherited it in my own woodland and it has been absent recently because of the current series of extreme droughty autumns that inhibit its germination.

LANDSCAPE USE The plant usually stays under 1 foot tall and makes an ornamental spring groundcover between other woodland garden plants.

ORNAMENTAL ATTRIBUTES The seedlings in fall have long cotyledons (seed leaves) and remind me of dill or fennel seedlings. The spring foliage is like delicate parsley and the white flowers are as dainty as tiny wild carrot flowers.

Chamaecrista fasciculata (partridge pea) blooms through late summer with ferny, pinnately compound leaves.

Chamaecrista fasciculata

Partridge pea

> Syn. *Cassia fasciculata*
> Caesalpiniaceae (senna family)

HOW TO GROW This species is native across the Lower Midwest northward to the southern shores of Lake Erie, southern tip of Lake Michigan, central Wisconsin, central Minnesota and southeastern South Dakota. It is found frequently in sand prairies and savannas but also in upland loamy soils from prairies to disturbed old fields and roadsides.

LANDSCAPE USE The plant reaches 2–3 feet tall and is a perfect nurse plant for a prairie reconstruction, natural landscape, or insectaries garden. The flowers are a favorite of bees. In my yard, the foliage is host to butterflies including cloudless sulphur, little yellow, gray hairstreak, and Horace's duskywing.

ORNAMENTAL ATTRIBUTES The delicate, pinnately compound leaves create a really fine, ferny texture. The flowers are produced for several weeks in mid to late summer and are true yellow followed by long, thin legume "pea pods." It's a plant to listen to: masses have an audible buzz of bees while in bloom and the fruits snap open in autumn, flinging the seeds a good distance.

Cirsium spp.
Thistles
Asteraceae (aster family)

The name thistle conjures up negative connotations, but it's the nonnative thistles that are the problems, most of which are legally listed as noxious weeds (and deservedly

so). These include the perennial Canada thistle (*Cirsium arvense*—which is not from Canada), and the biennials bull thistle (*Cirsium vulgare*), and musk thistle (*Carduus nutans*).

Native thistles are never as invasive as the alien species and I would have loved to include all the Midwest's native species here. When they are in bloom in my meadow/prairie restoration, I often pull up a chair to sit and watch all the hummingbirds, fluttering swallowtail butterflies, and other insects that visit. Who needs a television, computer, or smart phone when you have a stand of native thistles in bloom to entertain?

HOW TO GROW These are biennial species that produce a basal rosette of leaves their first season, then grow tall, flower, set seed, and die their second season. Their leaves are dense white hairy beneath, which is an easy way to distinguish them from the nonnative, invasive thistles.

LANDSCAPE USE Native thistles are premier plants for an insectaries garden and should be a part of every natural landscape. The flowers are exceedingly nectar rich and invite a wealth of pollinators from bees to butterflies and hummingbirds. Many songbirds relish the seeds, which are an integral food for fledgling American goldfinches that time their late-season nesting with this plant. Native thistles have so many creatures after their seeds that few survive and self-sowing is limited.

ORNAMENTAL ATTRIBUTES The first season rosette of leaves is actually quite attractive and not as spiny as nonnative thistles. The leaves are green above and whitened underneath. The showy flower heads are produced in late summer atop the plant and are light lavender purple or rarely white. When the fruits ripen, each head becomes whitened with the tuft of silky fibers that will carry the seed away.

411

A bumblebee is one of many visitors to the shaving brush lavender flowers of *Cirsium altissimum* (tall thistle).

Cirsium altissimum (tall thistle) is native throughout the Midwest and is found more in open woodlands, savannas, and woodland edges. The foliage is entire or lightly toothed.

Cirsium discolor (prairie thistle) is also native throughout the Midwest but found more often in prairie, meadow, and pasture settings away from trees. The foliage is deeply divided with sharply pointed lobes. This species can hybridize with tall thistle and intergrade so some plants are intermediate.

Cirsium muticum (swamp thistle) is another widespread biennial (or short-lived perennial) thistle but is restricted to growing wild in fens (where most prevalent), wet prairies, sedge meadows, and seeps. It's a choice for wetlands and is the sole host for a rare butterfly: the swamp metalmark. It has all the attributes of tall and prairie thistle and is easily identified by its flower head scales that lack spines and become sticky as blooming is about to begin.

A cloudless sulphur nectars from *Cirsium discolor* (prairie thistle).

Cirsium muticum (swamp thistle) blooms beneath blue skies at Chiwaukee Prairie, Wisconsin.

Collinsia verna
Blue-eyed Mary

Plantaginaceae (plantain family)

If you don't know blue-eyed Mary, get to know her by experiencing sparkling fragrant drifts where she grows wild. I once had a turkey hunter tell me that he came across a drift of a fragrant bi-colored white and blue wildflower (along the Blackwater River in western Missouri) that captured his attention to the point he had to sit for a spell in its presence. Blue-eyed Mary often is found near Virginia bluebells and the two of them are reasons to block out time on a busy calendar as one of those annual events as sacred as a holiday, birthday, or anniversary.

HOW TO GROW This wildflower is native in a scattered range across the Eastern and Lower Midwest northwestward to southern Wisconsin and southeastern Iowa. It grows in mesic floodplain forests and moist forested ravines in deep rich soils. It's a winter annual, germinating in fall and flowering by midspring.

LANDSCAPE USE If blue-eyed Mary were a perennial, it would be one of the most popular spring woodland wildflowers. Its annual nature means it thrives only in specific gardens that suit it—as it must self-sow. In gardens with moist, rich soils along a draw or stream terrace, it is a candidate for attempting to establish.

ORNAMENTAL ATTRIBUTES The bi-colored blue and white flowers have the slightest delicate perfume noticeable only when they are in abundance.

The bi-colored blue and white spring flowers of *Collinsia verna* (blue-eyed Mary) have a faint fragrance.

Coreopsis tinctoria (plains coreopsis) in full bloom is a remnant of its planting as a prairie reconstruction cover crop at Powell Gardens, Missouri.

Croton monanthogynus (goatweed) forms lovely planes of foliage clusters, each with tiny green flowers at their center.

Coreopsis tinctoria
Plains coreopsis, calliopsis
Asteraceae (aster family)

HOW TO GROW This species is native to the Great Plains originally ranging eastward specifically into the Dakotas, Nebraska, Missouri, and Arkansas but now naturalized in disturbed habitats throughout the Midwest. It is thought to have grown around prairie buffalo wallows, thriving in disturbed soils that are seasonally wet. It does well in almost any soil type and requires full sun.

LANDSCAPE USE The plant reaches 2–3 feet tall and makes a good choice for a cover crop in a prairie reconstruction, flowering the first season but quickly dying out as perennial plants take hold. Plains coreopsis has nectar-rich flowers and seeds that songbirds seek for food.

ORNAMENTAL ATTRIBUTES The plant is covered in showy flowers that are normally yellow but sometimes are bicolored yellow and brick red or are totally brick red.

Croton monanthogynus
Goatweed, prairie tea
Euphorbiaceae (spurge family)

HOW TO GROW Goatweed is native across the Lower Midwest and scattered or naturalized northward into Nebraska, Iowa, Wisconsin, and Michigan. It is found in dry upland prairie and savanna ridges, thriving in disturbed soil in full sun. It self-sows heavily in a garden setting deserving the name "weed."

LANDSCAPE USE This species is in my garden because it is the host plant for one of my favorite butterflies, the goatweed leafwing whose caterpillars create daytime chambers of rolled leaves ("leaf burritos" as described by some of my

The sparkling white flowers of *Erigeron annuus* (daisy fleabane) shine above golden *Rudbeckia hirta* (black-eyed Susan) flowers in a prairie restoration test plot at the University of Wisconsin Arboretum in Madison.

lepidopterist friends) on the plant. It is best planted in a site with poor, dry soil where little else will grow. Allegedly goats will eat it, but it is not browsed by other herbivores because of its fetid foliage.

ORNAMENTAL ATTRIBUTES The little gray-green leaves adorn what look like flat-topped, mini-acacia tree-looking plants. The plant exudes a very special, hard-to-describe aroma (described as "lemony" by one of my gardening friends) that gives it some additional charm.

Erigeron annuus
Daisy fleabane
Asteraceae (aster family)

HOW TO GROW This species is native across much of eastern North America including the entire Midwest. It is disturbance dependent; in other words, it needs open ground to germinate and grow in. It is ubiquitous in gardens where it thrives anywhere vegetation has been removed and the soil tilled. It is actually an annual or biennial.

LANDSCAPE USE I recently read about daisy fleabane as a wonderful new annual from America in the Royal Horticultural Society's *The Garden* magazine and I felt exonerated. I spare this "weed" where it emerges on the edges of natural landscaping beds and sometimes even when it comes up in more formal beds. It reaches 3–4 feet tall and blooms profusely for a few weeks in early summer. By the end of summer, it has gone to seed and consists of dry stalks that I remove. It also grows profusely in newly prepared and planted prairie reconstructions where it actually functions like a cover crop and may be a beautiful companion to black-eyed Susans. Daisy fleabane does attract a wealth of pollinators while in bloom but I've never tried it to deter fleas.

ORNAMENTAL ATTRIBUTES The flat-topped clusters of mini-daisylike flowers have yellow disc flowers and a fringe of thin white ray flowers atop tall stems. What's not to like except for the fact it is a native weed that can self-sow abundantly though it is very easy to remove and is stopped in its tracks by mulch.

RELATED PLANTS There are several similar species, none as widespread, with a couple of species more dependent on natural habitats.

Eryngium leavenworthii
Leavenworth's eryngo
Apiaceae (carrot family)

HOW TO GROW This species is native in the southern Great Plains from the Flint Hills and western Osage Plains of eastern Kansas southward to Texas. It is found in rocky prairies and can be cultivated in similar soils with perfect drainage—naturalizing once in pure gravel besides Powell

This patch of *Eryngium leavenworthii* (Leavenworth's eryngo) at Kansas State University Gardens in Manhattan is past peak of flowering but remains beautiful with only a few purple flowers still in bloom.

Gardens' greenhouses. The plant thrives in fast drainage, heat, drought, and intense sunlight.

LANDSCAPE USE It makes a strikingly showy conversation-piece plant for a gravel or rock garden in full sun. This is an annual for that "hell strip" where you thought nothing would grow.

ORNAMENTAL ATTRIBUTES The late-summer into fall flowers are almost shockingly beautiful and like no other with a flower head that looks like a purple pineapple. Gardeners will recognize that the spiny, bluish-green foliage looks like that of related perennial sea hollies.

Euphorbia marginata
Snow-on-the-mountain
Euphorbiaceae (spurge family)

HOW TO GROW Snow-on-the-mountain should really be called snow-on-the-prairie, as it is native to the Great Plains and eastward into the western Midwest. It has naturalized throughout the Midwest and is a popular self-sowing annual in gardens. It thrives in almost any well-drained soil in full sun, self-sowing best in bare soil.

LANDSCAPE USE This species can be added to traditional perennial borders where it may fill in between seasonally blooming plants. It grows about 3 feet tall and its fruits are eaten by some birds, especially mourning doves.

ORNAMENTAL ATTRIBUTES Despite its gray-green leaves, the plant really stands out in a garden in late summer as the

White-edged, leafy bracts surrounding small flowers show that *Euphorbia marginata* (snow-on-the-mountain) is related to the poinsettia.

Glandularia canadensis (rose verbena) makes a magnificent carpet of flowers during peak bloom in spring.

Helenium amarum (plains sneezeweed) readily self-sows in gravelly edges between beds and concrete walks.

upper foliage is edged in white. These whitened leafy bracts around the ornamentally insignificant flowers atop the plant are where the name "snow" comes from.

Glandularia canadensis

Rose verbena

> Syn. *Verbena canadensis*
> Verbenaceae (vervain family)

HOW TO GROW Rose verbena is native across the Lower Midwest, but is most widespread in the Ozark Highlands, Osage Plains, and Flint Hills. It has naturalized northward to Michigan, Wisconsin, and Iowa. It is found on open rock outcrops, glades or other rocky, gravelly, or sandy openings in prairies, savannas, or woodlands in full sun or partial shade. It is a very short-lived perennial best treated as a self-sowing annual.

LANDSCAPE USE The plant makes a fine edge to an annual flower border blooming most abundantly in spring but reblooming a bit through the entire growing season. It attracts many pollinating insects especially butterflies so should be included in an insectaries garden. It makes an ideal rock garden plant or planting in or above a rock wall where its trailing habit is enhanced. This habit lends itself as an ideal spiller plant in containers.

ORNAMENTAL ATTRIBUTES Bright fuchsia pink flowers are produced over a long period (recorded in bloom every month of the year at Powell Gardens) though peaking in spring and early summer, sometimes with a nice rebloom in early fall. Flowers may occasionally be truer pink or even purplish, very rarely white. The leaves are divided and create a fine texture along the ground-hugging stems.

Helenium amarum

Plains sneezeweed, bitterweed

> Asteraceae (aster family)

HOW TO GROW This sneezeweed is native to the Ozark Highlands and into adjacent areas of southeastern Kansas, southern Illinois, and western Kentucky. It has naturalized sparingly northward into Nebraska, Iowa, Wisconsin, and Michigan. It is found in sandy, gravelly, or rocky soils of dry prairies and glades but thrives in disturbed soils including railroad ballast. It will grow in any well-drained soil in full sun.

LANDSCAPE USE The plant ranges from 6 to 30 inches tall and is a super annual for a hot sunny rock garden, on or around rock walls, and between path stepping-stones.

ORNAMENTAL ATTRIBUTES The leaves are threadlike and so very finely textured all season. The showy yellow flowers are produced in late summer into fall.

RELATED PLANTS 'Dakota Gold' is a popular seed strain of the species available at many midwestern nurseries.

Helenium flexuosum

Purple-headed sneezeweed

> Asteraceae (aster family)

HOW TO GROW This species is native to the Ozark Highlands eastward across the Lower Midwest to Ohio, and naturalized northward to Wisconsin and Michigan. It grows in prairies, meadows, swales, and open floodplain woodlands and is often described as perennial but is more of a biennial or a very short-lived perennial—never more than a biennial in my personal garden or at Powell Gardens where it is

Helenium flexuosum (purple-headed sneezeweed) makes a striking midsummer wildflower at Paintbrush Prairie, Missouri.

Helianthus annuus (common sunflower) blooms in the wild adjacent to Cooley Lake, a backwater of the Missouri River near Kansas City.

grown from local, wild-collected seed. It requires soils that are at least moist in spring and full sun to partial shade.

LANDSCAPE USE The plant reaches around 30 inches tall. It is easy to grow and add for seasonal color to any flower border.

ORNAMENTAL ATTRIBUTES Purple-headed sneezeweed blooms in late summer into early fall. Its flower head is comprised of a lovely purplish-brown cone of disc flowers surrounded by a skirt of three-lobed, showy yellow ray flowers.

Helianthus annuus
Common sunflower

Asteraceae (aster family)

Helianthus annuus is the state flower of Kansas.

HOW TO GROW The common sunflower probably originated in Mexico and has been cultivated for almost 5000 years. Native Americans most certainly brought the plant northward and now it has naturalized throughout the Midwest. It grows in disturbed soils and is considered an agricultural pest in some areas. It is common on roadsides as a wild plant in the Lower Midwest and is much more rarely encountered in the Upper Midwest.

LANDSCAPE USE The wild form of this species is a stunning addition to a natural landscape valued for its beautiful flowers that attract many pollinators. It makes a phenomenal cut flower. Birds and wildlife also relish the edible seeds.

ORNAMENTAL ATTRIBUTES A close inspection of the dark brown center of disc flowers reveals that they are arranged in a marvelous pattern that efficiently packs in their fruits (the sunflower seeds). This mathematical arrangement is known as Fibonacci number (learn more at momath.org/home/fibonacci-numbers-of-sunflower-seed-spirals). The

The large flower umbels of *Heracleum maximum* (cow parsnip) atop sturdy stems look like gigantic versions of alien Queen Anne's lace.

showy ray flowers are golden yellow, rarely pale or reddish.

NOTES Many cultivars and hybrids of common sunflower have been made through history and it is an important crop grown for its edible seeds and their oil. Would Vincent Van Gogh be as popular without this American plant?

Heracleum maximum
Cow parsnip

Apiaceae (carrot family)

HOW TO GROW Cow parsnip is native across the Upper Midwest and sparingly southward to the Ohio and Lower Missouri Rivers. It's a biennial found in open floodplain woodlands, along rivers and streams, edges of moist forests and in disturbed habitats including moist meadows and roadsides. Propagation is easy by seed.

LANDSCAPE USE The plant reaches 4–6 feet tall, sometimes taller, and makes a spectacular wildflower for the edge of woodlands, moist meadows, or other natural landscapes.

ORNAMENTAL ATTRIBUTES This species blooms in early summer with huge flat umbels of white flowers that are a real standout in the greens of the season. The foliage is large and dramatic and the seed heads hold well into winter and remain highly ornamental.

Hydrophyllum appendiculatum
Appendaged waterleaf
Hydrophyllaceae (waterleaf family)

HOW TO GROW Appendaged waterleaf is native across the Eastern Midwest westward through the central Mississippi Valley as far north as central Minnesota, and up the Missouri River basin into the core of the Ozark Highlands and upstream to eastern Nebraska and western Iowa. It is found on mesic north- and east-facing slopes, ravines, and mesic floodplains in humus-rich, well-drained soil. Plants germinate in spring, flower the following spring, quickly set seed, and die.

LANDSCAPE USE This species is a good addition to a sheltered, moist woodland garden where it may self-sow. As with all waterleafs, its flowers are visited by many types of bees.

ORNAMENTAL ATTRIBUTES The spring seedlings are quite pretty with two quarter-sized cotyledon seed-leaves and an emerging water-spotted true leaf in between.

Second-spring plants produce a fresh basal tuft of divided, strikingly water-spangled leaves from which the flowering stem arises with leaves that mature green, are fuzzy and maple-shaped. Appendaged waterleaf produces the showiest light lilac flowers of any waterleaf in the characteristic scorpioid (spiral like a scorpion's abdomen) cymes characteristic of plants in the borage and waterleaf families (some resources now lump these into one family).

Impatiens capensis
Spotted jewelweed, touch-me-not
Balsaminaceae (balsam family)

HOW TO GROW Spotted jewelweed is native across the entire Midwest and is found in wetlands from seeps and springs to fens and sedge meadows and even disturbed wet meadows and ditches. It thrives in a garden in wet soils in full sun to light shade and self-sows abundantly to the point of being a nuisance. The spring seedlings are frost hardy but a flowering plant in fall will succumb to the first frost.

LANDSCAPE USE The plant usually reaches 3 feet tall and is suitable only to wet natural gardens where its aggressive nature may be embraced. It is a fine pollinator plant and in peak bloom during ruby-throated hummingbird migration a bird may guard a patch.

ORNAMENTAL ATTRIBUTES The flowers, which are yellow-orange with darker orange to orange-red spots and lip,

Hydrophyllum appendiculatum (appendaged waterleaf) produces fuzzy maple-shaped leaves.

The orange flowers of *Impatiens capensis* (spotted jewelweed) are often pollinated by hummingbirds.

Impatiens pallida (yellow jewelweed) has pale yellow flowers and luscious translucent foliage.

Silver bracts blushed with pink separate whorls of spotted, parchment-colored flowers on the stem of *Monarda punctata* (dotted horsemint).

produce fruits that burst with the slightest touch, propelling the seeds a great distance. That's how it gets the name "touch-me-not" and why it's a fun plant to engage children. The seed's coat may be scraped off to reveal a robin's egg blue surface underneath along with the aroma of wintergreen—and the seed is edible.

RELATED PLANTS Yellow jewelweed or pale touch-me-not (*Impatiens pallida*) is similar with yellow flowers and is found throughout the Midwest. It has an equally aggressive, self-sowing behavior in a garden. In the wild it is found in mesic forest in moist, humus-rich soils. It prefers shade though it may grow side-by-side with spotted jewelweed. Yellow jewelweed blooms in late summer into fall and is more often pollinated by bumblebees. It produces similar touch-me-not fruits.

Monarda punctata
Dotted horsemint, spotted beebalm
Lamiaceae (mint family)

The pointed leafy bracts whorled between these layers of flowers are often whitened or silvered and blushed with shades of pink for an extraordinary composition that shows well even after the flowers have gone to seed.

HOW TO GROW This species is native over much of eastern North America with a somewhat divergent range around the Great Lakes, southern Great Plains, and Mid-Atlantic to southeastern states. It's only absent from the Dakotas and

Nebraska in the Midwest. It is found in open, sandy prairies and sand savannas, sometimes colonizing similar sites along roadsides and railways. It requires well-drained soils, thriving and self-sowing in almost pure sand. It is a very short-lived perennial or biennial and should be treated as an annual in normal garden soils.

LANDSCAPE USE Dotted horsemint reaches about 2 feet tall and is valued as a long-blooming plant suitable for a container or annual flower border. It attracts an amazing array of pollinators making it one of the finest choices for an insectaries garden as it attracts so many beneficial/predatory insects.

ORNAMENTAL ATTRIBUTES The flowers are arranged in tiered whorls along the top of the stem and are the color of aged parchment with dark purple spots. The pointed leafy bracts whorled between these layers of flowers are often whitened or silvered and blushed with shades of pink for an extraordinary composition that shows well even after the flowers have gone to seed.

RELATED PLANTS Lemon bergamot (*Monarda citriodora*) is a similar species but has lavender to purplish flowers and bracts. It is native to the southern Great Plains northward into the Flint Hills and Osage Plains of eastern Kansas and Ozark Highlands of southwestern Missouri and northwestern Arkansas. The species has naturalized elsewhere into Nebraska, Illinois, and Kentucky and is an annual herb that is gaining in popularity. It may be used as a quick color nurse plant for a prairie reconstruction and will quickly fade out as perennials take hold.

Monarda citriodora (lemon bergamot) is a striking wildflower while in full bloom with tiers of lavender flowers above rosy-purple leafy bracts.

Oenothera biennis (common evening primrose) is a striking ornamental while in full bloom.

Oenothera biennis

Common evening primrose

Onagraceae (evening primrose family)

HOW TO GROW Common evening primrose is native throughout the Midwest, where it is found in disturbed prairies, meadows, abandoned fields, roadsides, and other ruderal habitats. It is easy to grow in full sun or partial shade in almost any well-drained soil.

LANDSCAPE USE This species is a good nurse crop for a prairie restoration but is valued in a natural landscape for its abundant seed that is an excellent attractor of winter songbirds and a favorite food of common redpolls. The plant appeals to many pollinating insects and is host plant for the white-lined sphinx. The plant's first winter root is edible along with the basal rosette of leaves.

ORNAMENTAL ATTRIBUTES The flowers are bright yellow and produced in late summer into fall. They open at dusk so are aptly named and the flowers are fun to watch unfurl rather quickly right before one's eyes. The plant grows lanky and weedy-looking after peak flowering. Common evening primrose's autumn fruits remain showy into winter when they split and peel open to release the nutritious seeds.

Oenothera gaura

Biennial gaura, biennial beeblossom

Syn. *Gaura biennis*
Onagraceae (evening primrose family)

Identifying this species and long-flower gaura (*Gaura longiflora*, syn. *Oenothera filiformis*) is literally splitting hairs as the two are separated by the hairs on their stems and they are otherwise virtually identical to the naked eye. Apparently another difference is that biennial gaura produces flowers that are self-compatible and inbred while long-flower gaura cannot self-fertilize and requires cross-pollination. Some references lump them into a single species, including my botanist friend Mark Leoschke who remarks, "When in doubt, lump is my motto."

HOW TO GROW The species is native in northeastern North America and is found in the Midwest from Michigan, Ohio, and Kentucky westward across Indiana and Illinois to southern Wisconsin and eastern Iowa. I like its common name "biennial beeblossom." The stems are covered in wide-spreading hairs. If plants have short hairs pressed against the stems (adpressed is the botanical term), then they are long-flower gaura found mainly in the center of North America from Illinois westward across southern Iowa, Missouri, and Arkansas to southeastern Nebraska, eastern Kansas, and eastern Oklahoma. Long-flower gaura has naturalized northward into Wisconsin and eastward to Michigan and Ohio. Both species can be found in native and disturbed sites from prairies to meadows and roadsides in full sun in well-drained soils.

LANDSCAPE USE Gardeners rarely cultivate the two Midwest native gauras as they do the more compact and fuller flowering Lindheimer's gaura (*Oenothera lindheimeri*), which is native to southeastern Texas and Louisiana. The

419

Pinkish buds and flowers of *Oenothera gaura* (biennial gaura) are nice color echoes to big bluestem's budded "turkey foot" inflorescences.

Palafoxia callosa produces an abundance of rosy-pink shaded flower heads.

midwestern gauras attain a lanky 4–6 feet and bloom in late summer into early fall along with the tall prairie grasses that make good companions. They make an integral addition to a prairie restoration or natural landscape because the flowers attract so many beneficial insects from bees to moths. Gauras are host to a couple of unique flower moths including the exquisitely colored moth and caterpillar of the clouded crimson (*Schinia gaurae*) and the spectacular white-lined sphinx that regularly visit gardens for nectar.

ORNAMENTAL ATTRIBUTES Gauras are valued for their delicate pink-budded, white flowers atop tall stems that dance with the grasses in the wind.

RELATED PLANTS Scarlet gaura (*Gaura coccinea*) is found in western North America eastward into the Midwest in western Minnesota and the Loess Hills of western Iowa and northwestern Missouri. It grows shorter (2 feet tall), is longer lived, and has pink flowers but otherwise fills the same garden niche in well-drained soils.

Palafoxia callosa
Palafoxia

Asteraceae (aster family)

HOW TO GROW Palafoxia is native from the Ozark Highlands southwestward into Texas and has naturalized, mainly along gravelly roadsides and railroad right-of-ways northward and eastward to Indiana. It is found in rocky glades in full sun and may self-sow abundantly in gravelly or rocky situations.

LANDSCAPE USE The plant reaches 24–30 inches tall and is a colorful annual for a "hell strip" or other seemingly inhospitable dry and rocky site. It works well in rock gardens or rock walls where it self-sows.

ORNAMENTAL ATTRIBUTES Seedling plants have long thin leaves that are quite handsome. In late summer into early fall, the plant looks like a rosy-pink baby's breath producing a dome of airy flower heads. The fruiting heads are topped with clusters of gray seeds.

The long reddish stamen filaments of *Polanisia dodecandra* subsp. *trachysperma* (large-flowered redwhisker clammyweed) look like cat whiskers.

Polanisia dodecandra
Clammyweed
Cleomaceae (cleome family)

Clammyweed has a marketing problem: its name. Being so closely related to cleomes or beeplant, it should also be called beeplant. With increasing understanding of the importance of bees in our gardens, clammyweed is sure to become a more popular annual in gardens.

HOW TO GROW This species is native throughout the Midwest and beyond in all directions to include much of the heart of North America. It grows on sandy or gravelly shores, eroding bluffs, open glades, rock ledges, or other disturbed sites.

LANDSCAPE USE Clammyweed is similar to Rocky Mountain beeplant (*Cleome serrulata*) growing about 2 feet tall with nectar-rich flowers that attract many bees, butterflies, and other pollinators. It is not edible.

ORNAMENTAL ATTRIBUTES The leaves, stems, and short fruits are covered in glandular hairs giving the whole plant (when touched) a clammy feel and rather rank odor; its scent is like that of green peppers. The plant produces small, ¼-inch, notched white-petaled flowers, short-stalked fruits, and trifoliate leaves. It is quite weedy looking for lack of a better term.

RELATED PLANTS Clammyweed's western subspecies (Illinois and west), the large-flowered clammyweed (*Polanisia dodecandra* subsp. *trachysperma*) produces larger, ½-inch flowers and is much showier and suitable to an ornamental garden. Large-flowered clammyweed has naturalized eastward sparingly across the entire Midwest.

The iconic golden domes of *Polytaenia nuttallii* (prairie parsley) flowers are becoming a rare sight. This species is a declining member of our prairie flora.

Polytaenia nuttallii
Prairie parsley
Apiaceae (carrot family)

HOW TO GROW Prairie parsley is found in the middle of North America from southeastern Nebraska, eastern Iowa, central Wisconsin, and northwestern Indiana southward nearly to the Gulf of Mexico. It is an upland prairie species that has not adapted to disturbed habitats so is extirpated from southeastern Minnesota and southwestern Michigan. Prairie parsley is a biennial or very short-lived perennial that dies after it flowers and sets fruit. It grows well in moist to dry, well-drained soils in full sun and it may be propagated from seed but is slow and usually self-sows only lightly despite abundant seed production.

LANDSCAPE USE The plant may reach 4 feet tall and is best used in a natural landscape or prairie reconstruction.

ORNAMENTAL ATTRIBUTES The basal leaves make the plant look like a flat-leaved parsley until it matures and bolts (which may take several years). The towering dome-shaped umbels of yellow flowers are quite colorful in midsummer and form an ornamental crown of fruits in late summer that begin green, age to a showy yellow, and mature a dry golden brown.

NOTES Do not confuse this native species with nonnative common parsnip (*Pastinaca sativa*), which has become a widespread invasive weed.

Rudbeckia hirta (black-eyed Susan) embellishes a private hay prairie near Sedalia, Missouri.

Self-sown *Rudbeckia triloba* (brown-eyed Susan) enlivens a moist space between a concrete walk and a rock wall.

Rudbeckia hirta

Black-eyed Susan

Asteraceae (aster family)

HOW TO GROW Black-eyed Susan is native from the Rocky Mountains eastward in North America and is found everywhere in the Midwest, including disturbed prairies, meadows, and roadsides in a wide range of soils from wet to dry. It will grow in partial shade, but is at its best in full sun. Black-eyed Susans are usually biennial but some plants may be annual or very short-lived perennials. Dead-heading (removing the spent flowers) will extend the bloom and lengthen the life of the plant but you also rob wildlife of its seeds and ruin the winter interest of the plant.

LANDSCAPE USE Plants grow 18 inches to rarely 3 feet tall. They make a perfect cover crop when establishing a prairie reconstruction, suppressing weeds and creating a quick splash of bloom, but diminishing in cover as other longer-lived plants take hold. They are best in natural landscapes where they are allowed to self-sow. They do make nice container plants and additions to annual flowerbeds though their flowers do not rebloom all growing season as many traditional annuals do.

ORNAMENTAL ATTRIBUTES The glowing golden-yellow ray flowers surrounding the dark brown disc-flowers create a classic wildflower beloved by all. The blooms appear in midsummer and occasionally into fall, and the dried seed heads remain attractive into the winter landscape.

NOTES The hairs on the stems and leaves of dried plants can be quite irritating (and can cause a rash) so always wear gloves when handling them.

RELATED PLANTS There are several ornamental selections with larger flowers, golden to chestnut flower colors, and even those without "black eyes" dark disc flowers. These selections gradually revert to the wild form if allowed to reseed in a garden.

Rudbeckia triloba

Brown-eyed Susan

Asteraceae (aster family)

HOW TO GROW Brown-eyed Susan is native to the Eastern and Lower Midwest and northwestward to central Minnesota and southeastern Nebraska. It is mainly biennial and is found in wet prairies, along streams, and in open low woodlands. It grows well in any rich moist to wet soil, self-sowing abundantly in bare soil areas so it is sometimes considered a garden thug.

LANDSCAPE USE This species is best in a natural style garden where its spontaneous nature can be celebrated. It often self-sows in moist cracks on the edges of walks, in the low sections of rain gardens, or wetland swales.

ORNAMENTAL ATTRIBUTES Brown-eyed Susans are somewhat like mini black-eyed Susans—the flowers are one-fourth the size but produced in abundance like a dome of brown and gold over the plant. The winter fruit seed heads are among my favorites as they are dark brown but have a warm orange inner glow to them like no other plant.

Towering spikes of *Zizania aquatica* (wild rice) provide food for waterfowl at the end of summer.

Zizania aquatica
Wild rice

Poaceae (grass family)

This spectacular annual is off the radar of gardeners but available as seed, usually planted as waterfowl food on hunting properties. Yes, it is the culinary wild rice that produces the delicious, long black grains.

HOW TO GROW This species is native across the Upper Midwest from central Ohio westward to central Illinois, Iowa, and Minnesota and is found in floodplain backwaters, shallow lakes, and marshes. The seed needs water that is 6–18 inches deep with a muck bottom to germinate. The plant requires full sun. A good way to plant it is by mixing the seed in muck then throwing your "seed bomb" into shallow water.

LANDSCAPE USE Wild rice makes a spectacular plant with inflorescences that grow 5–8 feet tall. It is most commonly used in shallow ponds and wetlands as food for waterfowl but can be harvested by boat for human consumption.

ORNAMENTAL ATTRIBUTES The plant blooms at the end of summer with its towering spikes of golden-yellow anthers—showy from a great distance. The grass leaves of young plants float on the surface of the water but do become spiky emergent aquatics as they mature through summer.

NOTES Want a close encounter with wild rice? You can see it up close just a short distance from the Mall of America at the Minnesota Valley National Wildlife Refuge's visitor center's trail into the wetlands below. It's also a stunner when in bloom around Labor Day along the Lower Wisconsin River backwaters between Spring Green and Prairie du Chien—its spectacular golden plumes easily enjoyed from Highway 60.

423

Hardiness Zone Charts

USDA Plant Hardiness Zones

Temp °F			Zone	Temp °C		
−60	to	−55	1a	−51	to	−48
−55	to	−50	1b	−48	to	−46
−50	to	−45	2a	−46	to	−43
−45	to	−40	2b	−43	to	−40
−40	to	−35	3a	−40	to	−37
−35	to	−30	3b	−37	to	−34
−30	to	−25	4a	−34	to	−32
−25	to	−20	4b	−32	to	−29
−20	to	−15	5a	−29	to	−26
−15	to	−10	5b	−26	to	−23
−10	to	−5	6a	−23	to	−21
−5	to	0	6b	−21	to	−18
0	to	5	7a	−18	to	−15
5	to	10	7b	−15	to	−12
10	to	15	8a	−12	to	−9
15	to	20	8b	−9	to	−7
20	to	25	9a	−7	to	−4
25	to	30	9b	−4	to	−1
30	to	35	10a	−1	to	2
35	to	40	10b	2	to	4
40	to	45	11a	4	to	7
45	to	50	11b	7	to	10
50	to	55	12a	10	to	13
55	to	60	12b	13	to	16
60	to	65	13a	16	to	18
65	to	70	13b	18	to	21

AHS Plant Heat Zones

Zone	Number of Days per Year Above 86°F (30°C)
1	<1
2	1–7
3	>7–14
4	>14–30
5	>30–45
6	>45-60
7	>60–90
8	>90–120
9	>120–150
10	>150–180
11	>180–210
12	>210

Bibliography

Allison, R. B. 2005. *Wisconsin's Champion Trees*. Verona, Wisconsin: Wisconsin Book Publishing.

Antonio, T. M., and S. Masi. 2001. *The Sunflower Family in the Upper Midwest: A Photographic Guide to the Asteraceae in Illinois, Indiana, Iowa, Michigan, Minnesota, and Wisconsin*. Indianapolis: Indiana Academy of Science.

Armitage, A. M. 1997. *Herbaceous Perennial Plants: A Treatise on Their Identification, Culture, and Garden Attributes*. 2nd ed. Champaign, Illinois: Stipes Publishing.

Armitage, A. M. 2006. *Armitage's Native Plants for North American Gardens*. Portland, Oregon: Timber Press.

Bare, J. E. 1979. *Wildflowers and Weeds of Kansas*. Lawrence, Kansas: University Press of Kansas.

Barnard, I. 2010. *A Pocket Guide to Kansas Flint Hills Wildflowers and Grasses*. Wichita, Kansas: Great Plains Nature Center.

Bir, R. E. 1992. *Growing and Propagating Showy Native Woody Plants*. Chapel Hill, North Carolina: University of North Carolina Press.

Birdseye, C., and E. G. Birdseye. 1972. *Growing Woodland Plants*. New York: Dover Publications.

Blanchan, N. 1922. *Nature's Garden: An Aid to Knowledge of Our Wild Flowers and Their Insect Visitors*. New York: Doubleday, Page and Company.

Boon, B., and H. Groe. 1990. *Nature's Heartland: Native Plant Communities of the Great Plains*. Ames, Iowa: Iowa State University Press.

Braun, E. L. 1961. *The Woody Plants of Ohio*. Columbus, Ohio: Ohio State University Press.

Britton, N. L., and H. A. Brown. 1970. *An Illustrated Flora of the Northern United States and Canada*. 2nd ed. Vols. 1, 2, and 3. New York: Dover Publications.

Brockman, C. F. 2001. *Trees of North America: A Guide to Field Identification* Revised and updated. Golden Field Guides. New York: St. Martin's Press.

Bruce, H. 1976. *How to Grow Wildflowers and Wild Shrubs and Trees in Your Own Garden*. New York: Knopf.

Callaway, D. J. 1994. *The World of Magnolias*. Portland, Oregon: Timber Press.

Cappiello, P., and D. Shadow. 2005. *Dogwoods: The Genus* Cornus. Portland, Oregon: Timber Press.

Case, W. C., Jr., and R. B. Case. 1997. *Trilliums*. Portland, Oregon: Timber Press.

Christopher, T., ed.. 2011. *The New American Landscape: Leading Voices on the Future of Sustainable Gardening*. Portland, Oregon: Timber Press.

Cochrane, T. S., K. Elliot, and C. S. Lipke. 2008. *Prairie Plants of the University of Wisconsin-Madison Arboretum*. Madison, Wisconsin: University of Wisconsin Press.

Coffin, B., and L. Pfannmuller, eds. 1988. *Minnesota's Endangered Flora and Fauna*. Minneapolis, Minnesota: University of Minnesota Press.

Cope, A. B. 1986. *Native and Cultivated Conifers of Northeastern North America*. Ithaca, New York: Cornell University Press.

Courtenay, B., and J. H. Zimmerman. 1972. *Wildflowers and Weeds*. New York: Van Nostrand Reinhold Company.

Creasy, R. 2010. *Edible Landscaping*. San Francisco: Sierra Club Books.

Crockett, J. U. 1977. *Wildflower Gardening*. The Time-Life Encyclopedia of Gardening. Alexandria, Virginia: Time-Life Books.

Cullina, W. 2000. *The New England Wildflower Society Guide to Growing and Propagating Wildflowers of the United States and Canada*. New York: Houghton Mifflin Company.

Cullina, W. 2002. *Native Trees, Shrubs, and Vines: A Guide to Using, Growing, and Propagating North American Woody Plants*. New York: Houghton Mifflin Company.

Cullina, W. 2008. *Native Ferns, Moss, and Grasses*. New York: Houghton Mifflin Company.

Cullina, W. 2009. *Understanding Perennials: A New Look at an Old Favorite*. New York: Houghton Mifflin Company.

Curtis, J. T. 1959. *The Vegetation of Wisconsin: An Ordination of Plant Communities*. Madison, Wisconsin: University of Wisconsin Press.

Czarapata, E. J. 2005. *Invasive Plants of the Upper Midwest*. Madison, Wisconsin: University of Wisconsin Press.

Daniel, G. 1984. *Dune Country: A Hiker's Guide to the Indiana Dunes*. Athens, Ohio: Ohio University Press.

Darke, R. 1999. *The Color Encyclopedia of Ornamental Grasses: Sedges, Rushes, Restios, Cat-tails, and Selected Bamboos*. Portland, Oregon: Timber Press.

Darke, R. 2002. *The American Woodland Garden: Capturing the Spirit of the Deciduous Forest*. Portland, Oregon: Timber Press.

Darke, R. 2007. *The Encyclopedia of Grasses for Livable Landscapes*. Portland, Oregon: Timber Press.

Den Boer, A. F. 1959. *Ornamental Crab Apples*. American Association of Nurserymen.

Denison, E. 2008. *Missouri Wild Flowers: A Field Guide to the Wildflowers of Missouri*. 6th ed. Jefferson City, Missouri: Missouri Department of Conservation.

Dennis, J. V. 1985. *The Wildlife Gardener*. New York: Alfred A. Knopf.

Diekelmann, J., and R. M. Schuster. 2002. *Natural Landscaping: Designing with Native Plant Communities*. 2nd ed. Madison, Wisconsin: University of Wisconsin Press.

Dirr, M. A. 2004. *Hydrangeas for American Gardens*. Portland, Oregon: Timber Press.

Dirr, M. A. 2007. *Viburnums: Flowering Shrubs for Every Season*. Portland, Oregon: Timber Press.

Dirr, M. A. 2009. *Manual of Woody Landscape Plants: Their Identification, Ornamental Characteristics, Culture, Propagation, and Uses*. Champaign, Illinois: Stipes Publishing.

Eilers, L. J., and D. M. Roosa. 1994. *The Vascular Plants of Iowa: An Annotated Checklist and Natural History*. Iowa City: University of Iowa Press.

Eisendrath, E. R. 1978. *Missouri Wildflowers of the St. Louis Area*. St. Louis: Missouri Botanical Garden.

Faldet, D. S. 2009. *Oneota Flow: The Upper Iowa River and Its People*. Iowa City: University of Iowa Press.

Farrar, J. 1990. *Field Guide to Wildflowers of Nebraska and the Great Plains*. 1st ed. Lincoln, Nebraska: Nebraskaland Magazine.

Fasset, N. C. 1976. *Spring Flora of Wisconsin*. 4th ed. Madison, Wisconsin: University of Wisconsin Press.

Fell, E. W. 1955. *Flora of Winnebago County, Illinois*. Washington D.C.: Nature Conservancy.

Ferreniea, V. 1993. *Wildflowers in Your Garden*. New York: Random House.

Fiala, J. F. 1994. *Flowering Crabapples: The Genus* Malus. Portland, Oregon: Timber Press.

Fowells, H. A. 1965. *Silvics of Forest Trees of the United States*. Agriculture Handbook No. 271. Washington, D.C.: United States Department of Agriculture.

Freeman, C. C., and E. K. Schofield. 1991. *Roadside Wildflowers of the Southern Great Plains*. Lawrence, Kansas: University Press of Kansas.

Galle, F. C. 1997. *Hollies: The Genus* Ilex. Portland, Oregon: Timber Press.

Geiger, B. 2011. *Low-Key Genius: The Life and Work of Landscape-Gardener O. C. Simonds*. Ferme Ornee Press / Urbpublisher.

Gracie, C. 2012. *Spring Wildflowers of the Northeast: A Natural History.* Princeton, New Jersey: Princeton University Press.

Gillman, J. 2009. *How Trees Die: The Past, Present, and Future of Our Forests.* Yardley, Pennsylvania: Westholme Publishing.

Gleason, H. A., and A. Cronquist. 1964. *The Natural Geography of Plants.* New York: Columbia University Press.

Greenberg, J. 2002. *A Natural History of the Chicago Region.* Chicago: University of Chicago Press.

Grese, R. E. 1992. *Jens Jensen: Maker of Natural Parks and Gardens.* Baltimore, Maryland: Johns Hopkins University Press.

Grese, R. E., ed. 2011. *The Native Landscape Reader.* Amherst, Massachusetts: University of Massachusetts Press.

Grissell, E. 2001. *Insects and Gardens: In Pursuit of a Garden Ecology.* Portland, Oregon: Timber Press.

Haddock, M. J. 2005. *Wildflowers and Grasses of Kansas: A Field Guide.* Lawrence, Kansas: University Press of Kansas.

Harlow, W. M. 1957. *Trees of the Eastern and Central United States and Canada.* New York: Dover Publications.

Harstad, C. 1999. *Go Native! Gardening with Native Plants and Wildflowers in the Lower Midwest.* Bloomington, Indiana: Indiana University Press.

Hawker, J. L. 1992. *Missouri Landscapes: A Tour Through Time.* Rolla, Missouri: Missouri Department of Natural Resources.

Heims, D., and G. Ware. 2005. *Heucheras and Heucherellas: Coral Bells and Foamy Bells.* Portland, Oregon: Timber Press.

Hellander, M. E. 1992. *The Wild Gardener: The Life and Selected Writings of Eloise Butler.* St. Cloud, Minnesota: North Star Press.

Hightshoe, G. L. 1988. *Native Trees, Shrubs, and Vines for Urban and Rural America: A Planting Design Manual for Environmental Designers.* New York: Van Nostrand Reinhold.

Hill, P. 2007. *Design Your Natural Midwest Garden.* Madison, Wisconsin: Trails Books.

Hipp, A. L. 2004. *Spring Woodland Wildflowers of the University of Wisconsin-Madison Arboretum.* Madison, Wisconsin: Regents of the University of Wisconsin.

Hitchcock, A. S., and A. Chase. 1971. *Manual of the Grasses of the United States.* 2nd ed. Vols. 1 and 2. New York: Dover Publications.

Hoag, D. G. 1965. *Trees and Shrubs for the Northern Great Plains.* Minneapolis, Minnesota: Lund Press.

Homan, B. 2008. *The Missouri Gardener's Companion: An Insider's Guide to Gardening in the Show-Me State.* Guilford, Connecticut: Globe Pequot Press.

Homoya, M. A. 2012. *Wildflowers and Ferns of Indiana Forests: A Field Guide.* Bloomington, Indiana: Indiana University Press.

Hugo, N. R., and R. Llewellyn. 2011. *Seeing Trees: Discover the Extraordinary Secrets of Everyday Trees.* Portland, Oregon: Timber Press.

Hunter, C. G. 1992. *Wildflowers of Arkansas.* 3rd ed. Little Rock, Arkansas: Ozark Society Foundation.

Hunter, C. G. 1995. *Trees, Shrubs, and Vines of Arkansas.* 2nd ed. Little Rock, Arkansas: Ozark Society Foundation.

Jacobs, D. L., and R. L. Jacobs. 1997. *Trilliums in Woodland and Garden: American Treasures.* Decatur, Georgia: Eco-Gardens.

Jensen, J. 1990. *Siftings.* Baltimore, Maryland: Johns Hopkins University Press.

Jones, M. D., and E. W. Fell. 1994. *The Flora and Vegetational History of Winnebago County, Illinois.* Rockford, Illinois: Winnebago County Forest Preserve District.

Keator, G. 1998. *The Life of an Oak: An Intimate Portrait.* Berkeley, California: Heyday Books.

Keeler, H. L. 1969. *Our Northern Shrubs and How to Identify Them.* New York: Dover Publications.

Kershner, B. S., D. Mathews, G. Nelson, and R. Spellenberg. 2008. *National Wildlife Federation Field Guide to Trees of North America.* New York: Sterling Publishing Company.

Key, J. S. 1982. *Field Guide to Missouri Ferns.* Jefferson City, Missouri: Missouri Department of Conservation.

Kingsbury, N. 2006. *Natural Garden Style: Gardening Inspired by Nature.* London: Merrell Publishers.

Kingsbury, N. 2014. *Gardening with Perennials: Lessons from Chicago's Lurie Garden.* Chicago: University of Chicago Press.

Klaber, D. 1976. *Violets of the United States.* Cranbury, New Jersey: A. S. Barnes and Company.

Korling, T. 1972. *The Prairie: Swell and Swale.* Dundee, Illinois: Torkel Korling.

Korling, T., and R. O. Petty. 1977. *Wild Plants in Flower 3: Eastern Deciduous Forest.* Evanston, Illinois: Torkel Korling.

Kramer, J. 1973. *Natural Gardens: Gardening with Native Plants.* New York: Charles Scribner's Sons.

Kruschke, E. P. 1965. *Contributions to the Taxonomy of Crataegus.* Publications in Botany, No. 3. Milwaukee, Wisconsin: Milwaukee Public Museum.

Kuccera, C. L. 1998. *The Grasses of Missouri.* Columbia, Missouri: University of Missouri Press.

Kurz, D. 1997. *Shrubs and Woody Vines of Missouri.* Jefferson City, Missouri: Missouri Department of Conservation.

Kurz, D. 1999. *Ozark Wildflowers: A Field Guide to Common Ozark Wildflowers.* Helena, Montana: Falcon Press.

Kurz, D. 2003. *Trees of Missouri.* Jefferson City, Missouri: Missouri Department of Conservation.

Kurz, D. 2004. *Illinois Wildflowers.* Pettigrew, Arkansas: Cloudland.net Publishing.

Ladd, D. 1995. *Tallgrass Prairie Wildflowers: A Field Guide.* Helena, Montana: Falcon Press.

Lane, C. 2005. *Witch Hazels.* Portland, Oregon: Timber Press.

Laughlin, K. 1956. *Manual of the Hawthorns of Cook and Du Page Counties of Illinois.* Chicago: Kendall Laughlin.

Lawrence, E. 1971. *Lob's Wood.* Cincinnati, Ohio: Cincinnati Nature Center.

Lawton, B. P. 2002. *Mints: A Family of Herbs and Ornamentals.* Portland, Oregon: Timber Press.

Leopold, A. 1966. *A Sand County Almanac with Essays on Conservation from Round River.* Oxford: Oxford University Press.

Leopold, D. J., W. C. McComb, and R. N. Muller. 1998. *Trees of the Central Hardwood Forests of North America: An Identification and Cultivation Guide.* Portland, Oregon: Timber Press.

Lerner, C. 1979. *Flowers of a Woodland Spring.* New York: William Morrow and Company.

Lerner, C. 1980. *Seasons of the Tallgrass Prairie.* New York: William Morrow and Company.

Little, E. L., Jr. 1996. *Forest Trees of Oklahoma: How to Know Them.* Oklahoma City: Oklahoma Department of Vocational and Technical Education.

Locklear, J. H. 2011. *Phlox: A Natural History and Gardener's Guide.* Portland, Oregon: Timber Press.

Loewer, P. 1991. *The Wild Gardener: On Flowers and Foliage for the Natural Border.* Harrisburg, Pennsylvania: Stackpole Books.

Logan, W. B. 2005. *Oak: The Frame of Civilization.* New York: W. W. Norton and Company.

Lommasson, R. C. 1973. *Nebraska Wild Flowers.* Lincoln, Nebraska: University of Nebraska Press.

Martin, A. C., H. S. Zim, and A. L. Nelson. 1951. *American Wildlife and Plants: A Guide to Wildlife Food Habits.* New York: Dover Publications.

McCoy, D. 1987. *Oklahoma Wildflowers.* Lindsay, Oklahoma: Doyle McCoy.

Mackenzie, K. K. 1902. *Manual of the Flora of Jackson County, Missouri.* Kansas City, Missouri: New Era Printing Company.

McLane, S. 1950. *Garden Guide by Months for the Midwest.* Kansas City, Missouri: Frank Glenn Publishing Company.

McRae, E. A. 1998. *Lilies: A Guide for Growers and Collectors*. Portland, Oregon: Timber Press.

Mickel, J. T. 1994. *Ferns for American Gardens*. New York: Macmillan Publishing Company.

Miller, D. 2010. *Life Afloat*. Rockford, Illinois: Rockford Litho Center.

Mohlenbrock, R. H. 1980. *Spring Woodland Wildflowers of Illinois*. Springfield, Illinois: Illinois Department of Conservation, Division of Forestry.

Mohlenbrock, R. H., ed. 1988. *A Field Guide to the Wetlands of Illinois*. Springfield, Illinois: Illinois Department of Conservation, Division of Planning.

Mohlenbrock, R. H. 1996. *Forest Trees of Illinois*. Springfield, Illinois: Illinois Department of Conservation, Division of Forestry.

Mohlenbrock, R. H., and J. W. Voigt. 1959. *A Flora of Southern Illinois*. Carbondale, Illinois: Southern Illinois University Press.

Moore, D. M. 1972. *Trees of Arkansas*. 3rd ed. Little Rock, Arkansas: Arkansas Forestry Commission.

Morton Arboretum. 1990. *Woody Plants of the Morton Arboretum: A Handlist of Living Plants in the Outdoor Woody Plant Collections*. Lisle, Illinois: Morton Arboretum.

Moyle, J. B., and E. W. Moyle. 2001. *Northland Wildflowers: The Comprehensive Guide to the Minnesota Region*. Minneapolis, Minnesota: University of Minnesota Press.

Nelson, P. W. 2005. *The Terrestrial Natural Communities of Missouri*. Jefferson City, Missouri: Missouri Department of Conservation.

Newcomb, L. 1977. *Newcomb's Wildflower Guide*. Boston: Little, Brown and Company.

Newsholme, C. 2002. *Willows: The Genus* Salix. Portland, Oregon: Timber Press.

Nichols, S., and L. Entine. 1978. *Prairie Primer*. Madison, Wisconsin: University of Wisconsin—Extension.

Nold, R. 1999. *Penstemons*. Portland, Oregon: Timber Press.

North American Rock Garden Society. 2001. *Bulbs of North America*. Portland, Oregon: Timber Press.

Nowak, M. 2007. *Birdscaping in the Midwest: A Guide to Gardening with Native Plants to Attract Birds*. Blue Mounds, Wisconsin: Itchy Cat Press.

Nuzzo, V. 1984. Our native plants: plants of Dane County and where to find them. *The Capital Times* (Madison, Wisconsin).

Olsen, S. 2007. *Encyclopedia of Garden Ferns*. Portland, Oregon: Timber Press.

Osler, M. 1989. *A Gentle Plea for Chaos*. New York: Arcade Publishing.

Oudolf, P., and H. Gerritsen. 2000. *Dream Plants for the Natural Garden*. Portland, Oregon: Timber Press.

Oudolf, P., and H. Gerritsen. 2003. *Planting the Natural Garden*. Portland, Oregon: Timber Press.

Oudolf, P., and N. Kingsbury. 2005. *Planting Design: Gardens in Time and Space*. Portland, Oregon: Timber Press.

Owensby, C. E. 1980. *Kansas Prairie Wildflowers*. Ames, Iowa: Iowa State University Press.

Peattie, D. C. 1966. *A Natural History of Trees of Eastern and Central North America*. 2nd ed. New York: Bonanza Books.

Peel, L. 1997. *The Ultimate Sunflower Book*. New York: Harper Collins Publishers.

Phillips, N. 1984. *The Root Book*. Grand Rapids, Minnesota: N. Phillips.

Phipps, J. B. 2003. *Hawthorns and Medlars*. Portland, Oregon: Timber Press.

Pohl, R. W. 1978. *How to Know the Grasses*. 3rd ed. Dubuque, Iowa: Wm. C. Brown Company Publishers.

Porter, F. W. 2013. *Back to Eden*. Wilmington, Ohio: Orange Frazer Press.

Price, S. D. 1995. *Minnesota Gardens: An Illustrated History*. Afton, Minnesota: Afton Historical Society Press.

Reichman, O. J. 1987. *Konza Prairie*. Lawrence, Kansas: University Press of Kansas.

Rizzo, L. 2001. *Kansas City WildLands*. Jefferson City, Missouri: Missouri Department of Conservation.

Rock, H. W. 1981. *Prairie Propagation Handbook*. 6th ed. Milwaukee, Wisconsin: Milwaukee Department of Parks, Recreation and Culture.

Rosendahl, C. O. 1928. *Trees and Shrubs of Minnesota*. Minneapolis, Minnesota: University of Minnesota Press.

Rowantree, L. 2006. *Hardy Californians: A Woman's Life with Native Plants*. New, expanded ed. Berkeley, California: University of California Press.

Runkel, S. T., and D. M. Roosa. 1989. *Wildflowers of the Tallgrass Prairie: The Upper Midwest*. Ames, Iowa: Iowa State University Press.

Runkel, S. T., and D. M. Roosa. 1999. *Wildflowers and Other Plants of Iowa Wetlands*. Ames, Iowa: Iowa State University Press.

Rutkow, E. 2012. *American Canopy: Trees, Forests, and the Making of a Nation*. New York: Scribner.

Sabuco, J. J. 1990. *The Best of the Hardiest*. 3rd ed. Flossmoor, Illinois: Plantsmen's Publications.

Sachse, N. D. 1965. *A Thousand Ages*. Madison, Wisconsin: Regents of the University of Wisconsin.

Sargent, C. S. 1965. *Manual of the Trees of North America*. 2nd ed. Vols. 1 and 2. New York: Dover Publications.

Sawyers, C. E. 2007. *The Authentic Garden: Five Principles for Cultivating a Sense of Place*. Portland, Oregon: Timber Press.

Schopmeyer, C. S. 1974. *Seeds of Woody Plants in the United States*: Agriculture Handbook No. 450. Washington, D.C.: Forest Service, United States Department of Agriculture.

Seymour, R. 1997. *Wildflowers of Mammoth Cave National Park*. Lexington, Kentucky: University Press of Kentucky.

Simo, M. L. 2003. *Forest and Garden: Traces of Wildness in a Modernizing Land, 1897-1949*. Charlottesville, Virginia: University of Virginia Press.

Smith, A. 1996. *Big Bluestem: Journey into the Tall Grass*. Tulsa, Oklahoma: Council Oak Books.

Smith, E. B. 1978. *An Atlas and Annotated List of the Vascular Plants of Arkansas*. Fayetteville, Arkansas: University of Arkansas at Fayetteville.

Smith, J. R. 1980. *The Prairie Garden: 70 Native Plants You Can Grow in Town or Country*. Madison, Wisconsin: University of Wisconsin Press.

Smith, W. R. 1993. *Orchids of Minnesota*. Minneapolis, Minnesota: University of Minnesota Press.

Smith, W. R. 2008. *Trees and Shrubs of Minnesota*. Minneapolis, Minnesota: University of Minnesota Press.

Snyder, L. C. 1978. *Gardening in the Upper Midwest*. Minneapolis, Minnesota: University of Minnesota Press.

Snyder, L. C. 1980. *Trees and Shrubs for Northern Gardens*. Minneapolis, Minnesota: University of Minnesota Press.

Snyder, L.C. 1983. *Flowers for Northern Gardens*. Minneapolis, Minnesota: University of Minnesota Press.

Snyder, L. C. 1991. *Native Plants for Northern Gardens*. Chanhassen, Minnesota: Andersen Horticultural Library.

Snyder, L. C. 2000. *Trees and Shrubs for Northern Gardens*. Chanhassen, Minnesota: Andersen Horticultural Library.

Stein, J., D. Binion, and R. Acciavatti. 2003. *Field Guide to Native Oak Species of Eastern North America*. FHTET-2003-01. Morgantown, West Virginia: U.S. Forest Service, Forest Health Technology Enterprise Team.

Stein, S. 1993. *Noah's Garden: Restoring the Ecology of Our Own Back Yards*. New York: Houghton-Mifflin.

Steiner, L. M. 2010. *Prairie-Style Gardens: Capturing the Essence of the American Prairie Wherever You Live*. Portland, Oregon: Timber Press.

Steiner, L. M. 2005. *Landscaping with Native Plants of Minnesota*. St. Paul, Minnesota: Voyageur Press.

Steiner, L. M. 2007. *Landscaping with Native Plants of Wisconsin*. St. Paul, Minnesota: Voyageur Press.

Stephens, H. A. 1969. *Trees, Shrubs, and Woody Vines in Kansas*. Lawrence, Kansas: University Press of Kansas.

Sternberg, G. 2004. *Native Trees for North American Landscapes*. Portland, Oregon: Timber Press.

Steyermark, J. A. 1940. *Spring Flora of Missouri*. St. Louis: Missouri Botanical Garden.

Steyermark, J. A. 1963. *Flora of Missouri*. Ames, Iowa: The Iowa State University Press.

Stokes, D. W. 1981. *The Natural History of Wild Shrubs and Vines: Eastern and Central North America*. New York: Harper and Row.

Summers, B. 1996. *Missouri Orchids*. Jefferson City, Missouri: Missouri Department of Conservation.

Sutton, J. 2001. *The Plantfinder's Guide to Daisies*. Portland, Oregon: Timber Press.

Swink, F., and G. Wilhelm. 1994. *Plants of the Chicago Region*. 4th ed. Indianapolis: Indiana Academy of Science.

Tallamy, D. W. 2007. *Bringing Nature Home*. Portland, Oregon: Timber Press.

Taylor, K. S., and S. F. Hamblin. 1963. *A Handbook of Wildflower Cultivation*. New York: Macmillan Publishing Company.

Taylor, P. A. 1996. *Easy Care Native Plants: A Guide to Selecting and Using Beautiful American Flowers, Shrubs, and Trees in Gardens and Landscapes*. New York: Henry Holt and Company.

Teale, E. W. 1957. *Dune Boy: The Early Years of a Naturalist*. Bloomington, Indiana: Indiana University Press.

Tebbitt, M., M. Lidén, and H. Zetterlund. 2008. *Bleeding Hearts, Corydalis, and Their Relatives*. Portland, Oregon: Timber Press.

Tehon, L. R. 1942. *Field Book of Native Illinois Shrubs*. Urbana, Illinois: State of Illinois, Natural History Survey Division.

Tester, J. R. 1995. *Minnesota's Natural Heritage*. Minneapolis, Minnesota: University of Minnesota Press.

Tishler, W. H. 2000. *Midwestern Landscape Architecture*. Urbana, Illinois: University of Illinois Press.

Tishler, W. H., ed. 2012. *Jens Jensen: Writings Inspired by Nature*. Madison, Wisconsin: Wisconsin Historical Society Press.

Trehane, J. 2004. *Blueberries, Cranberries, and Other Vacciniums*. Portland, Oregon: Timber Press.

Tryon, R. 1980. *Ferns of Minnesota*. 2nd ed. Minneapolis, Minnesota: University of Minnesota Press.

Tufts, C. 1993. *The Backyard Naturalist*. Vienna, Virginia: National Wildlife Federation.

Tullock, J. 2005. *Growing Hardy Orchids*. Portland, Oregon: Timber Press.

Tylka, D. 2002. *Native Landscaping for Wildlife and People*. Jefferson City, Missouri: Missouri Department of Conservation.

Valder, P. 1995. *Wisterias: A Comprehensive Guide*. Portland, Oregon: Timber Press.

Van Bruggen, T. 1985. *The Vascular Plants of South Dakota*. 2nd ed. Ames, Iowa: Iowa State University Press.

Van Bruggen, T. 1992. *Wildflowers, Grasses, and Other Plants of the Northern Plains and Black Hills*. Interior, South Dakota: Badlands Natural History Association.

Van Der Linden, P. J., and D. R. Farrar. 2011. *Forest and Shade Trees of Iowa*. 3rd ed. Iowa City: University of Iowa Press.

Van Sweden, J. 1997. *Gardening with Nature: How James van Sweden and Wolfgang Oehme Plant Slopes, Meadows, Outdoor Rooms, and Garden Screens*. New York: Random House.

Van Sweden, J. 2011. *The Artful Garden: Creative Inspiration for Landscape Design*. New York: Random House.

Vance, F. R., J. R. Jowsey, and J. S. McLean. 1984. *Wildflowers of the Northern Great Plains*. Minneapolis, Minnesota: University of Minnesota Press.

Voigt, J. W., and R. H. Mohlenbrock. 1979. *Prairie Plants of Illinois*. Springfield, Illinois: State of Illinois, Department of Conservation, Division of Forest Resources and Natural Heritage.

Voss, E. G. 1972. *Michigan Flora: Part 1, Gymnosperms and Monocots*. Bloomfield Hills, Michigan: Cranbrook Institute of Science.

Voss, E. G. 1985. *Michigan Flora: Part 2, Dicots (Saururaceae—Cornaceae)*. Ann Arbor, Michigan: Regents of the University of Michigan.

Voss, E. G., and A. A. Reznicek. 2012. *Field Manual of Michigan Flora*. Ann Arbor, Michigan: University of Michigan Press.

Wagner, D. L. 2005. *Caterpillars of Eastern North America: A Guide to Identification and Natural History*. Princeton, New Jersey: Princeton University Press.

Waldron, G. 2003. *Trees of the Carolinian Forest: A Guide to Species, Their Ecology and Uses*. Erin, Ontario, Canada: Boston Mills Press.

Wasowski, A. 2000. *The Landscaping Revolution: Garden with Mother Nature not Against Her*. Chicago: Contemporary Books.

Wasowski, S. 2002. *Gardening with Prairie Plants: How to Create Beautiful Native Landscapes*. Minneapolis, Minnesota: University of Minnesota Press.

Watts, M. T. 1957. *Reading the Landscape: An Adventure in Ecology*. New York: Macmillan Publishing Company.

Weeks, S. S., and H. P. Weeks Jr. 2012. *Shrubs and Woody Vines of Indiana and the Midwest: Identification, Wildlife Values, and Landscaping Use*. West Lafayette, Indiana: Purdue University Press.

Weeks, S. S., H. P. Weeks Jr., and G. R. Parker. 2005. *Native Trees of the Midwest: Identification, Wildlife Values, and Landscaping Use*. West Lafayette, Indiana: Purdue University Press.

Wetter, M. A., T. S. Cochrane, M. R. Black, H. H. Iltis, and P. E. Berry. 2001. Checklist of the vascular plants of Wisconsin. Technical Bulletin 192, Department of Natural Resources, Madison, Wisconsin.

Wharton, M. E., and R. W. Barbour. 1973. *Trees and Shrubs of Kentucky*. Lexington, Kentucky: University Press of Kentucky.

Wharton, M. E., and R. W. Barbour. 1979. *A Guide to the Wildflowers and Ferns of Kentucky*. Lexington, Kentucky: University Press of Kentucky.

Wherry, E. T. 1948. *Wild Flower Guide: Northeastern and Midland United States*. Garden City, New York: Doubleday and Company.

Whitley, J. R., B. Bassett, J. G. Dillard, and R. A. Hafner. 1990. *Water Plants for Missouri Ponds*. Jefferson City, Missouri: Missouri Department of Conservation.

Winterringer, G. S. 1967. *Wild Orchids of Illinois*. Springfield, Illinois: State of Illinois.

Winterringer, G. S., and A. C. Lopinot. 1977. *Aquatic Plants of Illinois*. Springfield, Illinois: State of Illinois.

Wovcha, D. S., B. C. Delaney, and G. E. Nordquist. *Minnesota's St. Croix River Valley and Anoka Sandplain: A Guide to Native Habitats*. Minneapolis, Minnesota: University of Minnesota Press.

Yatskievych, G. 1999. *Steyermark's Flora of Missouri*. Vol 1. Jefferson City, Missouri: Missouri Department of Conservation.

Yatskievych, G. 2006. *Steyermark's Flora of Missouri*. Vol. 2. Jefferson City, Missouri: Missouri Department of Conservation.

Yatskievych, G. 2013. *Steyermark's Flora of Missouri*. Vol. 3. St. Louis, Missouri: Missouri Botanical Garden Press.

Yatskievych, K. 2000. *Field Guide to Indiana Wildflowers*. Bloomington, Indiana: Indiana University Press.

Acknowledgments

I would like to thank colleague Scott Woodbury, manager of the Whitmire Wildflower Garden at Shaw Nature Reserve, for passing along information about the opportunity to write this book and for all his work promoting native plants in the landscape.

I am thankful for Eric Tschanz and the staff of Powell Gardens for their support, though I wrote the book on my personal time.

I thank Mark Loeschke for being a botanical buddy since we first met and keeping me informed of all the new flora (and fauna) discoveries in Iowa and throughout the Midwest. Mark reviewed all the botanical nomenclature and terminology for this book and I thank him for writing the foreword to this book.

I am indebted to friend Paula Koch as copy editor of my manuscripts.

I thank botany mentors Kay Lancaster at Iowa State and Lowell Urbatsch at Louisiana State University for their incredible botanical knowledge and talent as great teachers.

I thank Robert W. Dyas, professor emeritus at Iowa State University, for being my mentor and leading me and so many other landscape architecture students to natural areas in Iowa, Missouri, and Wisconsin, and to Iowa State University Professor Gary Hightshoe for all his inspiration, work with, and publications about native plants.

I want to thank lifelong friend Perry Halse, who leads the care of Luther College's campus, for his assistance in clarifying the story of Jens Jensen's historic cottonwood tree there.

To Elizabeth Lorentzen, school teacher and fellow nature lover who always inspires me to look at the beauty of the color schemes, forms, and textures of native plants in all lights through each season.

To JoAnn and Leroy Mercer, the most generous former neighbors and friends, who shared their Rockford, Illinois garden with me as if it were my own. I will never forget all our forays to gardens across northern Illinois and southern Wisconsin along with the many wonderful meals we shared.

To Dan Williams for researching Rockford area parks for me and along with his wife Barbara Williams for always providing a place to stay when in Rockford, Illinois. I have so many fond birding memories with you over the past three decades.

To friend Beth Goeppinger, long-time naturalist at Richard Bong State Recreation Area and nature friend, may you continue to inspire many with the importance of our natural world.

To Gary and Judy Anderson for all our excursions together as "prairie people" and to Don and Sue Miller for the same excursions and to Don for his extraordinary gift of inspiring children and adults as naturalist at the Severson Dells Environmental Education Center.

To Ruth Little and Brian Pruka for a power summer of stalking the wilds of northern Illinois and southern Wisconsin in 1988 and beyond.

To Marjory Rand for countless nature walks around the Kishwaukee River in Illinois during all seasons.

To Kevin Kaltenbach for many other excursions to see so many inspiring natural areas and for all his volunteer work at Nachusa Grasslands.

To John and Janet Alesandrini as fellow naturalists and generous neighbors and many evenings breaking bread and visiting in their home and sustainable garden.

To Verne and Ardith Koenig who took me as an eight-year-old on his first Audubon Christmas Bird Count in 1969 and to Bud Bahr who also was often in the birding party and shared so much of his botanical knowledge with me at such an early age.

To friends in the Idalia Society of Mid-American lepidopterists including Brett Budach, Jackie Goetz, Linda Williams, Betsy Betros, Lenora Larson, and Micky Louis who shared so much with me.

To Marvin Snyder, past president of the American Conifer Society for sharing his garden and love for conifers.

To Linda Hezel of Prairie Birthday Farm for all her work with an organic edible landscape that includes so many of our native plants.

To Neil Dieboll, Mervin Wallace, and Alan Wade for starting native plant nurseries well before they were cool and whose respective nurseries—Prairie Nursery, Missouri Wildflowers Nursery, and Prairie Moon Nursery—continue to supply Midwestern native plants and share how and why to grow them to gardener and restorationists alike.

To the Iowa Natural Heritage Foundation for saving some of my favorite sacred wild places along the Upper Iowa River and throughout Iowa.

To the Natural Land Institute for saving critical natural areas across north central and northwestern Illinois and for sharing their facilities for so many past programs and events. I will never forget their founder, the late George Fell, for his humble and unwavering pursuit of preserving natural lands and inspiring creation of Illinois Nature Preserves.

To the Missouri Prairie Foundation for protecting so much native prairie in Missouri and for carrying on the Grow Native! program that inspires the use of native plants across the Lower Midwest.

To all the other midwestern departments of conservation, departments of natural resources, parks departments, forest preserves, nongovernmental organization land trusts, and the Nature Conservancy for all of their work protecting the original homes of our native plants.

To the American Public Gardens Association and all their midwestern member gardens for conserving and displaying native plants, and educating visitors about their importance.

Lastly and most importantly I thank my parents and family for always embracing my pursuits and always being there to celebrate life's journey.

Photo Credits

All photos by author except as follows:

Elizabeth Betros, pages 335 top right, 351 top right, 382 top left, 417 bottom left

Becky Klukas-Brewer, courtesy of Prairie Moon Nursery, page 338 top right

Mount Cuba Center, pages 391 top left, 395 bottom left, 399 top right

Rob Routledge, Sault College, Bugwood.org, page 312 bottom right

David Schwaegler, page 301 top left

Scott Woodbury, page 395 top right

FLICKR

Steve Law, used under a Creative Commons Attribution-Share Alike 2.0 Generic license, page 195 bottom left

WIKIMEDIA

SEWilco, used under a Creative Commons Attribution-Share Alike 2.5 Generic license, page 197 bottom right

Halpaugh, used under a Creative Commons Attribution-Share Alike 3.0 Unported License, page 313 top right

Index